MATHEMATICS
for
CIRCUITS
and FILTERS

MATHEMATICS
for
CIRCUITS
and FILTERS

edited by
Wai-Kai Chen

CRC Press
Boca Raton London New York Washington, D.C.

TK
454
M3295
2000

Library of Congress Cataloging-in-Publication Data

Mathematics for circuits and filters / Wai-Kai Chen, editor.
        p.   cm.
    Includes bibliographical references.
    ISBN 0-8493-0052-5 (alk. paper)
    1. Electric circuits, Linear—Mathematical models. 2. Electric filters—Mathematical
models. 3. Electric engineering—Mathematics. I. Chen, Wai-Kai, 1936–
TK454. M3295 1999
621.3—dc21
                                                                        99-043798
                                                                              CIP

Preface

The purpose of Mathematics for Circuits and Filters is to provide in a single volume a comprehensive reference work covering the broad spectrum of mathematics and symbols that underlie numerous applications in electrical circuits and filters. It is written and developed for practicing electrical engineers in industry, government, and academia. Over the years, the mathematical fundamentals of electrical circuits and filters have evolved to include a wide range of topics and a broad range of practice. To encompass such a wide range of knowledge, the book focuses on the key concepts , models, and equations that enable the electrical engineer to analyze, design, and predict the behavior of large-scale circuits, devices, filters, and systems.

Wai Kai Chen
Editor-in-Chief

Contributors

John R. Deller, Jr.
Michigan State University
East Lansing, Michigan

Igor Djokovic
California Institute of Technology
Pasadena, California

W. Kenneth Jenkins
University of Illinois
Urbana, Illinois

Jelena Kovacevic
AT&T Bell Laboratories
Murray Hill, New Jersey

Michael K. Sain
Universtiy of Notre Dame
Notre Dame, Indiana

Cheryl B. Schrader
University of Texas
San Antonio, Texas

Krishnaiyan Thulasiraman
University of Oklahoma
Norman, Oklahoma

P. P. Vaidyanathan
California Institute of Technology
Pasadena, California

Contents

1

Linear Operators and Matrices

Cheryl B. Schrader
The University of Texas

Michael K. Sain
University of Notre Dame

1.1 Introduction

It is only after the engineer masters linear concepts—linear models, and circuit and filter theory—that the possibility of tackling nonlinear ideas becomes achievable. Students frequently encounter linear methodologies, and bits and pieces of mathematics that aid in problem solution are stored away. Unfortunately, in memorizing the process of finding the inverse of a matrix or of solving a system of equations, the essence of the problem or associated knowledge may be lost. For example, most engineers are fairly comfortable with the concept of a vector space, but have difficulty in generalizing these ideas to the module level. Therefore, it is the intention of this section to provide a unified view of key concepts in the theory of linear circuits and filters, to emphasize interrelated concepts, to provide a mathematical reference to the handbook itself, and to illustrate methodologies through the use of many and varied examples.

This chapter begins with a basic examination of vector spaces over fields. In relating vector spaces the key ideas of linear operators and matrix representations come to the fore. Standard matrix operations are examined as are the pivotal notions of determinant, inverse, and rank. Next, transformations are shown to determine similar representations, and matrix characteristics such as singular values and eigenvalues are defined. Finally, solutions to algebraic equations are presented in the context of matrices and are related to this introductory chapter on mathematics as a whole.

Standard algebraic notation is introduced first. To denote an element s in a set S, use $s \in S$. Consider two sets S and T. The set of all ordered pairs (s, t) where $s \in S$ and $t \in T$ is defined as the Cartesian product set $S \times T$. A function f from S into T, denoted by $f : S \rightarrow T$, is a subset U of ordered pairs $(s, t) \in S \times T$ such that for every $s \in S$ one and only one $t \in T$ exists such that $(s, t) \in U$. The function evaluated at the element s gives t as a solution ($f(s) = t$), and each $s \in S$ as a first element in U appears exactly once. A binary operation is a function acting on a Cartesian product set $S \times T$. When $T = S$, one speaks of a binary operation on S.

1.2 Vector Spaces Over Fields

A **field** F is a nonempty set F and two binary operations, sum (+) and product, such that the following properties are satisfied for all $a, b, c \in F$:

1. Associativity: $(a + b) + c = a + (b + c)$; $(ab)c = a(bc)$
2. Commutativity: $a + b = b + a$; $ab = ba$
3. Distributivity: $a(b + c) = (ab) + (ac)$
4. Identities: (Additive) $0 \in F$ exists such that $a + 0 = a$
  (Multiplicative) $1 \in F$ exists such that $a1 = a$
5. Inverses: (Additive) For every $a \in F$, $b \in F$ exists such that $a + b = 0$
  (Multiplicative) For every nonzero $a \in F$, $b \in F$ exists such that $ab = 1$

Examples

- Field of real numbers R
- Field of complex numbers C
- Field of rational functions with real coefficients $R(s)$
- Field of binary numbers

The set of integers Z with the standard notions of addition and multiplication does not form a field because a multiplicative inverse in Z exists only for ± 1. The integers form a **commutative ring**. Likewise, polynomials in the indeterminate s with coefficients from F form a commutative ring $F[s]$. If field property 2 also is not available, then one speaks simply of a **ring**. An **additive group** is a nonempty set G and one binary operation + satisfying field properties 1, 4, and 5 for addition; i.e., associativity and the existence of additive identity and inverse. Moreover, if the binary operation + is commutative (field property 2), then the additive group is said to be **abelian**. Common notation regarding inverses is that the additive inverse for $a \in F$ is $b = -a \in F$. In the multiplicative case $b = a^{-1} \in F$.

An F-**vector space** V is a nonempty set V and a field F together with binary operations $+ : V \times V \to V$ and $* : F \times V \to V$ subject to the following axioms for all elements $v, w \in V$ and $a, b \in F$:

1. V and + form an additive abelian group
2. $a * (v + w) = (a * v) + (a * w)$
3. $(a + b) * v = (a * v) + (b * v)$
4. $(ab) * v = a * (b * v)$
5. $1 * v = v$

Examples

- The set of all n-tuples $(v_1, v_2, \ldots, v_n)$ for $n > 0$ and $v_i \in F$
- The set of polynomials of degree less than n with real coefficients ($F = R$)

Elements of V are referred to as **vectors**, whereas elements of F are **scalars**. Note that the terminology **vector space** V **over the field** F is used often. A **module** differs from a vector space in only one aspect; the underlying field in a vector space is replaced by a ring. Thus, a module is a direct generalization of a vector space.

When considering vector spaces of n-tuples, + is vector addition defined by element using the scalar addition associated with F. Multiplication ($*$), which is termed scalar multiplication, also is defined by element using multiplication in F. The additive identity in this case is the zero vector

(n-tuple of zeros) or null vector, and F^n denotes the set of n-tuples with elements in F, a vector space over F. A nonempty subset $\tilde{V} \subset V$ is called a **subspace** of V if for each $v, w \in \tilde{V}$ and every $a \in F$, $v + w \in \tilde{V}$ and $a * v \in \tilde{V}$. When the context makes things clear, it is customary to suppress the $*$, and write av in place of $a * v$.

A set of vectors $\{v_1, v_2, \ldots, v_m\}$ belonging to an F-vector space V is said to **span** the vector space if any element $v \in V$ can be represented by a linear combination of the vectors v_i. That is, scalars $a_1, a_2, \ldots, a_m \in F$ are such that

$$v = a_1 v_1 + a_2 v_2 + \cdots + a_m v_m \tag{1.1}$$

A set of vectors $\{v_1, v_2, \ldots, v_p\}$ belonging to an F-vector space V is said to be **linearly dependent** over F if scalars $a_1, a_2, \ldots, a_p \in F$, not all zero, exist such that

$$a_1 v_1 + a_2 v_2 + \cdots + a_p v_p = 0 \tag{1.2}$$

If the only solution for (1.2) is that all $a_i = 0 \in F$, then the set of vectors is said to be **linearly independent**.

Examples

- $(1, 0)$ and $(0, 1)$ are linearly independent.
- $(1, 0, 0)$, $(0, 1, 0)$, and $(1, 1, 0)$ are linearly dependent over R. To see this, simply choose $a_1 = a_2 = 1$ and $a_3 = -1$.
- $s^2 + 2s$ and $2s + 4$ are linearly independent over R, but are linearly dependent over $R(s)$ by choosing $a_1 = -2$ and $a_2 = s$.

A set of vectors $\{v_1, v_2, \ldots, v_n\}$ belonging to an F-vector space V is said to form a **basis** for V if it both spans V and is linearly independent over F. The number of vectors in a basis is called the **dimension** of the vector space, and is denoted $\dim(V)$. If this number is not finite, then the vector space is said to be **infinite dimensional**.

Examples

- In an n-dimensional vector space, any n linearly independent vectors form a basis.
- The **natural (standard) basis**

$$e_1 = \begin{bmatrix} 1 \\ 0 \\ 0 \\ \vdots \\ 0 \\ 0 \end{bmatrix}, \quad e_2 = \begin{bmatrix} 0 \\ 1 \\ 0 \\ \vdots \\ 0 \\ 0 \end{bmatrix}, \quad e_3 = \begin{bmatrix} 0 \\ 0 \\ 1 \\ \vdots \\ 0 \\ 0 \end{bmatrix}, \cdots, e_{n-1} = \begin{bmatrix} 0 \\ 0 \\ 0 \\ \vdots \\ 1 \\ 0 \end{bmatrix}, \quad e_n = \begin{bmatrix} 0 \\ 0 \\ 0 \\ \vdots \\ 0 \\ 1 \end{bmatrix}$$

both spans F^n and is linearly independent over F.

Consider any basis $\{v_1, v_2, \ldots, v_n\}$ in an n-dimensional vector space. Every $v \in V$ can be represented uniquely by scalars $a_1, a_2, \ldots, a_n \in F$ as

$$v = a_1 v_1 + a_2 v_2 + \cdots + a_n v_n \tag{1.3}$$

$$= \begin{bmatrix} v_1 & v_2 & \cdots & v_n \end{bmatrix} \begin{bmatrix} a_1 \\ a_2 \\ \vdots \\ a_n \end{bmatrix} \tag{1.4}$$

$$= \begin{bmatrix} v_1 & v_2 & \cdots & v_n \end{bmatrix} a \tag{1.5}$$

Here, $a \in F^n$ is a coordinate representation of $v \in V$ with respect to the chosen basis. The reader will be able to discern that each choice of basis will result in another representation of the vector under consideration. Of course, in the applications some representations are more popular and useful than others.

1.3   Linear Operators and Matrix Representations

First, recall the definition of a function $f : S \to T$. Alternate terminology for a function is mapping, operator, or transformation. The set S is called the **domain** of f, denoted $D(f)$. The **range** of f, $R(f)$, is the set of all $t \in T$ such that $(s, t) \in U$ ($f(s) = t$) for some $s \in D(f)$.

Examples

Use $S = \{1, 2, 3, 4\}$ and $T = \{5, 6, 7, 8\}$.

- $\tilde{U} = \{(1, 5), (2, 5), (3, 7), (4, 8)\}$ is a function. The domain is $\{1, 2, 3, 4\}$ and the range is $\{5, 7, 8\}$.
- $\hat{U} = \{(1, 5), (1, 6), (2, 5), (3, 7), (4, 8)\}$ is not a function.
- $\overline{U} = \{(1, 5), (2, 6), (3, 7), (4, 8)\}$ is a function. The domain is $\{1, 2, 3, 4\}$ and the range is $\{5, 6, 7, 8\}$.

If $R(f) = T$, then f is said to be **surjective (onto)**. Loosely speaking, all elements in T are used up. If $f : S \to T$ has the property that $f(s_1) = f(s_2)$ implies $s_1 = s_2$, then f is said to be **injective (one-to-one)**. This means that any element in $R(f)$ comes from a unique element in $D(f)$ under the action of f. If a function is both injective and surjective, then it is said to be **bijective (one-to-one and onto)**.

Examples

- $\tilde{U}$ is not onto because $6 \in T$ is not in $R(f)$. Also $\tilde{U}$ is not one-to-one because $f(1) = 5 = f(2)$, but $1 \neq 2$.
- $\overline{U}$ is bijective.

Now consider an operator $L : V \to W$, where V and W are vector spaces over the same field F. L is said to be a **linear operator** if the following two properties are satisfied for all $v, w \in V$ and for all $a \in F$:

$$L(av) \quad = \quad aL(v) \tag{1.6}$$
$$L(v + w) \quad = \quad L(v) + L(w) \tag{1.7}$$

Equation (1.6) is the property of homogeneity and (1.7) is the property of additivity. Together they imply the principle of **superposition**, which may be written as

$$L(a_1 v_1 + a_2 v_2) = a_1 L(v_1) + a_2 L(v_2) \tag{1.8}$$

for all $v_1, v_2 \in V$ and $a_1, a_2 \in F$. If (1.8) is not satisfied, then L is called a **nonlinear operator**.

Examples

- Consider $V = C$ and $F = C$. Let $L : V \to V$ be the operator that takes the complex conjugate: $L(v) = \overline{v}$ for $v, \overline{v} \in V$. Certainly

$$L(v_1 + v_2) = \overline{v_1 + v_2} = \overline{v_1} + \overline{v_2} = L(v_1) + L(v_2)$$

However,

$$L(a_1 v_1) = \overline{a_1 v_1} = \overline{a_1} \overline{v_1} = \overline{a_1} L(v_1) \neq a_1 L(v_1)$$

Then L is a nonlinear operator because homogeneity fails.

- For F-vector spaces V and W, let V be F^n and W be F^{n-1}. Examine $L : V \to W$, the operator that truncates the last element of the n-tuple in V; that is,

$$L((v_1, v_2, \ldots, v_{n-1}, v_n)) = (v_1, v_2, \ldots, v_{n-1})$$

Such an operator is linear.

The **null space (kernel)** of a linear operator $L : V \to W$ is the set

$$\ker L = \{v \in V \text{ such that } L(v) = 0\} \tag{1.9}$$

Equation (1.9) defines a vector space. In fact, $\ker L$ is a subspace of V. The mapping L is injective if and only if $\ker L = 0$; that is, the only solution in the right member of (1.9) is the trivial solution. In this case L is also called **monic**.

The **image** of a linear operator $L : V \to W$ is the set

$$\operatorname{im} L = \{w \in W \text{ such that } L(v) = w \text{ for some } v \in V\} \tag{1.10}$$

Clearly, $\operatorname{im} L$ is a subspace of W, and L is surjective if and only if $\operatorname{im} L$ is all of W. In this case L is also called **epic**.

A method of relating specific properties of linear mappings is the **exact sequence**. Consider a sequence of linear mappings

$$\cdots V \xrightarrow{L} W \xrightarrow{\tilde{L}} U \to \cdots \tag{1.11}$$

This sequence is said to be exact at W if $\operatorname{im} L = \ker \tilde{L}$. A sequence is called exact if it is exact at each vector space in the sequence. Examine the following special cases:

$$0 \quad \to \quad V \xrightarrow{L} W \tag{1.12}$$

$$W \quad \xrightarrow{\tilde{L}} \quad U \to 0 \tag{1.13}$$

Sequence (1.12) is exact if and only if L is monic, whereas (1.13) is exact if and only if $\tilde{L}$ is epic.

Further, let $L : V \to W$ be a linear mapping between finite-dimensional vector spaces. The **rank** of L, $\rho(L)$, is the dimension of the image of L. In such a case

$$\rho(L) + \dim(\ker L) = \dim V \tag{1.14}$$

Linear operators commonly are represented by **matrices**. It is quite natural to interchange these two ideas, because a matrix with respect to the standard bases is indistinguishable from the linear operator it represents. However, insight may be gained by examining these ideas separately. For V and W n- and m-dimensional vector spaces over F, respectively, consider a linear operator $L : V \to W$. Moreover, let $\{v_1, v_2, \ldots, v_n\}$ and $\{w_1, w_2, \ldots, w_m\}$ be respective bases for V and W. Then $L : V \to W$ can be represented uniquely by the matrix $M \in F^{m \times n}$ where

$$M = \begin{bmatrix} m_{11} & m_{12} & \cdots & m_{1n} \\ m_{21} & m_{22} & \cdots & m_{2n} \\ \vdots & \vdots & \ddots & \vdots \\ m_{m1} & m_{m2} & \cdots & m_{mn} \end{bmatrix} \tag{1.15}$$

The ith column of M is the representation of $L(v_i)$ with respect to $\{w_1, w_2, \ldots, w_m\}$. Element $m_{ij} \in F$ of (1.15) occurs in row i and column j.

Matrices have a number of properties. A matrix is said to be **square** if $m = n$. The **main diagonal** of a square matrix consists of the elements m_{ii}. If $m_{ij} = 0$ for all $i > j (i < j)$ a square matrix is said to be **upper (lower) triangular**. A square matrix with $m_{ij} = 0$ for all $i \neq j$ is **diagonal**. Additionally, if all $m_{ii} = 1$, a diagonal M is an **identity matrix**. A **row vector (column vector)** is a special case in which $m = 1$ ($n = 1$). Also, $m = n = 1$ results essentially in a scalar.

Matrices arise naturally as a means to represent sets of simultaneous linear equations. For example, in the case of Kirchhoff equations, a later section on graph theory shows how incidence, circuit, and cut matrices arise. Or consider a π network having node voltages v_i, $i = 1, 2$ and current sources i_i, $i = 1, 2$ connected across the resistors R_i, $i = 1, 2$ in the two legs of the π. The bridge resistor is R_3. Thus, the unknown node voltages can be expressed in terms of the known source currents in the manner

$$\frac{(R_1 + R_3)}{R_1 R_3} v_1 - \frac{1}{R_3} v_2 = i_1 \qquad (1.16)$$

$$\frac{(R_2 + R_3)}{R_2 R_3} v_2 - \frac{1}{R_3} v_1 = i_2 \qquad (1.17)$$

If the voltages, v_i, and the currents, i_i, are placed into a voltage vector $v \in R^2$ and current vector $i \in R^2$, respectively, then (1.16) and (1.17) may be rewritten in matrix form as

$$\begin{bmatrix} i_1 \\ i_2 \end{bmatrix} = \begin{bmatrix} \dfrac{(R_1 + R_3)}{R_1 R_3} & -\dfrac{1}{R_3} \\ -\dfrac{1}{R_3} & \dfrac{(R_2 + R_3)}{R_2 R_3} \end{bmatrix} \begin{bmatrix} v_1 \\ v_2 \end{bmatrix} \qquad (1.18)$$

A conductance matrix G may then be defined so that $i = Gv$, a concise representation of the original pair of circuit equations.

1.4 Matrix Operations

Vector addition in F^n was defined previously as element-wise scalar addition. Similarly, two matrices M and N, both in $F^{m \times n}$, can be added (subtracted) to form the resultant matrix $P \in F^{m \times n}$ by

$$m_{ij} \pm n_{ij} = p_{ij} \qquad i = 1, 2, \ldots, m \quad j = 1, 2, \ldots, n \qquad (1.19)$$

Matrix addition, thus, is defined using addition in the field over which the matrix lies. Accordingly, the matrix, each of whose entries is $0 \in F$, is an additive identity for the family. One can set up additive inverses along similar lines, which, of course, turn out to be the matrices each of whose elements is the negative of that of the original matrix.

Recall how scalar multiplication was defined in the example of the vector space of n-tuples. Scalar multiplication can also be defined between a field element $a \in F$ and a matrix $M \in F^{m \times n}$ in such a way that the product aM is calculated element-wise:

$$aM = P \iff am_{ij} = p_{ij} \qquad i = 1, 2, \ldots, m \quad j = 1, 2, \ldots, n \qquad (1.20)$$

Examples

$$(F = R): \qquad M = \begin{bmatrix} 4 & 3 \\ 2 & 1 \end{bmatrix} \qquad N = \begin{bmatrix} 2 & -3 \\ 1 & 6 \end{bmatrix} \qquad a = -0.5$$

- $M + N = P = \begin{bmatrix} 4+2 & 3-3 \\ 2+1 & 1+6 \end{bmatrix} = \begin{bmatrix} 6 & 0 \\ 3 & 7 \end{bmatrix}$

- $M - N = \tilde{P} = \begin{bmatrix} 4-2 & 3+3 \\ 2-1 & 1-6 \end{bmatrix} = \begin{bmatrix} 2 & 6 \\ 1 & -5 \end{bmatrix}$

- $aM = \hat{P} = \begin{bmatrix} (-0.5)4 & (-0.5)3 \\ (-0.5)2 & (-0.5)1 \end{bmatrix} = \begin{bmatrix} -2 & -1.5 \\ -1 & -0.5 \end{bmatrix}$

To multiply two matrices M and N to form the product MN requires that the number of columns of M equal the number of rows of N. In this case the matrices are said to be **conformable**. Although vector multiplication cannot be defined here because of this constraint, Chapter 2 examines this operation in detail using the tensor product. The focus here is on matrix multiplication. The resulting matrix will have its number of rows equal to the number of rows in M and its number of columns equal to the number of columns of N. Thus, for $M \in F^{m \times n}$ and $N \in F^{n \times p}$, $MN = P \in F^{m \times p}$. Elements in the resulting matrix P may be determined by

$$p_{ij} = \sum_{k=1}^{n} m_{ik} n_{kj} \tag{1.21}$$

Matrix multiplication involves one row and one column at a time. To compute the p_{ij} term in P, choose the ith row of M and the jth column of N. Multiply each element in the row vector by the corresponding element in the column vector and sum the result. Notice that in general, matrix multiplication is not commutative, and the matrices in the reverse order may not even be conformable. Matrix multiplication is, however, associative and distributive with respect to matrix addition. Under certain conditions, the field F of scalars, the set of matrices over F, and these three operations combine to form an **algebra**. Chapter 2 examines algebras in greater detail.

Examples

$$(F = R): \qquad M = \begin{bmatrix} 4 & 3 \\ 2 & 1 \end{bmatrix} \qquad N = \begin{bmatrix} 1 & 3 & 5 \\ 2 & 4 & 6 \end{bmatrix}$$

- $MN = P = \begin{bmatrix} 10 & 24 & 38 \\ 4 & 10 & 16 \end{bmatrix}$

  To find p_{11}, take the first row of M, [4 3], and the first column of N, $\begin{bmatrix} 1 \\ 2 \end{bmatrix}$, and evaluate (1.21): $4(1) + 3(2) = 10$. Continue for all i and j.

- NM does not exist because that product is not conformable.

- Any matrix $M \in F^{m \times n}$ multiplied by an identity matrix $I \in F^{n \times n}$ such that $MI \in F^{m \times n}$ results in the original matrix M. Similarly, $IM = M$ for I an $m \times m$ identity matrix over F. It is common to interpret I as an identity matrix of appropriate size, without explicitly denoting the number of its rows and columns.

The **transpose** $M^T \in F^{n \times m}$ of a matrix $M \in F^{m \times n}$ is found by interchanging the rows and columns. The first column of M becomes the first row of M^T, the second column of M becomes the second row of M^T, and so on. The notations M^t and M' are used also. If $M = M^T$ the matrix is called **symmetric**. Note that two matrices $M, N \in F^{m \times n}$ are equal if and only if all respective elements are equal: $m_{ij} = n_{ij}$ for all i, j. The **Hermitian transpose** $M^* \in C^{n \times m}$ of $M \in C^{m \times n}$ is also termed the **complex conjugate transpose**. To compute M^*, form M^T and take the complex conjugate of every element in M^T. The following properties also hold for matrix transposition for all $M, N \in F^{m \times n}$, $P \in F^{n \times p}$, and $a \in F$: $(M^T)^T = M$, $(M+N)^T = M^T + N^T$, $(aM)^T = aM^T$, and $(MP)^T = P^T M^T$.

Examples

$$(F = C): \quad M = \begin{bmatrix} j & 1 - j \\ 4 & 2 + j3 \end{bmatrix}$$

- $M^T = \begin{bmatrix} j & 4 \\ 1 - j & 2 + j3 \end{bmatrix}$
- $M^* = \begin{bmatrix} -j & 4 \\ 1 + j & 2 - j3 \end{bmatrix}$

1.5 Determinant, Inverse, and Rank

Consider square matrices of the form $[m_{11}] \in F^{1 \times 1}$. For these matrices, define the **determinant** as m_{11} and establish the notation $\det([m_{11}])$ for this construction. This definition can be used to establish the meaning of $\det(M)$, often denoted $|M|$, for $M \in F^{2 \times 2}$. Consider

$$M = \begin{bmatrix} m_{11} & m_{12} \\ m_{21} & m_{22} \end{bmatrix} \tag{1.22}$$

The **minor** of m_{ij} is defined to be the determinant of the submatrix which results from the removal of row i and column j. Thus, the minors of m_{11}, m_{12}, m_{21}, and m_{22} are m_{22}, m_{21}, m_{12}, and m_{11}, respectively. To calculate the determinant of this M, (1) choose any row i (or column j), (2) multiply each element m_{ik} (or m_{kj}) in that row (or column) by its minor and by $(-1)^{i+k}$ (or $(-1)^{k+j}$), and (3) add these results. Note that the product of the minor with the sign $(-1)^{i+k}$ (or $(-1)^{k+j}$) is called the **cofactor** of the element in question. If row 1 is chosen, the determinant of M is found to be $m_{11}(+m_{22}) + m_{12}(-m_{21})$, a well-known result. The determinant of 2×2 matrices is relatively easy to remember: multiply the two elements along the main diagonal and subtract the product of the other two elements. Note that it makes no difference which row or column is chosen in step 1.

A similar procedure is followed for larger matrices. Consider

$$\det(M) = \begin{vmatrix} m_{11} & m_{12} & m_{13} \\ m_{21} & m_{22} & m_{23} \\ m_{31} & m_{32} & m_{33} \end{vmatrix} \tag{1.23}$$

Expanding about column 1 produces

$$
\begin{aligned}
\det(M) &= m_{11} \begin{vmatrix} m_{22} & m_{23} \\ m_{32} & m_{33} \end{vmatrix} - m_{21} \begin{vmatrix} m_{12} & m_{13} \\ m_{32} & m_{33} \end{vmatrix} + m_{31} \begin{vmatrix} m_{12} & m_{13} \\ m_{22} & m_{23} \end{vmatrix} \tag{1.24} \\
&= m_{11}(m_{22}m_{33} - m_{23}m_{32}) - m_{21}(m_{12}m_{33} - m_{13}m_{32}) \\
&\quad + m_{31}(m_{12}m_{23} - m_{13}m_{22}) \tag{1.25} \\
&= m_{11}m_{22}m_{33} + m_{12}m_{23}m_{31} + m_{13}m_{21}m_{32} - m_{13}m_{22}m_{31} \\
&\quad - m_{11}m_{23}m_{32} - m_{12}m_{21}m_{33} \tag{1.26}
\end{aligned}
$$

An identical result may be achieved by repeating the first two columns next to the original matrix:

$$
\begin{array}{ccccc}
m_{11} & m_{12} & m_{13} & m_{11} & m_{12} \\
m_{21} & m_{22} & m_{23} & m_{21} & m_{22} \\
m_{31} & m_{32} & m_{33} & m_{31} & m_{32}
\end{array} \tag{1.27}
$$

Then, form the first three products of (1.26) by starting at the upper left corner of (1.27) with m_{11}, forming a diagonal to the right, and then repeating with m_{12} and m_{13}. The last three products are subtracted in (1.26) and are formed by starting in the upper right corner of (1.27) with m_{12} and taking a diagonal to the left, repeating for m_{11} and m_{13}. Note the similarity to the 2×2 case. Unfortunately, such simple schemes fail above the 3×3 case.

Determinants of $n \times n$ matrices for $n > 3$ are computed in a similar vein. As in the earlier cases the determinant of an $n \times n$ matrix may be expressed in terms of the determinants of $(n-1) \times (n-1)$ submatrices; this is termed **Laplace's expansion**. To expand along row i or column j in $M \in F^{n \times n}$, write

$$\det(M) = \sum_{k=1}^{n} m_{ik} \tilde{m}_{ik} = \sum_{k=1}^{n} m_{kj} \tilde{m}_{kj} \tag{1.28}$$

where the m_{ik} (m_{kj}) are elements of M. The $\tilde{m}_{ik}$ ($\tilde{m}_{kj}$) are cofactors formed by deleting the ith (kth) row and the kth (jth) column of M, forming the determinant of the $(n-1) \times (n-1)$ resulting submatrix, and multiplying by $(-1)^{i+k}$ ($(-1)^{k+j}$). Notice that minors and their corresponding cofactors are related by ± 1.

Examples

$$(F = R): \qquad M = \begin{bmatrix} 0 & 1 & 2 \\ 3 & 4 & 5 \\ 2 & 3 & 6 \end{bmatrix}$$

- Expanding about row 1 produces

$$\begin{aligned} \det(M) &= 0 \begin{vmatrix} 4 & 5 \\ 3 & 6 \end{vmatrix} - 1 \begin{vmatrix} 3 & 5 \\ 2 & 6 \end{vmatrix} + 2 \begin{vmatrix} 3 & 4 \\ 2 & 3 \end{vmatrix} \\ &= -(18 - 10) + 2(9 - 8) = -6 \end{aligned}$$

- Expanding about column 2 yields

$$\begin{aligned} \det(M) &= -1 \begin{vmatrix} 3 & 5 \\ 2 & 6 \end{vmatrix} + 4 \begin{vmatrix} 0 & 2 \\ 2 & 6 \end{vmatrix} - 3 \begin{vmatrix} 0 & 2 \\ 3 & 5 \end{vmatrix} \\ &= -(18 - 10) + 4(0 - 4) - 3(0 - 6) = -6 \end{aligned}$$

- Repeating the first two columns to form (1.27) gives

$$\begin{array}{ccccc} 0 & 1 & 2 & 0 & 1 \\ 3 & 4 & 5 & 3 & 4 \\ 2 & 3 & 6 & 2 & 3 \end{array}$$

Taking the appropriate products,

$$0 \cdot 4 \cdot 6 + 1 \cdot 5 \cdot 2 + 2 \cdot 3 \cdot 3 - 1 \cdot 3 \cdot 6 - 0 \cdot 5 \cdot 3 - 2 \cdot 4 \cdot 2$$

results in -6 as the determinant of M.

- Any square matrix with a zero row and/or zero column will have zero determinant. Likewise, any square matrix with two or more identical rows and/or two or more identical columns will have determinant equal to zero.

Determinants satisfy many interesting relationships. For any $n \times n$ matrix, the determinant may be expressed in terms of determinants of $(n-1) \times (n-1)$ matrices or first-order minors. In turn, determinants of $(n-1) \times (n-1)$ matrices may be expressed in terms of determinants of

$(n-2) \times (n-2)$ matrices or second-order minors, etc. Also, the determinant of the product of two square matrices is equal to the product of the determinants:

$$\det(MN) = \det(M)\det(N) \tag{1.29}$$

For any $M \in F^{n \times n}$ such that $|M| \neq 0$, a unique inverse $M^{-1} \in F^{n \times n}$ satisfies

$$MM^{-1} = M^{-1}M = I \tag{1.30}$$

For (1.29) one may observe the special case in which $N = M^{-1}$, then $(\det(M))^{-1} = \det(M^{-1})$. The **inverse** M^{-1} may be expressed using determinants and cofactors in the following manner. Form the matrix of cofactors

$$\tilde{M} = \begin{bmatrix} \tilde{m}_{11} & \tilde{m}_{12} & \cdots & \tilde{m}_{1n} \\ \tilde{m}_{21} & \tilde{m}_{22} & \cdots & \tilde{m}_{2n} \\ \vdots & \vdots & \ddots & \vdots \\ \tilde{m}_{n1} & \tilde{m}_{n2} & \cdots & \tilde{m}_{nn} \end{bmatrix} \tag{1.31}$$

The transpose of (1.31) is referred to as the **adjoint** matrix or $\text{adj}(M)$. Then,

$$M^{-1} = \frac{\tilde{M}^T}{|M|} = \frac{\text{adj}(M)}{|M|} \tag{1.32}$$

Examples

- Choose M of the previous set of examples. The cofactor matrix is

$$\begin{bmatrix} 9 & -8 & 1 \\ 0 & -4 & 2 \\ -3 & 6 & -3 \end{bmatrix}$$

Because $|M| = -6$, M^{-1} is

$$\begin{bmatrix} -\frac{3}{2} & 0 & \frac{1}{2} \\ \frac{4}{3} & \frac{2}{3} & -1 \\ -\frac{1}{6} & -\frac{1}{3} & \frac{1}{2} \end{bmatrix}$$

Note that (1.32) is satisfied.

- $M = \begin{bmatrix} 2 & 1 \\ 4 & 3 \end{bmatrix}$ $\text{adj}(M) = \begin{bmatrix} 3 & -1 \\ -4 & 2 \end{bmatrix}$ $M^{-1} = \begin{bmatrix} \frac{3}{2} & -\frac{1}{2} \\ -2 & 1 \end{bmatrix}$

In the 2×2 case this method reduces to interchanging the elements on the main diagonal, changing the sign on the remaining elements, and dividing by the determinant.

- Consider the matrix equation in (1.18). Because $\det(G) \neq 0$, whenever the resistances are nonzero, with R_1 and R_2 having the same sign, the node voltages may be determined in terms of the current sources by multiplying on the left of both members of the equation using G^{-1}. Then, $G^{-1}i = v$.

The **rank** of a matrix M, $\rho(M)$, is the number of linearly independent columns of M over F, or using other terminology, the dimension of the image of M. For $M \in F^{m \times n}$ the number of linearly independent rows and columns is the same, and is less than or equal to the minimum of m and n. If $\rho(M) = n$, M is of **full column rank**; similarly, if $\rho(M) = m$, M is of **full row rank**. A square matrix with all rows (and all columns) linearly independent is said to be **nonsingular**. In this case $\det(M) \neq 0$. The rank of M also may be found from the size of the largest square submatrix with a nonzero determinant. A full-rank matrix has a full-size minor with a nonzero determinant.

The **null space (kernel)** of a matrix $M \in F^{m \times n}$ is the set

$$\ker M = \{v \in F^n \text{ such that } Mv = 0\} \tag{1.33}$$

Over F, ker M is a vector space with dimension defined as the **nullity** of M, $\nu(M)$. The fundamental theorem of linear equations relates the rank and nullity of a matrix $M \in F^{m \times n}$ by

$$\rho(M) + \nu(M) = n \tag{1.34}$$

If $\rho(M) < n$, then M has a nontrivial null space.

Examples

- $M = \begin{bmatrix} 0 & 1 \\ 0 & 0 \end{bmatrix}$

  The rank of M is 1 because only one linearly independent column of M is found. To examine the null space of M, solve $Mv = 0$. Any element in ker M is of the form $\begin{bmatrix} f_1 \\ 0 \end{bmatrix}$ for $f_1 \in F$. Therefore, $\nu(M) = 1$.

- $M = \begin{bmatrix} 1 & 4 & 5 & 2 \\ 2 & 5 & 7 & 1 \\ 3 & 6 & 9 & 0 \end{bmatrix}$

  The rank of M is 2 and the nullity is 2.

1.6 Basis Transformations

This section describes a change of basis as a linear operator. Because the choice of basis affects the matrix of a linear operator, it would be most useful if such a basis change could be understood within the context of matrix operations. Thus, the new matrix could be determined from the old matrix by matrix operations. This is indeed possible. This question is examined in two phases. In the first phase the linear operator maps from a vector space to itself. Then a basis change will be called a **similarity transformation**. In the second phase the linear operator maps from one vector space to another, which is not necessarily the same as the first. Then a basis change will be called an **equivalence transformation**. Of course, the first situation is a special case of the second, but it is customary to make the distinction and to recognize the different terminologies. Philosophically, a fascinating special situation exists in which the second vector space, which receives the result of the operation, is an identical copy of the first vector space, from which the operation proceeds. However, in order to avoid confusion, this section does not delve into such issues.

For the first phase of the discussion, consider a linear operator that maps a vector space into itself, such as $L : V \rightarrow V$, where V is n-dimensional. Once a basis is chosen in V, L will have a unique matrix representation. Choose $\{v_1, v_2, \ldots, v_n\}$ and $\{\bar{v}_1, \bar{v}_2, \ldots, \bar{v}_n\}$ as two such bases. A matrix $M \in F^{n \times n}$ may be determined using the first basis, whereas another matrix $\bar{M} \in F^{n \times n}$ will result in the latter choice. According to the discussion following (1.15), the ith column of M is the representation of $L(v_i)$ with respect to $\{v_1, v_2, \ldots, v_n\}$, and the ith column of $\bar{M}$ is the representation of $L(\bar{v}_i)$ with respect to $\{\bar{v}_1, \bar{v}_2, \ldots, \bar{v}_n\}$. As in (1.4), any basis element v_i has a unique representation in terms of the basis $\{\bar{v}_1, \bar{v}_2, \ldots, \bar{v}_n\}$. Define a matrix $P \in F^{n \times n}$ using the ith column as this representation. Likewise, $Q \in F^{n \times n}$ may have as its ith column the unique representation of $\bar{v}_i$ with respect to $\{v_1, v_2, \ldots, v_n\}$. Either represents a basis change which is a linear operator. By construction, both P and Q are nonsingular. Such matrices and linear operators are sometimes called **basis transformations**. Notice that $P = Q^{-1}$.

If two matrices M and $\bar{M}$ represent the same linear operator L they must somehow carry essentially the same information. Indeed, a relationship between M and $\bar{M}$ may be established. Consider $a_v, a_w, \bar{a}_v, \bar{a}_w \in F^n$ such that $Ma_v = a_w$ and $\bar{M}\bar{a}_v = \bar{a}_w$. Here, a_v denotes the representation of v with respect to the basis v_i, $\bar{a}_v$ denotes the representation of v with respect to the basis $\bar{v}_i$, and so forth. In order to involve P and Q in these equations it is possible to make use of a sketch:

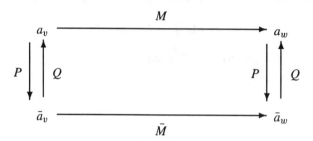

In this sketch a vector at a given corner can be multiplied by a matrix on an arrow leaving the corner and set equal to the vector that appears at the corner at which that arrow arrives. Thus, for example, $a_w = Ma_v$ may be deduced from the top edge of the sketch. It is interesting to perform "chases" around such sketches. By way of illustration, consider the lower right corner. Progress around the sketch counterclockwise so as to reach the lower left corner and set the result equal to that obtained by progressing clockwise to the lower left corner. In equations this is carried out as follows:

$$\bar{a}_w = Pa_w = PMa_v = PMQ\bar{a}_v = \bar{M}\bar{a}_v \tag{1.35}$$

Inasmuch as $\bar{a}_v \in F^n$ is arbitrary it follows that

$$\bar{M} = PMP^{-1} \tag{1.36}$$

Sketches that have this type of property, namely the same result when the sketch is traversed from a starting corner to an finishing corner by two paths, are said to be **commutative**. It is perhaps more traditional to show the vector space F^n instead of the vectors at the corners. Thus, the sketch would be called a **commutative diagram of vector spaces and linear operators**. M and $\bar{M}$ are said to be **similar** because a nonsingular matrix $P \in F^{n \times n}$ is such that (1.36) is true. The matrix P is then called a **similarity transformation**. Note that all matrix representations associated with the same linear operator, from a vector space to itself, are similar. Certain choices of bases lead to special forms for the matrices of the operator, as are apparent in the following examples.

Examples

- Choose $L : R^2 \to R^2$ as the linear operator that rotates a vector by $90°$. Pick $\{v_1, v_2\}$ as the natural basis $\{[1\ 0]^T, [0\ 1]^T\}$ and $\{\bar{v}_1, \bar{v}_2\}$ as $\{[1\ 1]^T, [1\ 0]^T\}$. Then,

$$L(v_1) = \begin{bmatrix} v_1 & v_2 \end{bmatrix} \begin{bmatrix} 0 \\ 1 \end{bmatrix}$$

$$L(v_2) = \begin{bmatrix} v_1 & v_2 \end{bmatrix} \begin{bmatrix} -1 \\ 0 \end{bmatrix}$$

$$L(\bar{v}_1) = \begin{bmatrix} \bar{v}_1 & \bar{v}_2 \end{bmatrix} \begin{bmatrix} 1 \\ -2 \end{bmatrix}$$

$$L(\bar{v}_2) = \begin{bmatrix} \bar{v}_1 & \bar{v}_2 \end{bmatrix} \begin{bmatrix} 1 \\ -1 \end{bmatrix}$$

so that

$$M = \begin{bmatrix} 0 & -1 \\ 1 & 0 \end{bmatrix} \qquad \bar{M} = \begin{bmatrix} 1 & 1 \\ -2 & -1 \end{bmatrix}$$

To find the similarity transformation P, determine the representations of the basis vectors $\mathbf{v}_i$ in terms of the basis $\{\bar{v}_1, \bar{v}_2\}$. Then

$$P = \begin{bmatrix} 0 & 1 \\ 1 & -1 \end{bmatrix} \qquad P^{-1} = \begin{bmatrix} 1 & 1 \\ 1 & 0 \end{bmatrix}$$

so that $PMP^{-1} = \bar{M}$.

- Suppose that $M \in F^{n \times n}$ is a representation of $L : V \to V$. Assume a $v \in F^n$ exists such that the vectors $\{v, Mv, \ldots, M^{n-1}v\}$ are linearly independent. Thus, these n vectors can be chosen as an alternate basis. (Section 1.7 discusses the characteristic equation of a matrix M.) Using the Cayley-Hamilton theorem, which states that every matrix satisfies its own characteristic equation, it is always possible to write $M^n v$ as a linear combination of these alternate basis vectors

$$-\alpha_n v - \alpha_{n-1} Mv - \cdots - \alpha_1 M^{n-1} v$$

for $\alpha_i \in F$. The matrix representation of L with respect to the alternate basis given by this set of linearly independent vectors is

$$\bar{M} = \begin{bmatrix} 0 & 0 & \cdots & 0 & -\alpha_n \\ 1 & 0 & \cdots & 0 & -\alpha_{n-1} \\ 0 & 1 & \cdots & 0 & -\alpha_{n-2} \\ \vdots & \vdots & \ddots & \vdots & \vdots \\ 0 & 0 & \cdots & 0 & -\alpha_2 \\ 0 & 0 & \cdots & 1 & -\alpha_1 \end{bmatrix}$$

which is the transpose of what is known as the **companion form**.

For the second phase of the discussion, select a pair of bases, one for the vector space V and one for the vector space W, and construct the resulting matrix representation M of $L : V \to W$. Another choice of bases exists for V and W, with the property that the resulting matrix $\bar{M}$ representing L is of the form

$$\begin{bmatrix} I & 0 \\ 0 & 0 \end{bmatrix} \tag{1.37}$$

where I is $\rho(M) \times \rho(M)$. Such a matrix is said to be in **normal form**. It is possible to transform M into $\bar{M}$ with the assistance of three types of transformations which are called **elementary**: (1) interchange any two rows or any two columns, (2) scale any row or column by a nonzero element in F, and (3) add any F-multiple of any row (column) to any other row (column). It is apparent that each of the three transformations involving rows may be accomplished by multiplying M on the left by a nonsingular matrix. Column operations require a corresponding multiplication on the right. The secret to understanding elementary transformations is to recognize that each of them will carry one basis into another. Not as easy to see, but equally true, is that any transformation that carries one basis into another basis must be a product of such elementary transformations. The elementary column transformations are interpreted as changing the basis in the vector space from which the operator takes vectors, whereas the elementary row transformations correspond to changing the basis in the vector space into which the operator places its result. It stands to reason that a simultaneous adjustment of both sets of basis vectors could lead to some quite special forms for the matrices of an operator. Of course, a great deal of linear algebra and its applications is concerned with just such constructions and forms. Space does not permit a complete treatment here.

If a matrix M has been transformed into normal form, certain types of key information become available. For example, one knows the rank of M because $\rho(M)$ is the number of rows and columns of the identity in (1.37). Perhaps more importantly, the normal form is easily factored in a fundamental way, and so such a construction is a natural means to construct two factors of minimal rank for a given matrix. The reader is cautioned, however, to be aware that computational linear algebra is quite a different subject than theoretical linear algebra. One common saying is that "if an algorithm is straightforward, then it is not numerically desirable." This may be an exaggeration, but it is well to recognize the implications of finite precision on the computer. Space limitations prevent addressing numerical issues.

Many other thoughts can be expressed in terms of elementary basis transformations. By way of illustration, elementary basis transformations offer an alternative in finding the inverse of a matrix. For a nonsingular matrix $M \in F^{n \times n}$, append to M an $n \times n$ identity I to form the $n \times 2n$ matrix

$$\hat{M} = \begin{bmatrix} M & I \end{bmatrix} \tag{1.38}$$

Perform elementary row transformations on (1.38) to transform M into normal form. Then M^{-1} will appear in the last n columns of the transformed matrix.

Examples

$$(F = R): \qquad M = \begin{bmatrix} 2 & 1 \\ 4 & 3 \end{bmatrix}$$

- Transform M into normal form. The process can be carried out in many ways. For instance, begin by scaling row 1 by $\frac{1}{2}$,

$$L_1 M = \begin{bmatrix} \frac{1}{2} & 0 \\ 0 & 1 \end{bmatrix} \begin{bmatrix} 2 & 1 \\ 4 & 3 \end{bmatrix} = \begin{bmatrix} 1 & \frac{1}{2} \\ 4 & 3 \end{bmatrix}$$

Clear the first element in row 2 by

$$L_2 L_1 M = \begin{bmatrix} 1 & 0 \\ -4 & 1 \end{bmatrix} \begin{bmatrix} 1 & \frac{1}{2} \\ 4 & 3 \end{bmatrix} = \begin{bmatrix} 1 & \frac{1}{2} \\ 0 & 1 \end{bmatrix}$$

Finally, perform a column operation to produce $\bar{M}$:

$$L_2 L_1 M R_1 = \begin{bmatrix} 1 & \frac{1}{2} \\ 0 & 1 \end{bmatrix} \begin{bmatrix} 1 & -\frac{1}{2} \\ 0 & 1 \end{bmatrix} = \begin{bmatrix} 1 & 0 \\ 0 & 1 \end{bmatrix}$$

The rank of M is 2.

- Recall M^{-1} from previous examples. Form $\hat{M}$ and transform M into normal form by the following row operations:

$$L_1 \hat{M} = \begin{bmatrix} \frac{1}{2} & 0 \\ 0 & 1 \end{bmatrix} \begin{bmatrix} 2 & 1 & 1 & 0 \\ 4 & 3 & 0 & 1 \end{bmatrix} = \begin{bmatrix} 1 & \frac{1}{2} & \frac{1}{2} & 0 \\ 4 & 3 & 0 & 1 \end{bmatrix}$$

$$L_2 L_1 \hat{M} = \begin{bmatrix} 1 & 0 \\ -4 & 1 \end{bmatrix} \begin{bmatrix} 1 & \frac{1}{2} & \frac{1}{2} & 0 \\ 4 & 3 & 0 & 1 \end{bmatrix} = \begin{bmatrix} 1 & \frac{1}{2} & \frac{1}{2} & 0 \\ 0 & 1 & -2 & 1 \end{bmatrix}$$

$$L_3 L_2 L_1 \hat{M} = \begin{bmatrix} 1 & -\frac{1}{2} \\ 0 & 1 \end{bmatrix} \begin{bmatrix} 1 & \frac{1}{2} & \frac{1}{2} & 0 \\ 0 & 1 & -2 & 1 \end{bmatrix} = \begin{bmatrix} 1 & 0 & \frac{3}{2} & -\frac{1}{2} \\ 0 & 1 & -2 & 1 \end{bmatrix}$$

1.7 Characteristics: Eigenvalues, Eigenvectors, and Singular Values

A matrix has certain characteristics associated with it. Of these, characteristic values or eigenvalues may be determined through the use of matrix pencils. In general a matrix pencil may be formed from two matrices M and $N \in F^{m \times n}$ and an indeterminate λ in the manner

$$\begin{bmatrix} \lambda N - M \end{bmatrix} \in F[\lambda]^{m \times n} \tag{1.39}$$

In determining eigenvalues of a square matrix $M \in F^{n \times n}$ one assumes the special case in which $N = I \in F^{n \times n}$.

Assume that M is a square matrix over the complex numbers. Then, $\lambda \in C$ is called an **eigenvalue** of M if some nonzero vector $v \in C^n$ exists such that

$$Mv = \lambda v \tag{1.40}$$

Any such $v \neq 0$ satisfying (1.40) is said to be an **eigenvector** of M associated with λ. It is easy to see that (1.40) can be rewritten as

$$(\lambda I - M)v = 0 \tag{1.41}$$

Because (1.41) is a set of n linear homogeneous equations, a nontrivial solution ($v \neq 0$) exists if and only if

$$\Delta(\lambda) = \det(\lambda I - M) = 0 \tag{1.42}$$

In other words, $(\lambda I - M)$ is singular. Therefore, λ is an eigenvalue of M if and only if it is a solution of (1.42). The polynomial $\Delta(\lambda)$ is the **characteristic polynomial** and $\Delta(\lambda) = 0$ is the **characteristic equation**. Moreover, every $n \times n$ matrix has n eigenvalues that may be real, complex or both, where complex eigenvalues occur in complex-conjugate pairs. If two or more eigenvalues are equal they are said to be repeated (not distinct). It is interesting to observe that although eigenvalues are unique, eigenvectors are not. Indeed, an eigenvector can be multiplied by any nonzero element of C and still maintain its essential features. Sometimes this lack of uniqueness is resolved by selecting unit length for the eigenvectors with the aid of a suitable norm.

Recall that matrices representing the same operator are similar. One may question if these matrices indeed contain the same characteristic information. To answer this question, examine

$$\det(\lambda I - \bar{M}) = \det(\lambda P P^{-1} - P M P^{-1}) = \det(P(\lambda I - M) P^{-1}) \tag{1.43}$$

$$= \det(P) \det(\lambda I - M) \det(P^{-1}) = \det(\lambda I - M) \tag{1.44}$$

From (1.44) one may deduce that similar matrices have the same eigenvalues because their characteristic polynomials are equal.

For every square matrix M with distinct eigenvalues, a similar matrix $\bar{M}$ is diagonal. In particular, the eigenvalues of M, and hence $\bar{M}$, appear along the main diagonal. Let $\lambda_1, \lambda_2, \ldots, \lambda_n$ be the eigenvalues (all distinct) of M and let $v_1, v_2, \ldots, v_n$ be corresponding eigenvectors. Then, the vectors $\{v_1, v_2, \ldots, v_n\}$ are linearly independent over C. Choose $P^{-1} = Q = [v_1 \, v_2 \cdots v_n]$ as the **modal matrix**. Because $Mv_i = \lambda_i v_i$, $\bar{M} = PMP^{-1}$ as before.

For matrices with repeated eigenvalues, a similar approach may be followed wherein $\bar{M}$ is block diagonal, which means that matrices occur along the diagonal with zeros everywhere else. Each

matrix along the diagonal is associated with an eigenvalue and takes a specific form depending upon
the characteristics of the matrix itself. The modal matrix consists of generalized eigenvectors, of
which the aforementioned eigenvector is a special case; thus the modal matrix is nonsingular. The
matrix $\bar{M}$ is the **Jordan canonical form**. Space limitations preclude a detailed analysis of such
topics here; the reader is directed to Chen (1984) for further development.

Examples

$$(F = C): \qquad M = \begin{bmatrix} 1 & 4 \\ 2 & 3 \end{bmatrix}$$

- The characteristic polynomial is $\Delta(\lambda) = (\lambda - 1)(\lambda - 3) - 8 = (\lambda - 5)(\lambda + 1)$. The
  eigenvalues are $\lambda_1 = 5, \lambda_2 = -1$. To find the associated eigenvectors recall that for each
  λ_i, $(\lambda_i I - M)$ is singular, and write (1.41)

$$(\lambda_1 I - M)v_1 = \begin{bmatrix} 4 & -4 \\ -2 & 2 \end{bmatrix} \begin{bmatrix} v_{11} \\ v_{12} \end{bmatrix} = \begin{bmatrix} 0 \\ 0 \end{bmatrix}$$

$$(\lambda_2 I - M)v_2 = \begin{bmatrix} -2 & -4 \\ -2 & -4 \end{bmatrix} \begin{bmatrix} v_{21} \\ v_{22} \end{bmatrix} = \begin{bmatrix} 0 \\ 0 \end{bmatrix}$$

Then, $v_{11} = v_{12}$ and $v_{21} = -2v_{22}$ so that $v_1 = [1 \ 1]^T$ and $v_2 = [-2 \ 1]^T$ are eigenvectors associated with λ_1 and λ_2, respectively.

- Because the eigenvalues of M are distinct, M may be diagonalized. For verification,
  choose $P^{-1} = [v_1 \ v_2]$. Then

$$\bar{M} = PMP^{-1} = \frac{1}{3} \begin{bmatrix} 1 & 2 \\ -1 & 1 \end{bmatrix} \begin{bmatrix} 1 & 4 \\ 2 & 3 \end{bmatrix} \begin{bmatrix} 1 & -2 \\ 1 & 1 \end{bmatrix}$$

$$= \frac{1}{3} \begin{bmatrix} 15 & 0 \\ 0 & -3 \end{bmatrix} = \begin{bmatrix} 5 & 0 \\ 0 & -1 \end{bmatrix} = \begin{bmatrix} \lambda_1 & 0 \\ 0 & \lambda_2 \end{bmatrix}$$

In general a matrix $M \in F^{m \times n}$ of rank r can be written in terms of its **singular-value decomposition** (SVD),

$$M = U \Sigma V^* \tag{1.45}$$

For any M, **unitary** matrices U and V of dimension $m \times m$ and $n \times n$, respectively, form the
decomposition; that is, $UU^* = U^*U = I$ and $VV^* = V^*V = I$. The matrix $\Sigma \in F^{m \times n}$ is of the
form

$$\begin{bmatrix} \Sigma_r & 0 \\ 0 & 0 \end{bmatrix} \tag{1.46}$$

for $\Sigma_r \in F^{r \times r}$, a diagonal matrix represented by

$$\begin{bmatrix} \sigma_1 & 0 & \cdots & 0 \\ 0 & \sigma_2 & \cdots & 0 \\ \vdots & \vdots & \ddots & \vdots \\ 0 & 0 & \cdots & \sigma_r \end{bmatrix} \tag{1.47}$$

The elements σ_i, called **singular values**, are related by $\sigma_1 \geq \sigma_2 \geq \cdots \geq \sigma_r > 0$, and the columns
of U (V) are referred to as **left (right) singular vectors**. Although the unitary matrices U and V are
not unique for a given M, the singular values are unique.

Singular-value decomposition is useful in the numerical calculation of rank. After performing a SVD, the size of the matrix Σ_r may be decided. Additionally, the **generalized inverse** of a matrix M may be found by

$$M^\dagger = V \begin{bmatrix} \Sigma_r^{-1} & 0 \\ 0 & 0 \end{bmatrix} U^* \tag{1.48}$$

It can be verified easily that $MM^\dagger M = M$ and $M^\dagger M M^\dagger = M^\dagger$. In the special case in which M is square and nonsingular,

$$M^\dagger = M^{-1} = V\Sigma^{-1}U^* \tag{1.49}$$

1.8 On Linear Systems

Consider a set of n simultaneous algebraic equations, in m unknowns, written in the customary matrix form $w = Mv$ where

$$\begin{bmatrix} w_1 \\ w_2 \\ \vdots \\ w_n \end{bmatrix} = \begin{bmatrix} m_{11} & m_{12} & \cdots & m_{1m} \\ m_{21} & m_{22} & \cdots & m_{2m} \\ \vdots & \vdots & \ddots & \vdots \\ m_{n1} & m_{n2} & \cdots & m_{nm} \end{bmatrix} \begin{bmatrix} v_1 \\ v_2 \\ \vdots \\ v_m \end{bmatrix} \tag{1.50}$$

In the context of the foregoing discussion (1.50) represents the action of a linear operator. If the left member is a given vector, in the usual manner, then a first basic issue concerns whether the vector represented by the left member is in the image of the operator or not. If it is in the image, the equation has at least one solution; otherwise the equation has no solution. A second basic issue concerns the kernel of the operator. If the kernel contains only the zero vector, then the equation has at most one solution; otherwise more than one solution can occur, provided that at least one solution exists.

When one thinks of a set of simultaneous equations as a "system" of equations, the intuitive transition to the idea of a **linear system** is quite natural. In this case the vector in the left member becomes the **input** to the system, and the solution to (1.50), when it exists and is unique, is the **output** of the system.

Other than being a description in terms of inputs and outputs, as above, linear systems may also be described in terms of sets of other types of equations, such as differential equations or difference equations. When that is the situation, the familiar notion of initial condition becomes an instance of the idea of **state**, and one must examine the intertwining of states and inputs to give outputs. Then, the idea of (1.50), when each input yields a unique output, is said to define a **system function**.

If the differential (difference) equations are linear and have constant coefficients, the possibility exists of describing the system in terms of transforms, for example, in the s-domain or z-domain. This leads to fascinating new interpretations of the ideas of the foregoing sections, this time, for example, over fields of rational functions. Colloquially, such functions are best known as **transfer functions**.

Associated with systems described in the time domain, s-domain, or z-domain some characteristics of the system also aid in analysis techniques. Among the most basic of these are the entities termed **poles** and **zeros**, which have been linked to the various concepts of system **stability**. Both poles and zeros may be associated with matrices of transfer functions, and with the original differential or difference equations themselves. A complete and in-depth treatment of the myriad meanings of poles and zeros is a challenging undertaking, particularly in matrix cases. For a recent survey of the ideas, see Schrader and Sain [3]. However, a great many of the definitions involve such concepts as rank, pencils, eigenvalues, eigenvectors, special matrix forms, vector spaces, and modules—the very ideas sketched out in the sections preceding.

One very commonly known idea for representing solutions to (1.50) is **Cramer's rule**. When $m = n$, and when M has an inverse, the use of Cramer's rule expresses each unknown variable individually by using a ratio of determinants. Choose the ith unknown v_i. Define the determinant M_i as the determinant of a matrix formed by replacing column i in M with w. Then,

$$v_i = \frac{M_i}{\det(M)} \tag{1.51}$$

It turns out that this very interesting idea makes fundamental use of the notion of *multiplying* vectors, which is not part of the axiomatic framework of the vector space. The reader may want to reflect further on this observation, in respect to the foregoing treatment of determinants. When the framework of vector spaces is expanded to include vector multiplication, as in the case of determinants, one gets to the technical subject of **algebras**. The next chapter returns to this concept.

The concepts presented above allow for more detailed considerations in the solution of circuit and filter problems, using various approaches outlined in the remainder of this text. The following chapter provides for the multiplication of vectors by means of the foundational idea of bilinear operators and matrices. The next chapters on transforms—Fourier, z, and Laplace—provide the tools for analysis by allowing a set of differential or difference equations describing a circuit to be written as a system of linear algebraic equations. Moreover, each transform itself can be viewed as a linear operator, and thus becomes a prime example of the ideas of this chapter. The remaining chapters focus on graph-theoretical approaches to the solution of systems of algebraic equations.

A brief treatment cannot deal with all the interesting questions and answers associated with linear operators and matrices. For a more detailed treatment of these standard concepts, see any basic algebra text, for example, Greub [2].

References

[1] Chen, C.-T., *Linear System Theory and Design,* New York, CBS College Publishing, 1984.

[2] Greub, W. H., *Linear Algebra,* New York, Springer-Verlag, 1967.

[3] Schrader, C.B. and Sain, M.K., "Research on system zeros: a survey," *Int. J. Control,* 50(4), 1407–1433, October 1989.

2

Bilinear Operators and Matrices

Michael K. Sain
University of Notre Dame

Cheryl B. Schrader
The University of Texas

2.1 Introduction

The key player in Chapter 1 was the F-vector space V, together with its associated notion of bases, when they exist, and linear operators taking one vector to another. The idea of basis is, of course, quite central to the applications because it permits a vector v in V to be represented by a list of scalars from F. Such lists of scalars are the quantities with which one computes. No doubt the idea of an F-vector space V is the most common and widely encountered notion in applied linear algebra. It is typically visualized, on the one hand, by long lists of axioms, most of which seem quite reasonable, but none of which is particularly exciting, and on the other hand by images of classical addition of force vectors, velocity vectors, and so forth. The notion seems to do no harm, and helps one to keep his or her assumptions straight. As such it is accepted by most engineers as a plausible background for their work, even if the ideas of matrix algebra are more immediately useful. Perhaps some of the least appreciated but most crucial of the vector space axioms are the four governing the scalar multiplication of vectors. These link the abelian group of vectors to the field of scalars. Along with the familiar distributive covenants, these four agreements intertwine the vectors with the scalars in much the same way that the marriage vows bring about the union of man and woman. This section brings forth a new addition to the marriage.

As useful as it is, the notion of an F-vector space V fails to provide for one of the most important ideas in the applications—the concept of multiplication of vectors. In a vector space one can add vectors and multiply vectors by scalars, but one cannot multiply vectors by vectors. Yet there are numerous situations in which one faces exactly these operations. Consider, for instance, the cross and dot products from field theory. Even in the case of matrices, the ubiquitous and crucial matrix multiplication is available, when it is defined. The key to the missing element in the discussion lies in the terminology for matrix operations, which will be familiar to the reader as the **matrix algebra**.

What must occur in order for vector-to-vector multiplication to be available is for the vector space to be extended into an algebra.

Unfortunately, the word "algebra" carries a rather imprecise meaning from the most elementary and early exposures from which it came to signify the collection of operations done in arithmetic, at the time when the operations are generalized to include symbols or literals such as a, b, and c or x, y, and z. Such a notion generally corresponds closely with the idea of a field, F, as defined in Chapter 1, and is not much off the target for an environment of scalars. It may, however, come as a bit of a surprise to the reader that algebra is a technical term, in the same spirit as fields, vector spaces, rings, etc. Therefore, if one is to have available a notion of multiplication of vectors, then it is appropriate to introduce the precise notion of an **algebra**, which captures the desired idea in an axiomatic sense.

2.2   Algebras

Chapter 1 mentioned that the integers I and the polynomials $F[s]$ in s with coefficients in a field F were instances of a ring. In this section it is necessary to carry the concept of ring beyond the example stage. Of course, a long list of axioms could be provided, but it may be more direct just to cite the changes necessary to the field axioms already provided in Section 1.2. To be precise, the axioms of a **commutative ring** differ from the axioms of a field only by the removal of the multiplicative inverse. Intuitively, this means that one cannot always divide, even if the element in question is nonzero. Many important commutative rings are found in the applications; however, this chapter is centered on **rings**, wherein one more axiom is removed—the commutativity of multiplication. The ring of $n \times n$ matrices with elements from a field is a classic and familiar example of such a definition. It may be remarked that in some references a distinction is made between **rings** and **rings with identity**, the latter having a multiplicative identity and the former not being so equipped. This treatment has no need for such a distinction, and hereafter the term "ring" is understood to mean ring with identity, or, as described above, a field with the specified two axioms removed.

It is probably true that the field is the most comfortable of axiomatic systems for most persons because it corresponds to the earliest and most persistent of calculation notions. However, it is also true that the ring has an intuitive and immediate understanding as well, which can be expressed in terms of the well-known phrase "playing with one arm behind one's back". Indeed, each time an axiom is removed, it is similar to removing one of the options in a game. This adds to the challenge of a game, and leads to all sorts of new strategies. Such is the case for **algebras**, as is clear from the next definition. What follows is not the most general of possible definitions, but probably that which is most common.

An **algebra** A is an F-vector space A which is equipped with a multiplication $a_1 a_2$ of vectors a_1 and a_2 in such a manner that it is also a ring. First, addition in the ring is simply addition of vectors in the vector space. Second, a special relationship exists between multiplication of vectors and scalar multiplication in the vector space. If a_1 and a_2 are vectors in A, and if f is a scalar in F, then the following identity holds:

$$f(a_1 a_2) = (f a_1) a_2 = a_1 (f a_2) \qquad (2.1)$$

Note that the order of a_1 and a_2 does not change in the above equalities. This must be true because no axiom of commutativity exists for multiplication. The urge to define a symbol for vector multiplication is resisted here so as to keep things as simple as possible. In the same way the notation for scalar multiplication, as introduced in Chapter 1, is suppressed here in the interest of simplicity. Thus, the scalar multiplication can be associated either with the vector product, which lies in A, or with one or other of the vector factors. This is exactly the familiar situation with the matrix algebra.

Hidden in the definition of the **algebra** A above is the precise detail arising from the statement that A is a ring. Associated with that detail is the nature of the vector multiplication represented above with the juxtaposition a_1a_2. Because all readers are familiar with several notions of vector multiplication, the question arises as to just what constitutes such a multiplication. It turns out that a precise notion for multiplication can be found in the idea of a bilinear operator. Thus, an alternative description of Section 2.3 is that of vector spaces equipped with vector multiplication. Moreover, one is tempted to inquire whether a vector multiplication exists that is so general in nature that all other vector multiplications can be derived from it. In fact, this is the case, and the following section sets the stage for introducing such a multiplication.

2.3 Bilinear Operators

Suppose that there are three F-vector spaces: U, V, and W. Recall that $U \times V$ is the Cartesian product of U with V, and denotes the set of all ordered pairs, the first from U and the second from V. Now, consider a mapping b from $U \times V$ into W. For brevity of notation, this can be written b: $U \times V \to W$. The mapping b is a **bilinear operator** if it satisfies the pair of conditions

$$b(f_1u_1 + f_2u_2, v) = f_1b(u_1, v) + f_2b(u_2, v) \tag{2.2}$$

$$b(u, f_1v_1 + f_2v_2) = f_1b(u, v_1) + f_2b(u, v_2) \tag{2.3}$$

for all f_1 and f_2 in F, for all u, u_1, and u_2 in U, and for all v, v_1, and v_2 in V. The basic idea of the bilinear operator is apparent from this definition. It is an operator with two arguments, having the property that if either of the two arguments is fixed, the operator becomes linear in the remaining argument. A moment's reflection will show that the intuitive operation of multiplication is of this type.

One of the important features of a bilinear operator is that its image need not be a subspace of W. This is in marked contrast with the image of a linear operator, whose image is always a subspace. This property leads to great interest in the manipulations associated with vector products. At the same time, it brings about a great deal of nontriviality. The best way to illustrate the point is with an example.

Example. Suppose that U, V, and W have bases $\{u_1, u_2\}$, $\{v_1, v_2\}$, and $\{w_1, w_2, w_3, w_4\}$, respectively. Then, vectors u in U and v in V can be represented in the manner

$$u = f_1u_1 + f_2u_2 \tag{2.4}$$

$$v = g_1v_1 + g_2v_2 \tag{2.5}$$

where f_i and g_i are elements of F for $i = 1, 2$. Define a bilinear map by the action

$$b(u, v) = 2f_1g_1w_1 + 3f_1g_2w_2 + 3f_2g_1w_3 + 2f_2g_2w_4 \tag{2.6}$$

It is clear that every vector

$$h_1w_1 + h_2w_2 + h_3w_3 + h_4w_4 \tag{2.7}$$

in the image of b has the property that $9h_1h_4 = 4h_2h_3$. If the $\{h_i, i = 1, 2, 3, 4\}$ are given so as to satisfy the latter condition, consider the task of showing that this vector in W is a vector in the image of b. Suppose that $h_1 = 0$. Then either h_2 or h_3 is zero, or both are zero. If h_2 is zero, one may choose $f_1 = 0$, $f_2 = 1$, $g_1 = h_3/3$, and $g_2 = h_4/2$. If $h_3 = 0$, one may choose $g_1 = 0$, $g_2 = 1$, $f_1 = h_2/3$, and $f_2 = h_4/2$. An analogous set of constructions is available when $h_4 = 0$. For the remainder of the argument, it is assumed that neither h_1 nor h_2 is zero. Accordingly, none of the coordinates $\{h_i, i = 1, 2, 3, 4\}$ is zero. Without loss, assume that $f_1 = 1$. Then, g_1 is given by $h_1/2$, g_2 is found from $h_2/3$, and f_2 is constructed from $h_3/3g_1$, which is then $2h_3/3h_1$. It is easy to check that these choices produce the correct first three coordinates; the last coordinate is $4h_3h_2/9h_1$, which by virtue of the property $9h_1h_4 = 4h_2h_3$ is equal to h_4 as desired. Thus, a vector in W is in the image of b if and only if the relation $9h_1h_4 = 4h_2h_3$ is satisfied. Next, it is shown that the vectors in this class are not closed under addition. For this purpose, simply select a pair of vectors represented by $(1, 1, 9, 4)$ and $(4, 9, 1, 1)$. The sum, $(5, 10, 10, 5)$, does not satisfy the condition.

It is perhaps not so surprising that the image of b in this example is not a subspace of W. After all, the operator b is nonlinear, when both of its arguments are considered. What may be surprising is that a natural and classical way can be used to circumvent this difficulty, at least to a remarkable degree. The mechanism that is introduced in order to address such a question is the **tensor**. The reader should bear in mind that many technical personnel have prior notions and insights on this subject emanating from areas such as the theory of mechanics and related bodies of knowledge. For these persons, the authors wish to emphasize that the following treatment is **algebraic** in character and may exhibit, at least initially, a flavor different from that to which they may be accustomed. This difference is quite typical of the distinctive points of view that often can be found between the mathematical areas of algebra and analysis. Such differences are fortunate insofar as they promote progress in understanding.

2.4   Tensor Product

The notions of tensors and tensor product, as presented in this treatment, have the intuitive meaning of a very general sort of bilinear operator, in fact, the *most* general such operator. Once again, F-vector spaces U, V, and W are assumed. Suppose that $b\colon U \times V \to W$ is a bilinear operator. Then the pair (b, W) is said to be a **tensor product** of U and V if two conditions are met. The first condition is that W is the smallest F-vector space that contains the image of b. Using alternative terminology this could be expressed as W being the **vector space generated by the image of b**. The term **generated** in this expression refers to the formation of all possible linear combinations of elements in the image of b. The second condition relates b to an arbitrary bilinear operator $\breve{b}\colon U \times V \to X$, in which X is another F-vector space. To be precise, the second condition states that for every such $\breve{b}$, a **linear** operator $\breve{B}\colon W \to X$ exists with the property that

$$\breve{b}(u, v) = \breve{B}(b(u, v)) \tag{2.8}$$

for all pairs (u, v) in $U \times V$. Intuitively, this means that the arbitrary bilinear operator $\breve{b}$ can be factored in terms of the given bilinear operator b, which does not depend upon $\breve{b}$, and a linear operator $\breve{B}$ which does depend upon $\breve{b}$.

The idea of the tensor product is truly remarkable. Moreover, for any bilinear operator $\breve{b}$, the induced linear operator $\breve{B}$ is unique. The latter result is easy to see. Suppose that there are two such induced linear operators, e.g., $\breve{B}_1$ and $\breve{B}_2$. It follows immediately that

$$(\breve{B}_1 - \breve{B}_2)(b(u, v)) = 0 \tag{2.9}$$

for all pairs (u, v). However, the first condition of the tensor product assures that the image of b contains a set of generators for W, and thus that $(\breve{B}_1 - \breve{B}_2)$ must in fact be the zero operator. Therefore, once the tensor product of U and V is put into place, bilinear operations are in a one-to-one correspondence with linear operations. This is the essence of the tensor idea, and a very significant way to parameterize product operations in terms of matrices. In a certain sense, then, the idea of Chapter 2 is to relate the fundamentally nonlinear product operation to the linear ideas of Chapter 1. That this is possible is, of course, classical; nonetheless, it remains a relatively novel idea for numerous workers in the applications. Intuitively what happens here is that the idea of *product* is abstracted in the bilinear operator b, with all the remaining details placed in the realm of the induced linear operator $\breve{B}$.

When a pair (b, W) satisfies the two conditions above, and is therefore a tensor product for U and V, it is customary to replace the symbol b with the more traditional symbol $\otimes$. However, in keeping with the notion that $\otimes$ represents a product and not just a general mapping it is common to write $u \otimes v$ in place of the more correct, but also more cumbersome, $\otimes(u, v)$. Along the same lines, the space W is generally denoted $U \otimes V$. Thus, a tensor product is a pair $(U \otimes V, \otimes)$. The former is called the **tensor product of U with V**, and $\otimes$ is loosely termed the **tensor product**. Clearly, $\otimes$ is the most general sort of product possible in the present situation because all other products can be expressed in terms of it by means of linear operators $\check{B}$. Once again, the colloquial use of the word "product" is to be identified with the more precise algebraic notion of bilinear operation. In this way the tensor product becomes a sort of "grandfather" for all vector products.

Tensor products can be constructed for arbitrary vector spaces. They are not, however, unique. For instance, if $U \otimes V$ has finite dimension, then W obviously can be replaced by any other F-vector space of the same dimension, and $\otimes$ can be adjusted by a vector space isomorphism. Here, the term **isomorphism** denotes an invertible linear operator between the two spaces in question. It can also be said that the two tensor product spaces $U \otimes V$ and W are isomorphic to each other. Whatever the terminology chosen, the basic idea is that the two spaces are essentially the same within the axiomatic framework in use.

2.5 Basis Tensors

Attention is now focused on the case in which U and V are finite-dimensional vector spaces over the field F. Suppose that $\{u_1, u_2, \ldots, u_m\}$ is a basis for U and $\{v_1, v_2, \ldots, v_n\}$ is a basis for V. Consider the vectors

$$u_1 \otimes v_1 \quad u_1 \otimes v_2, \ldots, u_1 \otimes v_n \quad u_2 \otimes v_1, \ldots, u_m \otimes v_n \tag{2.10}$$

which can be represented in the manner $\{u_i \otimes v_j, \ i = 1, 2, \ldots, m; \ j = 1, 2, \ldots, n\}$. These vectors form a basis for the vector space $U \otimes V$. To understand the motivation for this, note that vectors in U and V, respectively, can be written uniquely in the forms

$$u = \sum_{i=1}^{m} f_i u_i \tag{2.11}$$

$$v = \sum_{j=1}^{n} g_j v_j \tag{2.12}$$

Recall that $\otimes$, which is an alternate notation for b, is a bilinear operator. It follows then that

$$u \otimes v = \sum_{i=1}^{m} \sum_{j=1}^{n} f_i g_j u_i \otimes v_j \tag{2.13}$$

which establishes that the proposed basis vectors certainly span the image of $\otimes$, and thus that they span the tensor product space $U \otimes V$. It also can be shown that the proposed set of basis vectors is linearly independent. However, in the interest of brevity for this summary exposition, the details are omitted.

From this point onward, inasmuch as the symbol $\otimes$ has replaced b, it will be convenient to use b in place of $\check{b}$ and B in place of $\check{B}$. It is hoped that this leads to negligible confusion. Thus, in the sequel b refers simply to a bilinear operator and B to its induced linear counterpart.

Example. Consider the bilinear form $b: R^2 \times R^3 \to R$ with action defined by

$$b(f_1, f_2, g_1, g_2, g_3) = 2f_2 g_3 \tag{2.14}$$

Observe that this can be put into the more transparent form

$$\begin{bmatrix} f_1 & f_2 \end{bmatrix} \begin{bmatrix} 0 & 0 & 0 \\ 0 & 0 & 2 \end{bmatrix} \begin{bmatrix} g_1 \\ g_2 \\ g_3 \end{bmatrix} \tag{2.15}$$

which, in turn, can be written in the compact notation $u^T M v$. Clearly, U has dimension two, and V has dimension three. Thus, $U \otimes V$ has a basis with six elements. The operator b maps into R, which has a basis with one element. All bases are chosen to be standard. Thus, an ith basis vector contains the multiplicative field element 1 in its ith row, and the additive field element 0 in its other rows. Therefore, the matrix of B has one row and six columns. To compute the entries, it is necessary to agree upon the order of the basis elements in $R^2 \otimes R^3$. It is customary to choose the natural ordering as introduced above:

$$u_1 \otimes v_1 \quad u_1 \otimes v_2 \quad u_1 \otimes v_3 \quad u_2 \otimes v_1 \quad u_2 \otimes v_2 \quad u_2 \otimes v_3 \tag{2.16}$$

The coordinate h_1 associated with the basis vector [1] in R, considered to be a vector space, is given by

$$h_1 = b(1, 0, 1, 0, 0) = 0 \tag{2.17}$$

when u and v are given by the respective first basis vectors in R^2 and R^3, respectively:

$$u = \begin{bmatrix} 1 \\ 0 \end{bmatrix} \tag{2.18}$$

$$v = \begin{bmatrix} 1 \\ 0 \\ 0 \end{bmatrix} \tag{2.19}$$

Similarly, for the other five pairings in order, one obtains

$$h_1 = b(1, 0, 0, 1, 0) = 0 \tag{2.20}$$
$$h_1 = b(1, 0, 0, 0, 1) = 0 \tag{2.21}$$
$$h_1 = b(0, 1, 1, 0, 0) = 0 \tag{2.22}$$
$$h_1 = b(0, 1, 0, 1, 0) = 0 \tag{2.23}$$
$$h_1 = b(0, 1, 0, 0, 1) = 2 \tag{2.24}$$

in order. In view of these calculations, together with the definitions of matrices in Chapter 1, it follows that the matrix description of B: $R^2 \otimes R^3 \to R$ is given by

$$[B] = \begin{bmatrix} 0 & 0 & 0 & 0 & 0 & 2 \end{bmatrix} \tag{2.25}$$

Observe that all the numerical information concerning B has been arrayed in $[B]$. It becomes increasingly clear then that such numerical entries define all possible bilinear forms of this type.

Example. In order to generalize the preceding example, one has only to be more general in describing the matrix of M. Suppose that

$$[M] = \begin{bmatrix} m_{11} & m_{12} & m_{13} \\ m_{21} & m_{22} & m_{23} \end{bmatrix} \tag{2.26}$$

so that the bilinear operator b has action

$$b(f_1, f_2, g_1, g_2, g_3) = \begin{bmatrix} f_1 & f_2 \end{bmatrix} \begin{bmatrix} m_{11} & m_{12} & m_{13} \\ m_{21} & m_{22} & m_{23} \end{bmatrix} \begin{bmatrix} g_1 \\ g_2 \\ g_3 \end{bmatrix} \tag{2.27}$$

Thus, it is easy to determine that

$$[B] = \begin{bmatrix} m_{11} & m_{12} & m_{13} & m_{21} & m_{22} & m_{23} \end{bmatrix} \tag{2.28}$$

The two examples preceding help in visualizing the linear operator B by means of its matrix. They do not, however, contribute to the understanding of the nature of the tensor product of two vectors. For that purpose, it is appropriate to carry the examples a bit further.

Example. The foregoing example presents the representations

$$\begin{bmatrix} f_1 \\ f_2 \end{bmatrix} \tag{2.29}$$

for **u** and

$$\begin{bmatrix} g_1 \\ g_2 \\ g_3 \end{bmatrix} \tag{2.30}$$

for v. From the development of the ideas of the tensor product, it was established that $b(u, v) = B(u \otimes v)$. The construction of $u \otimes v$ proceeds according to definition in the manner

$$u \otimes v = \left(\sum_{i=1}^{2} f_i u_i \right) \otimes \left(\sum_{j=1}^{3} g_j v_j \right) \tag{2.31}$$

$$= \sum_{i=1}^{2} \sum_{j=1}^{3} f_i g_j u_i v_j \tag{2.32}$$

From this and the basis ordering chosen above, it is clear that the representation of $u \otimes v$ is given by

$$[u \otimes v] = \begin{bmatrix} f_1 g_1 & f_1 g_2 & f_1 g_3 & f_2 g_1 & f_2 g_2 & f_2 g_3 \end{bmatrix}^{\mathrm{T}} \tag{2.33}$$

The total picture for the tensor representation of $b(u, v)$, then, is

$$b(f_1, f_2, g_1, g_2, g_3)$$

$$= \begin{bmatrix} f_1 \\ f_2 \end{bmatrix}^{\mathrm{T}} \begin{bmatrix} m_{11} & m_{12} & m_{13} \\ m_{21} & m_{22} & m_{23} \end{bmatrix} \begin{bmatrix} g_1 \\ g_2 \\ g_3 \end{bmatrix} \tag{2.34}$$

$$= \begin{bmatrix} m_{11} & m_{12} & m_{13} & m_{21} & m_{22} & m_{23} \end{bmatrix} \left(\begin{bmatrix} f_1 \\ f_2 \end{bmatrix} \otimes \begin{bmatrix} g_1 \\ g_2 \\ g_3 \end{bmatrix} \right) \tag{2.35}$$

$$= \begin{bmatrix} m_{11} & m_{12} & m_{13} & m_{21} & m_{22} & m_{23} \end{bmatrix} \begin{bmatrix} f_1 g_1 \\ f_1 g_2 \\ f_1 g_3 \\ f_2 g_1 \\ f_2 g_2 \\ f_2 g_3 \end{bmatrix} \tag{2.36}$$

The reader should have no difficulty extending the notions of these examples to cases in which the dimensions of U and V differ from those used here. The extension to an X with dimension larger than 1 is similar in nature, and can be carried out row by row.

Example. Another sort of example, which is likely to be familiar to most readers, is the formation of the ordinary matrix product $m(P, Q) = PQ$ for compatible matrices P and Q over the field F. Clearly, the matrix product m is a bilinear operator. Thus, a linear operator M exists that has the property

$$m(P, Q) = M(P \otimes Q) \tag{2.37}$$

The matrix $P \otimes Q$ is known in the applications as the **Kronecker product**. If the basis vectors are chosen in the usual way, then its computation has the classical form. Thus, the Kronecker product of two matrices is seen to be the most general of all such products. Indeed, any other product, including the usual matrix product, can be found from the Kronecker product by multiplication with a matrix.

2.6 Multiple Products

It may happen that more than two vectors are multiplied together. Thus, certain famous and well-known field formulas include both crosses and dots. While the notion of multiple product is part and parcel of the concept of ring, so that no further adjustments need be made there, one must undertake the question of how these multiple products are reflected back into the tensor concept. The purpose of this section, therefore, is to sketch the major ideas concerning such questions. A basic and natural step is the introduction of a generalization of bilinear operators.

For obvious reasons, not the least of which is the finite number of characters in the alphabet, it is now necessary to modify notation so as to avoid the proliferation of symbols. With regard to the foregoing discussion, the modification, which is straightforward, is to regard a bilinear operator in the manner $b\colon U_1 \times U_2 \to V$ in place of the previous $U \times V \to W$. Generalizing, consider p F-vector spaces $U_i, i = 1, 2, \ldots, p$. Let $m\colon U_1 \times U_2 \times \cdots \times U_p \to V$ be an operator which satisfies the condition

$$m\left(u_1, u_2, \ldots, u_{i-1}, f_i u_i + \check{f}_i \check{u}_i, u_{i+1}, \ldots, u_p\right)$$
$$= f_i m\left(u_1, u_2, \ldots, u_{i-1}, u_i, u_{i+1}, \ldots, u_p\right) \tag{2.38}$$
$$+ \check{f}_i m\left(u_1, u_2, \ldots, u_{i-1}, \check{u}_i, u_{i+1}, \ldots, u_p\right)$$

for $i = 1, 2, \ldots, p$, for all f_i and $\check{f}_i$ in F, and for all u_i and $\check{u}_i$ in U_i. Thus, m is said to be a p-**linear** operator. Observe in this definition that when $p - 1$ of the arguments of m are fixed, m becomes a linear operator in the remaining argument. Clearly, the bilinear operator is a special case of this definition, when $p = 2$. Moreover, the definition captures the intuitive concept of multiplication in a precise algebraic sense.

Next, the notion of tensor product is extended in a corresponding way. To do this, suppose that m and V satisfy two conditions. The first condition is that V is the smallest F-vector space that contains the image of m. Equivalently, V is the F-vector space generated by the image of m. Recall that the image of m is not equal to V, even in the bilinear case $p = 2$. The second condition is that

$$\check{m}\left(u_1, u_2, \ldots, u_p\right) = \check{M}\left(m\left(u_1, u_2, \ldots, u_p\right)\right), \tag{2.39}$$

where $\check{M}\colon V \to W$ is a linear operator, W is an F-vector space, and $\check{m}\colon U_1 \times U_2 \times \cdots \times U_p \to W$ is an arbitrary p-linear operator. If m satisfies these two conditions, the action of m is more traditionally written in the manner

$$m\left(u_1, u_2, \ldots, u_p\right) = u_1 \otimes u_2 \otimes \cdots \otimes u_p \tag{2.40}$$

and the space V is given the notation

$$V = U_1 \otimes U_2 \otimes \cdots \otimes U_p \tag{2.41}$$

Once again, existence of the tensor product pair (m, V) is not a problem, and the same sort of uniqueness holds, that is, up to isomorphism.

It is now possible to give a major example of the multiple product idea. The general import of this example far exceeds the interest attached to more elementary illustrations. Therefore, it is accorded its own section. The reason for this will shortly become obvious.

2.7 Determinants

The body of knowledge associated with the theory of determinants tends to occupy a separate and special part of the memory which one reserves for mathematical knowledge. This theory is encountered somewhat indirectly during matrix inversion, and thus is felt to be related to the matrix algebra. However, this association can be somewhat misleading. Multiplication in the matrix algebra is really a multiplication of linear operators, but determinants are more naturally seen in terms of multiplication of vectors. The purpose of this section is to make this idea apparent, and to suggest that a natural way to correlate this body of knowledge is with the concept of an algebra constructed upon a given F-vector space. As such, it becomes a special case of the ideas previously introduced. Fitting determinants into the larger picture is then much less of a challenge than is usually the case, which can save precious human memory.

Consider at the outset a square array of field elements from F, denoted customarily by

$$D = \begin{pmatrix} d_{11} & d_{12} & \cdots & d_{1p} \\ d_{21} & d_{22} & \cdots & d_{2p} \\ \vdots & \vdots & \ddots & \vdots \\ d_{p1} & d_{p2} & \cdots & d_{pp} \end{pmatrix} \tag{2.42}$$

The **determinant** of D will be denoted by $\det(D)$. It is assumed that all readers are comfortable with at least one of the algorithms for computing $\det(D)$. The key idea about $\det(D)$ is that it is a p-linear operator on its columns or upon its rows. In fact, two of the three classical properties of determinants are tantamount precisely to this statement. The third property, which concerns interchanging columns or rows, is also of great interest here (see below).

Without loss of generality, suppose that $\det(D)$ is regarded as a p-linear function of its columns. If the columns, in order, are denoted by $d_1, d_2, \ldots, d_p$, then it is possible to set up a p-linear operator

$$m\left(d_1, d_2, \ldots, d_p\right) = \det(D) \tag{2.43}$$

Accordingly, tensor theory indicates that this operator can be expressed in the manner

$$m\left(d_1, d_2, \ldots, d_p\right) = M\left(d_1 \otimes d_2 \otimes \cdots \otimes d_p\right) \tag{2.44}$$

It is interesting to inquire about the nature of the matrix $[M]$.

In order to calculate $[M]$, it is necessary to select bases for $U_i, i = 1, 2, \ldots, p$. In this case it is possible to identify U_i for each i with a fixed space U of dimension p. Let $\{u_1, u_2, \ldots, u_p\}$ be a basis for U and represent this basis by the standard basis vectors $\{e_1, e_2, \ldots, e_p\}$ in F^p. Moreover, select a basis for F and represent it by the multiplicative unit 1 in F. Then the elements of $[M]$ are found by calculating

$$\det\left(\begin{array}{cccc} e_{i_1} & e_{i_2} & \cdots & e_{i_p} \end{array} \right) \tag{2.45}$$

for all sequences $i_1 i_2 \cdots i_p$ in the increasing numerical order introduced earlier. Thus, if $p = 3$, this set of sequences is 111, 112, 113, 121, 122, 123, 131, 132, 133, 211, 212, 213, 221, 222, 223, 231, 232, 233, 311, 312, 313, 321, 322, 323, 331, 332, 333.

Example. For the case $p = 3$ described above, it is desired to calculate $[M]$. The first few calculations are given by

$$\det \begin{pmatrix} e_1 & e_1 & e_1 \end{pmatrix} = 0 \tag{2.46}$$

$$\det \begin{pmatrix} e_1 & e_1 & e_2 \end{pmatrix} = 0 \tag{2.47}$$

$$\det \begin{pmatrix} e_1 & e_1 & e_3 \end{pmatrix} = 0 \tag{2.48}$$

$$\det \begin{pmatrix} e_1 & e_2 & e_1 \end{pmatrix} = 0 \tag{2.49}$$

$$\det \begin{pmatrix} e_1 & e_2 & e_2 \end{pmatrix} = 0 \tag{2.50}$$

$$\det \begin{pmatrix} e_1 & e_2 & e_3 \end{pmatrix} = +1 \tag{2.51}$$

$$\det \begin{pmatrix} e_1 & e_3 & e_1 \end{pmatrix} = 0 \tag{2.52}$$

$$\det \begin{pmatrix} e_1 & e_3 & e_2 \end{pmatrix} = -1 \tag{2.53}$$

$$\det \begin{pmatrix} e_1 & e_3 & e_3 \end{pmatrix} = 0 \tag{2.54}$$

$$\det \begin{pmatrix} e_2 & e_1 & e_1 \end{pmatrix} = 0 \tag{2.55}$$

$$\det \begin{pmatrix} e_2 & e_1 & e_2 \end{pmatrix} = 0 \tag{2.56}$$

$$\det \begin{pmatrix} e_2 & e_1 & e_3 \end{pmatrix} = -1 \tag{2.57}$$

$$\vdots$$

Rather than provide the entire list in this form, it is easier to give the elements in the right members of the equations. Employing determinant theory, it follows that those sequences with repeated subscripts correspond to 0. Moreover, interchanging two columns changes the sign of the determinant, the third property mentioned previously. Thus, the desired results are

$$0, 0, 0, 0, 0, +1, 0, -1, 0, 0, 0, -1, 0, 0, 0, +1, 0, 0, 0, +1, 0, -1, 0, 0, 0, 0, 0 \tag{2.58}$$

Then $[M]$ is a row matrix having these numerical entries. It is 1×27.

Example. The preceding example indicates that the formation of the determinant in tensor notation results in the appearance of numerous multiplications by zero. This is inefficient. Moreover, if all the zero entries in $[M]$ are dropped, the result is a product of the form

$$\begin{bmatrix} +1 & -1 & -1 & +1 & +1 & -1 \end{bmatrix} \begin{bmatrix} d_{11}d_{22}d_{33} \\ d_{11}d_{32}d_{23} \\ d_{21}d_{12}d_{33} \\ d_{21}d_{32}d_{13} \\ d_{31}d_{12}d_{23} \\ d_{31}d_{22}d_{13} \end{bmatrix} \tag{2.59}$$

easily seen to be the standard formula for classical calculation of the determinant. In view of this result, one immediately wonders what to do about all the dropped zeros. The following section shows how to do away with all the zeros. In the process, however, more things happen than might have been anticipated; as a result, an entirely new concept appears.

2.8   Skew Symmetric Products

The determinant is an instance of skew symmetry in products. Consider a p-linear operator m: $U_1 \times U_2 \times \cdots \times U_p \to V$ with the property that each interchange of two arguments changes the sign of the result produced by m. Thus, for example,

$$m\left(u_1, \ldots, u_{i-1}, u_i, \ldots, u_{j-1}, u_j, \ldots, u_p\right)$$
$$= -m\left(u_1, \ldots, u_{i-1}, u_j, \ldots, u_{j-1}, u_i, \ldots, u_p\right) \tag{2.60}$$

If a list of k interchanges is performed, the sign is changed k times. Such an operator is described as **skew symmetric**.

Provided that only skew symmetric multiplications are of interest, the tensor construction can be streamlined. Let (m_{skewsym}, V) be a pair consisting of a skew symmetric p-linear operator and an F-vector space V. This pair is said to constitute a **skew symmetric tensor product** for the F-vector spaces $U_1, U_2, \ldots, U_p$, if two conditions hold. The reader can probably guess what these two conditions are. Condition one is that V is the F-vector space generated by the image of m_{skewsym}. Condition two is the property that for every skew symmetric p-linear operator $\check{m}_{\text{skewsym}}$: $U_1 \times U_2 \times \cdots \times U_p \to W$, a linear operator $\check{M}_{\text{skewsym}}: V \to W$ exists having the feature

$$\check{m}_{\text{skewsym}}\left(u_1, u_2, \ldots, u_p\right) = \check{M}_{\text{skewsym}}\left(m_{\text{skewsym}}\left(u_1, u_2, \ldots, u_p\right)\right) \tag{2.61}$$

If these two conditions hold for the pair (m_{skewsym}, V), then the custom is to write

$$m_{\text{skewsym}}\left(u_1, u_2, \ldots, u_p\right) = u_1 \wedge u_2 \wedge \cdots \wedge u_p \tag{2.62}$$

for the action of m_{skewsym} and

$$V = U_1 \wedge U_2 \wedge \cdots \wedge U_p \tag{2.63}$$

for the product of the vector spaces involved. Once again, this skew symmetric tensor product exists, and is unique in the usual way.

Now suppose that $U_i = U, i = 1, 2, \ldots, p$, and that $\{u_1, u_2, \ldots, u_p\}$ is a basis for U. It is straightforward to show that $u_{i_1} \wedge u_{i_2} \wedge \cdots \wedge u_{i_p}$ vanishes whenever two of its arguments are equal. Without loss, assume that $u_{i_j} = u_{i_k} = u$. If u_{i_j} and u_{i_k} are switched, the sign of the product must change. However, after the switch, the argument list is identical to the previous list. Because the only number whose value is unchanged after negation is zero, the conclusion follows. Accordingly, the basis for $U_1 \wedge U_2 \wedge \cdots \wedge U_p$ is the family

$$\left\{u_{i_1} \wedge u_{i_2} \wedge \cdots \wedge u_{i_p}\right\} \tag{2.64}$$

where each $i_1 i_2 \cdots i_p$ consists of p distinct nonzero natural numbers, and where the ordinary convention is to arrange the $\{i_k\}$ so as to increase from left to right. A moment's reflection shows that only one such basis element can exist, which is $u_1 \wedge u_2 \wedge \cdots \wedge u_p$. Thus, if $p = 4$, the basis element in question is $u_1 \wedge u_2 \wedge u_3 \wedge u_4$. If we return to the example of the determinant, and regard it as a skew symmetric p-linear operator, then the representation

$$m_{\text{skewsym}}\left(d_1, d_2, \ldots, d_p\right) = M_{\text{skewsym}}\left(d_1 \wedge d_2 \wedge \cdots \wedge d_p\right) \tag{2.65}$$

is obtained. Next observe that each of the p columns of the array D can be written as a unique linear combination of the basis vectors $\{u_1, u_2, \ldots, u_p\}$ in the manner

$$d_j = \sum_{i=1}^{p} d_{ij} u_i \tag{2.66}$$

for $j = 1, 2, \ldots, p$. Then it follows that $d_1 \wedge d_2 \wedge \cdots \wedge d_p$ is given by

$$\sum_{i_1=1}^{p} \sum_{i_2=1}^{p} \cdots \sum_{i_p=1}^{p} d_{i_1 1} d_{i_2 2} \cdots d_{i_p p} u_{i_1} \wedge u_{i_2} \wedge \cdots \wedge u_{i_p} \tag{2.67}$$

which is a consequence of the fact that $\wedge$ is a p-linear operator. The only nonzero terms in this p-fold summation are those for which the indices $\{i_k, k = 1, 2, \ldots, p\}$ are distinct. The reader will correctly surmise that these terms are the building blocks of $\det(D)$. Indeed,

$$d_1 \wedge d_2 \wedge \cdots \wedge d_p = \det(D) u_1 \wedge u_2 \wedge \cdots \wedge u_p \tag{2.68}$$

and, if U is R^p with the $\{u_i\}$ chosen as standard basis elements, then

$$d_1 \wedge d_2 \wedge \cdots \wedge d_p = \det(D) \tag{2.69}$$

because $u_1 \wedge u_2 \wedge \cdots \wedge u_p$ becomes 1 in F. Moreover, it is seen from this example how the usual formula for $\det(D)$ is altered if the columns of D are representations with respect to a basis other than the standard basis. In turn, this shows how the determinant changes when the basis of a space is changed. The main idea is that it changes by a constant which is constructed from the determinant whose columns are the corresponding vectors of the alternate basis. Finally, because this new basis is given by an invertible linear transformation from the old basis, it follows that the determinant of the transformation is the relating factor.

It can now be observed that the change from a tensor product based upon $\otimes$ to a tensor product based upon $\wedge$ has indeed eliminated the zero multiplications associated with skew symmetry. However, and this could possibly be a surprise, it has reduced everything to one term, which is the coordinate relative to the singleton basis element in a tensor product space of dimension one. This may be considered almost a tautology, except for the fact that it produces a natural generalization of the determinant to arrays in which the number of rows is not equal to the number of columns. Without loss, assume that the number of columns is less than the number of rows.

Example. Consider, then, an array of field elements from F, with less columns than rows, denoted by

$$D = \begin{pmatrix} d_{11} & d_{12} & \cdots & d_{1p} \\ d_{21} & d_{22} & \cdots & d_{2p} \\ \vdots & \vdots & \ddots & \vdots \\ d_{q1} & d_{q2} & \cdots & d_{qp} \end{pmatrix} \tag{2.70}$$

where $p < q$. The apparatus introduced in this section still permits the formation of a skew symmetric p-linear operation in the manner $d_1 \wedge d_2 \wedge \cdots \wedge d_p$. This is a natural generalization in the sense that the ordinary determinant is recovered when $p = q$. Moreover, the procedure of calculation is along the same lines as before, with the representations

$$d_j = \sum_{i=1}^{q} d_{ij} u_i \tag{2.71}$$

for $j = 1, 2, \ldots, p$. Note that the upper limit on the summation has changed from p to q. The reader will observe, then, that $d_1 \wedge d_2 \wedge \cdots \wedge d_p$ can be found once again by the familiar step

$$\sum_{i_1=1}^{q} \sum_{i_2=1}^{q} \cdots \sum_{i_p=1}^{q} d_{i_1 1} d_{i_2 2} \cdots d_{i_p p} u_{i_1} \wedge u_{i_2} \wedge \cdots \wedge u_{i_p} \tag{2.72}$$

which is a consequence once again of the fact that $\wedge$ is a p-linear operator. In this case, however, there is more than one way to form nonzero products in the family

$$\left\{ u_{i_1} \wedge u_{i_2} \wedge \cdots \wedge u_{i_p} \right\} \tag{2.73}$$

in which $i_1 i_2 \cdots i_p$ contains p distinct numbers, and where the traditional convention is to arrange the $\{i_k\}$ so that the numbers in each list are increasing, while the numbers which these sequences represent are also increasing. This verbiage is best illustrated quickly.

Example. Suppose that $p = 3$ and $q = 4$. Thus, the sequences $i_1 i_2 i_3$ of interest are 123, 124, 134, 234. It can be observed that each of these sequences describes a 3×3 subarray of D in which the indices are associated with rows. In a sense these subarrays lead to all the interesting 3×3 minors of D, inasmuch as all the others are either zero or negatives of these four. Some of these designated four minors could also be zero, but that would be an accident of the particular problem instead of a general feature.

Example. This example investigates in greater detail the idea of $q > p$. Suppose that $p = 2$ and $q = 3$. Further, let the given array be

$$D = \begin{pmatrix} d_{11} & d_{12} \\ d_{21} & d_{22} \\ d_{31} & d_{32} \end{pmatrix} \tag{2.74}$$

Choose the standard basis $\{e_1, e_2, e_3\}$ for $U = F^3$. Then

$$d_1 = d_{11}e_1 + d_{21}e_2 + d_{31}e_3 \tag{2.75}$$
$$d_2 = d_{12}e_1 + d_{22}e_2 + d_{32}e_3 \tag{2.76}$$

from which one computes that

$$d_1 \wedge d_2 = (d_{11}e_1 + d_{21}e_2 + d_{31}e_3) \wedge (d_{12}e_1 + d_{22}e_2 + d_{32}e_3) \tag{2.77}$$
$$= (d_{11}d_{22} - d_{21}d_{12})\, e_1 \wedge e_2$$
$$+ (d_{11}d_{32} - d_{31}d_{12})\, e_1 \wedge e_3 \tag{2.78}$$
$$+ (d_{21}d_{32} - d_{31}d_{22})\, e_2 \wedge e_3$$

that evidently can be rewritten in the form

$$d_1 \wedge d_2 = \det\begin{pmatrix} d_{11} & d_{12} \\ d_{21} & d_{22} \end{pmatrix} e_1 \wedge e_2$$
$$+ \det\begin{pmatrix} d_{11} & d_{12} \\ d_{31} & d_{32} \end{pmatrix} e_1 \wedge e_3 \tag{2.79}$$
$$+ \det\begin{pmatrix} d_{21} & d_{22} \\ d_{31} & d_{32} \end{pmatrix} e_2 \wedge e_3$$

making clear the idea that the 2×2 minors of D become the coordinates of the expansion in terms of the basis $\{e_1 \wedge e_2, e_1 \wedge e_3, e_2 \wedge e_3\}$ for $R^3 \wedge R^3$.

2.9 Solving Linear Equations

An important application of the previous section is to relate the skew symmetric tensor algebra to one's intuitive idea of matrix inversion. Consider the linear equation

$$\begin{bmatrix} d_{11} & d_{12} & \cdots & d_{1p} \\ d_{21} & d_{22} & \cdots & d_{2p} \\ \vdots & \vdots & \ddots & \vdots \\ d_{p1} & d_{p2} & \cdots & d_{pp} \end{bmatrix} \begin{bmatrix} x_1 \\ x_2 \\ \vdots \\ x_p \end{bmatrix} = \begin{bmatrix} c_1 \\ c_2 \\ \vdots \\ c_p \end{bmatrix} \tag{2.80}$$

With the aid of the usual notation for columns of D, rewrite this equation in the manner

$$\sum_{i=1}^{p} x_i d_i = c \tag{2.81}$$

where c is a vector whose ith element is c_i. To solve for x_k, multiply both members of this equation by the quantity

$$d_1 \wedge d_2 \wedge \cdots \wedge d_{k-1} \wedge d_{k+1} \wedge \cdots \wedge d_p \tag{2.82}$$

which will be denoted by d_{k-}. This multiplication can be done either on the left or the right, and the vector product which is used is $\wedge$. Multiplying on the left provides the result

$$x_k d_{k-} \wedge d_k = d_{k-} \wedge c \tag{2.83}$$

Now if $\det(D)$ is not zero, then this equation solves for

$$x_k = (d_{k-} \wedge c)/(d_{k-} \wedge d_k) \tag{2.84}$$

and this is essentially Cramer's rule. Using this rule conventionally one performs enough interchanges so as to move c to the kth column of the array. If these interchanges are performed in an analogous way with regard to d_k in the denominator, the traditional form of the rule results. This approach to the result shows that the solution proceeds by selecting multiplication by a factor which annihilates all but one of the terms in the equation, where each term is concerned with one column of D.

This is simply a new way of viewing an already known result. However, the treatment of Section 2.8 suggests the possibility of extending the construction to the case in which are found more unknowns than equations. The latter procedure follows via a minor adjustment of the foregoing discussion, and thus it seems instructive to illustrate the steps by means of an example.

Example. Consider the problem corresponding to $q = 3$ and $p = 2$ and given by three equations in two unknowns as follows:

$$\begin{bmatrix} 1 & 2 \\ 3 & 4 \\ 5 & 6 \end{bmatrix} \begin{bmatrix} x_1 \\ x_2 \end{bmatrix} = \begin{bmatrix} 7 \\ 8 \\ 9 \end{bmatrix} \tag{2.85}$$

Begin by rewriting the equation in the form

$$x_1 \begin{bmatrix} 1 \\ 3 \\ 5 \end{bmatrix} + x_2 \begin{bmatrix} 2 \\ 4 \\ 6 \end{bmatrix} = \begin{bmatrix} 7 \\ 8 \\ 9 \end{bmatrix} \tag{2.86}$$

To solve for x_2, multiply from the left with $\begin{bmatrix} 1 & 3 & 5 \end{bmatrix}^T$. This gives

$$x_2 \begin{bmatrix} 1 \\ 3 \\ 5 \end{bmatrix} \wedge \begin{bmatrix} 2 \\ 4 \\ 6 \end{bmatrix} = \begin{bmatrix} 1 \\ 3 \\ 5 \end{bmatrix} \wedge \begin{bmatrix} 7 \\ 8 \\ 9 \end{bmatrix} \tag{2.87}$$

which then implies

$$x_2 \begin{bmatrix} -2 \\ -4 \\ -2 \end{bmatrix} = \begin{bmatrix} -13 \\ -26 \\ -13 \end{bmatrix} \tag{2.88}$$

which implies that $x_2 = 13/2$. Next, consider a left multiplication by $[2 \ 4 \ 6]^T$. Then

$$x_1 \begin{bmatrix} 2 \\ 4 \\ 6 \end{bmatrix} \wedge \begin{bmatrix} 1 \\ 3 \\ 5 \end{bmatrix} = \begin{bmatrix} 2 \\ 4 \\ 6 \end{bmatrix} \wedge \begin{bmatrix} 7 \\ 8 \\ 9 \end{bmatrix} \tag{2.89}$$

which then implies

$$x_1 \begin{bmatrix} 2 \\ 4 \\ 2 \end{bmatrix} = \begin{bmatrix} -12 \\ -24 \\ -12 \end{bmatrix} \tag{2.90}$$

which implies that $x_1 = -6$. It is easy to check that these values of x_1 and x_2 are the unique solution of the equation under study.

The reader is cautioned that the construction of the example above produces necessary conditions on the solution. If any of these conditions cannot be satisfied, no solution can be found. On the other hand, if solutions to the necessary conditions are found, we must check that these solutions satisfy the original equation. Space limitations prevent any further discussion of this quite fascinating point.

2.10 Symmetric Products

The treatment presented in Section 2.8 has a corresponding version for this section. Consider a p-linear operator $m \colon U_1 \times U_2 \times \cdots \times U_p \to V$ with the property that each interchange of two arguments leaves the result produced by m unchanged. Symbolically, this is expressed by

$$
\begin{aligned}
m &\left(u_1, \ldots, u_{i-1}, u_i, \ldots, u_{j-1}, u_j, \ldots, u_p\right) \\
&= m\left(u_1, \ldots, u_{i-1}, u_j, \ldots, u_{j-1}, u_i, \ldots, u_p\right)
\end{aligned}
\tag{2.91}
$$

Such an operator is said to be **symmetric**.

If only symmetric multiplications are of interest, the tensor construction can once again be trimmed to fit. Let (m_{sym}, V) be a pair consisting of a symmetric p-linear operator and an F-vector space V. This pair is said to constitute a **symmetric tensor product** for the F-vector spaces $U_1, U_2, \ldots, U_p$ if two conditions hold. First, V is the F-vector space generated by the image of m_{sym}; and, second, for every symmetric p-linear operator $\check{m}_{\text{sym}} \colon U_1 \times U_2 \times \cdots \times U_p \to W$, there is a linear operator $\check{M}_{\text{sym}} \colon V \to W$ such that

$$
\check{m}_{\text{sym}}\left(u_1, u_2, \ldots, u_p\right) = \check{M}_{\text{sym}}\left(m_{\text{sym}}\left(u_1, u_2, \ldots, u_p\right)\right)
\tag{2.92}
$$

In such a case, one writes

$$
m_{\text{sym}}\left(u_1, u_2, \ldots, u_p\right) = u_1 \vee u_2 \vee \cdots \vee u_p
\tag{2.93}
$$

to describe the action of m_{sym} and

$$
V = U_1 \vee U_2 \vee \cdots \vee U_p
\tag{2.94}
$$

for the symmetric tensor product of the vector spaces involved. As before, this symmetric tensor product exists and is essentially unique.

Next, let $U_i = U$, $i = 1, 2, \ldots, p$, and $\{u_1, u_2, \ldots, u_p\}$ be a basis for U. Because the interchange of two arguments does not change the symmetric p-linear operator, the basis elements are characterized by the family

$$
\left\{u_{i_1} \vee u_{i_2} \vee \cdots \vee u_{i_p}\right\},
\tag{2.95}
$$

where each $i_1 i_2 \cdots i_p$ consists of all combinations of p nonzero natural numbers, written in increasing order, and where the ordinary convention is to arrange the basis vectors so that the numbers $i_1 i_2 \cdots i_p$ increase. Unlike the skew symmetric situation, quite a few such basis vectors can, in general, exist. For instance, the first basis element is $u_1 \vee u_1 \vee \cdots \vee u_1$, with p factors, and the last one is $u_p \vee u_p \vee \cdots \vee u_p$, again with p factors.

Example. Suppose that $p = 3$ and that the dimension of U is 4. The sequences $i_1 i_2 i_3$ of interest in the representation of symmetric p-linear products are 111, 112, 113, 114, 122, 123, 124, 133, 134, 144, 222, 223, 224, 233, 234, 244, 333, 334, 344, 444. Section 2.7 showed a related example which produced 27 basis elements for the tensor product space built upon $\otimes$. In this case, it would be 64. The same situation in Section 2.8 produced four basis elements for a tensor product space constructed with $\wedge$. Twenty basis elements are found for the tensor product space produced with $\vee$. Notice that $20 + 4 \neq 64$. This means that the most general space based on $\otimes$ is not just a direct sum of those based on $\wedge$ and $\vee$.

Example. For an illustration of the symmetric product idea, choose $U = R^2$ and form a symmetric bilinear form in the arrangement of a quadratic form $u^T M u$:

$$m_{\text{sym}}(f_1, f_2) = \begin{bmatrix} f_1 \\ f_2 \end{bmatrix}^T \begin{bmatrix} m_{11} & m_{12} \\ m_{21} & m_{22} \end{bmatrix} \begin{bmatrix} f_1 \\ f_2 \end{bmatrix} \tag{2.96}$$

A word about the matrix M is in order. Because of the relationship

$$M = \tfrac{1}{2}\left(M + M^T\right) + \tfrac{1}{2}\left(M - M^T\right) \tag{2.97}$$

it is easy to see that M may be assumed to be symmetric without loss of generality, as the remaining term in this representation of $m_{\text{sym}}(f_1, f_2)$ leads to a zero contribution in the result. Thus, one is concerned with a natural candidate for the symmetric tensor mechanism. The tensor construction begins by considering

$$\begin{bmatrix} f_1 \\ f_2 \end{bmatrix} \vee \begin{bmatrix} f_1 \\ f_2 \end{bmatrix} \tag{2.98}$$

Choose a standard basis $\{e_1, e_2\}$ for R^2. Then the expression introduced above becomes

$$\sum_{i=1}^{2}\sum_{j=1}^{2} f_i f_j e_i \vee e_j \tag{2.99}$$

which becomes

$$f_1^2 e_1 \vee e_1 + 2 f_1 f_2 e_1 \vee e_2 + f_2^2 e_2 \vee e_2 \tag{2.100}$$

The result may be represented with the matrix

$$\begin{bmatrix} f_1^2 & 2 f_1 f_2 & f_2^2 \end{bmatrix}^T \tag{2.101}$$

Because $p = 2$, the basis vectors of interest are seen to be $\{e_1 \vee e_1, e_1 \vee e_2, e_2 \vee e_2\}$. Inserted into the expression for m_{sym}, these produce the results $m_{11}, m_{12} = m_{21}, m_{22}$, respectively. Thus,

$$\begin{bmatrix} M_{\text{sym}} \end{bmatrix} = \begin{bmatrix} m_{11} & m_{12} & m_{22} \end{bmatrix} \tag{2.102}$$

Finally, the symmetric tensor expression for m_{sym} is

$$m_{\text{sym}}(f_1, f_2) = \begin{bmatrix} m_{11} & m_{12} & m_{22} \end{bmatrix} \begin{bmatrix} f_1^2 \\ 2 f_1 f_2 \\ f_2^2 \end{bmatrix} \tag{2.103}$$

Example. If M, as in the previous example, is real and symmetric, then it is known to satisfy the equation

$$ME = E\Lambda \tag{2.104}$$

where E is a matrix of eigenvectors of M, satisfying $E^T E = I$, and Λ is a diagonal matrix of eigenvalues $\{\lambda_i\}$, which are real. Then

$$M = E\Lambda E^T \tag{2.105}$$

and the quadratic form $u^T M u$ becomes

$$u^T M u = \left[E^T u \right]^T \Lambda \left[E^T u \right] \tag{2.106}$$

$$= \sum_{i=1}^{p} \lambda_i \left[E^T u \right]_i \tag{2.107}$$

When one considers u to be an arbitrary vector in R^p, this quadratic form is non-negative for all u if and only if the $\{\lambda_i\}$ are non-negative and is positive for all nonzero u if and only if the $\{\lambda_i\}$ are positive. M is then **non-negative definite** and **positive definite**, respectively.

Example. With reference to the preceding example, it is sometimes of interest to choose $U = C^p$, where the reader is reminded that C denotes the complex numbers. In this case a similar discussion can be carried out, with M as a complex matrix. The quadratic form must be set up a bit differently, with the structure $u^* M u$, in which superscript $*$ denotes a combined transposition and conjugation. Also, without loss one can assume that $M^* = M$. A common instance of this sort of situation occurs when M is a function $M(s)$ of the Laplace variable s, which is under consideration on the axis $s = j\omega$.

2.11 Summary

The basic idea of this chapter was to examine the axiomatic framework for equipping an F-vector space V with a vector multiplication, and thus develop it into an algebra. Despite the fact that vector multiplication is manifestly a nonlinear operation, it was shown that a very useful matrix theory can be developed for such multiplications. The treatment is based upon notions of algebraic tensor products from which all other multiplications can be derived. The authors showed that the determinant is nothing but a product of vectors, with a special skew symmetric character attached to it. Specializations of the tensor product to such cases, and to the analogous case of symmetric products, were discussed.

As a final remark, it should be mentioned that tensor products themselves develop into a complete algebra of their own. Although space does not permit treatment here, note that the key idea is the definition

$$\left(v_1 \otimes \cdots \otimes v_p \right) \otimes \left(v_{p+1} \otimes \cdots \otimes v_q \right) = v_1 \otimes \cdots \otimes v_q \tag{2.108}$$

References

[1] Greub, W.H., *Multilinear Algebra*, New York, Springer-Verlag, 1967.

3

The Laplace Transform

John R. Deller, Jr.
Michigan State University

3.1 Introduction

The Laplace transform (LT) is the cornerstone of classical circuits, systems, and control theory. Developed as a means of rendering cumbersome differential equation solutions simple algebraic problems, the engineer has transcended this original motivation and has developed an extensive toolbag of analysis and design methods based on the "s-plane." After a motivating differential equation (circuit) problem is presented, this chapter introduces the formal principles of the LT, including its properties and methods for forward and inverse computation. The transform is then applied to the analysis of circuits and systems, exploring such topics as the system function and stability analysis. Two appendices conclude the chapter, one of which relates the LT to other signal transforms to be covered in this volume.

3.2   Motivational Example

Series RLC Circuit

Let us motivate our study of the LT with a simple circuit example. Consider the series RLC circuit shown in Fig. 3.1, in which we leave the component values unspecified for the moment.

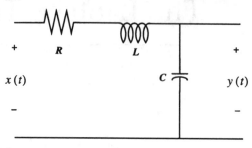

3.1   Series RLC circuit example.

With the input and output to the circuit taken to be the voltages x and y, respectively, the input–output dynamics of this circuit are found to be governed by a linear, constant-coefficient differential equation

$$x(t) = LC\frac{d^2y}{dt^2} + RC\frac{dy}{dt} + y(t) \qquad (3.1)$$

This equation arises from a circuit example, but is typical of many second-order systems arising in mechanical, fluid, acoustic, biomedical, chemical, and other engineering models. We can, therefore, view the circuit as a "system," and the theory explored here as having broad applicability in modeling and analysis of systems.

Suppose we are asked to find the complete solution for the output voltage y, given input

$$x(t) = M_x e^{\sigma_x t} \cos(\omega_x t + \theta_x) u(t) \qquad (3.2)$$

in which u denotes the **unit step function**,

$$u(t) \stackrel{\text{def}}{=} \begin{bmatrix} 0, \ t < 0 & 1/2, \ t = 0 & 1, \ t > 0 \end{bmatrix} \qquad (3.3)$$

For convenience and without loss of generality we assume that $M_x > 0$. The initial conditions [at time $t(0^-)$] on the circuit are given to be[1] $i(0^-) = i_0$ and $y(0^-) = y_0$, where y_0 and i_0 are known quantities.

Homogeneous Solution and the Natural Response

The **homogeneous** or **complementary solution** of the differential equation, say $\tilde{y}$, represents the **natural**, or **unforced**, **response** of the system. The natural response occurs because of the inequity between the initial conditions and the conditions imposed upon the system by the input at the instant it is applied. The system must adjust to these new conditions and will do so in accordance with its own physical properties (e.g., circuit values). For stable systems (see "Poles and Zeroes—Part II"),

---

[1] The notations 0^+ and 0^- indicate limits at $t = 0$ from the right and left, respectively. That $y(0^+) = y(0^-)$, for example, indicates continuity of $y(t)$ at $t = 0$.

the natural response will consist of **transients**, signals which decay exponentially with time. The unforced response will always be present unless the system is stable, and either the input was applied at time $t = -\infty$ (transient will have diminished in the infinite time interval prior to time zero) or the initial conditions on the system exactly nullify the "shock" of the input so that a transient adjustment is not necessary. The form of the natural response (homogeneous solution) is not dependent on the input (except for its use in determining changes around time $t = 0$), but rather on the inherent properties of the system.

The homogeneous solution is found by initially seeking the input-free response which may be written as

$$0 = \left(LC\mathcal{D}^2 + RC\mathcal{D} + 1\right) y(t) \quad \text{with } \mathcal{D}^i \stackrel{\text{def}}{=} \frac{d^i}{dt^i} \tag{3.4}$$

The characteristic equation is therefore

$$0 = \left(LCp^2 + RCp + 1\right) \tag{3.5}$$

Solving for the roots, we find

$$p_1, p_2 = \frac{-RC \pm \sqrt{R^2C^2 - 4LC}}{2LC} = -\frac{R}{2L} \pm \sqrt{\frac{R^2}{4L^2} - \frac{1}{LC}} \tag{3.6}$$

In general, p_1 and p_2 can be real and unequal (overdamped case), equal and real (critically damped case), or complex conjugates (underdamped case). Except in the critically damped case (in which $R^2C^2 = 4LC$ so that two identical real roots are found), the homogeneous solution takes the form

$$\tilde{y}(t) = Ae^{p_1 t} + Be^{p_2 t} \tag{3.7}$$

with A and B to be specified at the end of the solution.

For the sake of discussion, assume the underdamped case in which the natural response will be oscillatory, corresponding to a complex-conjugate pair of roots of (3.5). In this case we have $R^2C^2 < 4LC$ and $p_2 = p_1^*$. Let us define $p_h \stackrel{\text{def}}{=} p_1$, so the two roots are p_h and p_h^*. The meaning of the subscript "h" will become clear later. Some manipulation of (3.7) will simplify our future work. We have

$$\tilde{y}(t) = Ae^{p_h t} + Be^{p_h^* t} \tag{3.8}$$

We observe that A and B must be complex conjugates if $\tilde{y}$ is to be a *real* signal. Thus, we write

$$\tilde{y}(t) = Ae^{p_h t} + A^* e^{p_h^* t} \tag{3.9}$$

After some manipulation using Euler's relation, $e^{j\alpha} = \cos(\alpha) + j\sin(\alpha)$, we have

$$\begin{aligned} \tilde{y}(t) &= 2A_{\text{re}}e^{\sigma_h t}\cos(\omega_h t) - 2A_{\text{im}}e^{\sigma_h t}\sin(\omega_h t) \\ &= 2|A|e^{\sigma_h t}\cos(\omega_h t + \theta_A) \end{aligned} \tag{3.10}$$

where $A = A_{\text{re}} + jA_{\text{im}} = |A|e^{j\theta_A}$ and $p_h = \sigma_h + j\omega_h$. The numbers A_{re} and A_{im}, or, equivalently, $|A|$ and θ_A, are to be determined later.[2] Note that the number p_h is often called the **complex frequency** associated with the damped[3] sinusoid. The complex frequency is simply a convenient mathematical way to hold the damping and frequency information in one quantity. Later in the chapter we see that p_h is also a "pole" of the system being analyzed.

---

[2]In fact, because A_{re} and A_{im} are unknowns, we could replace $2A_{\text{re}}$, $2A_{\text{im}}$, and $2|A|$ with some simpler notations if desired.

[3]We use the term "damping" to refer to the real part of any complex frequency, "σ", with the understanding that two other cases are actually possible: If $\sigma = 0$ the signal is undamped, and if $\sigma > 0$, the signal is exponentially increasing.

Nonhomogeneous Solution and the Forced Response

The **nonhomogeneous**, or **particular**, **solution**, say $\tilde{y}$, represents the system's **forced**, or **driven**, **response**. For certain prevalent types of inputs, the forced response represents an attempt to "track" the forcing function in some sense. If the natural response in a particular problem is transient and the forced response is not, then a "long time" (theoretically, $t \to \infty$) after the input is applied, only this "tracking" response will remain. In the present circuit example the forced response is uniquely present as $t \to \infty$ if $\sigma_h < 0$ (natural response exponentially decays) and $\sigma_x \geq 0$ (forcing function persists in driving the circuit for all time). Further, in the special case in which the forcing function is a periodic or constant signal [an **undamped sinusoid**, **undamped complex exponential**, or a **constant** (in each case $\sigma_x = 0$)] this "tracking" response, in the absence of any transients, is called the **steady-state response** of the system. Note that the forced response will be present for all $t > 0$, but may become uniquely evident only after the transient natural response dies out. Also, note that the forced response might never become evident if it is itself a transient ($\sigma_x < 0$), even though in this case the forced response will still represent an attempt to track the input.

For mathematical convenience in finding $\tilde{y}$, we let x be replaced by

$$
\begin{aligned}
x(t) &= M_x e^{\sigma_x t} e^{j(\omega_x t + \theta_x)} u(t) \\
&= \left[M_x e^{\sigma_x t} \cos(\omega_x t + \theta_x) + j M_x e^{\sigma_x t} \sin(\omega_x t + \theta_x) \right] u(t)
\end{aligned}
\tag{3.11}
$$

Because of the linearity of the system, the real and imaginary parts of the solution $\tilde{y}$ will correspond to the real and imaginary parts of the complex x. Because we want the response to the real part (the cosine), we simply take the real part of the solution at the end.[4] It is extremely useful to rewrite (3.11) as

$$
x(t) = M_x e^{j\theta_x} e^{(\sigma_x + j\omega_x)t} u(t) = M_x e^{j\theta_x} e^{p_x t} u(t)
\tag{3.12}
$$

We shall call the complex number $\bar{X} = M_x e^{j\theta_x}$ a **generalized phasor** for the sinusoid, noting that the quantity is a conventional phasor (see, e.g., [5]) when $\sigma_x = 0$ (i.e., when x represents an undamped sinusoid). The complex frequency associated with the signal x is $p_x \overset{\text{def}}{=} \sigma_x + j\omega_x$.

Any signal that can be written as a sum of exponentials will be an **eigensignal** of a linear, time-invariant (LTI) system such as the present circuit. The forced response of a system to an eigensignal is a scaled, time-shifted version of the eigensignal. This means that an eigensignal generally has its amplitude and phase altered by the system, but *never its frequency*! Many signals used in engineering analysis are eigensignals. This is the case with the present input x.

Because x is an eigensignal, the nonhomogeneous solution will be of the form

$$
\tilde{y}(t) = |H(p_x)| M_x e^{\sigma_x t} e^{j(\omega_x t + \theta_x + \arg\{H(p_x)\})} = M_y e^{j\theta_y} e^{p_x t}
\tag{3.13}
$$

where $|H(p_x)|$ represents the amplitude scaling imposed upon a signal of complex frequency p_x, and $\arg\{H(p_x)\}$ is the phase shift. For the moment, do not be concerned about the seemingly excessive notation $|H(p_x)|$ and $\arg\{H(p_x)\}$. The number $H(p_x) = |H(p_x)|e^{j \arg\{H(p_x)\}}$, called the **eigenvalue** of the system for complex frequency p_x, is a package containing the scaling and phase change induced upon a sinusoid or exponential input of complex frequency p_x. In (3.13) we have implicitly defined $M_y \overset{\text{def}}{=} |H(p_x)| M_x$ and $\theta_y = \theta_x + \arg\{H(p_x)\}$, noting that

$$
\bar{Y} = M_y e^{j\theta_y} = H(p_x) M_x e^{j\theta_x} = H(p_x) \bar{X}
\tag{3.14}
$$

is the generalized phasor for the forced response $\tilde{y}$.

---

[4]Alternatively, we could also find the response to $x^*(t)$, and average the two responses at each t.

Let us now put expressions (3.12) and (3.13) into the differential equation (3.1) [ignoring the $u(t)$ because we are seeking a solution for $t > 0$],

$$M_x e^{j\theta_x} e^{p_x t} = p_x^2 LC M_y e^{j\theta_y} e^{p_x t} + p_x RC M_y e^{j\theta_y} e^{p_x t} + M_y e^{j\theta_y} e^{p_x t} \qquad (3.15)$$

Dividing through by $e^{p_x t}$ (note this critical step),

$$M_x e^{j\theta_x} = p_x^2 LC M_y e^{j\theta_y} + p_x RC M_y e^{j\theta_y} + M_y e^{j\theta_y} \qquad (3.16)$$

Isolating $M_y e^{j\theta_y}$ on the left side of (3.16), we have

$$M_y e^{j\theta_y} = \frac{M_x e^{j\theta_x}}{LC p_x^2 + RC p_x + 1} \qquad (3.17)$$

Because all quantities on the right are known, we can now solve for M_y and θ_y. For example (so that we can compare this result with future work), suppose we have system parameters

$$L = 2 \text{ H}, \quad C = 1 \text{ F}, \quad R = 1 \text{ } \Omega, \quad y_0 = \tfrac{1}{2} \text{ V}, \quad i_0 = 0 \text{ A} \qquad (3.18)$$

and signal parameters

$$M_x = 3 \text{ V}, \quad \sigma_x = -0.1/\text{s}, \quad \omega_x = 1 \text{ rd/s}, \quad \theta_x = \pi/4 \text{ rd} \qquad (3.19)$$

Then we find $M_y e^{j\theta_y} = 2.076 e^{-j(0.519\pi)}$

Whatever the specific numbers, *let us now assume that M_y and θ_y are known.* We have

$$\tilde{\tilde{y}}(t) = M_y e^{j\theta_y} e^{(\sigma_x + j\omega_x)t} \qquad (3.20)$$

Taking the real part,

$$\tilde{\tilde{y}}(t) = M_y e^{\sigma_x t} \cos(\omega_x t + \theta_y) \qquad (3.21)$$

This nonhomogeneous solution is valid for $t > 0$.

Total Solution

Combining the above results, we obtain the complete solution for $t > 0$,

$$y(t) = \tilde{y}(t) + \tilde{\tilde{y}}(t) = 2|A| e^{\sigma_h t} \cos(\omega_h t + \theta_A) + M_y e^{\sigma_x t} \cos\left(\omega_x t + \theta_y\right) \qquad (3.22)$$

We must apply the initial conditions to find the unknown numbers $|A|$ and θ_A. By physical considerations we know that $y(0^+) = y(0^-) = y_0$ and $i(0^+) = i(0^-) = i_0$, so

$$y(0) = y_0 = 2|A| \cos(\theta_A) + M_y \cos\left(\theta_y\right) \qquad (3.23)$$

and

$$i(0) = i_0 = C \left. \frac{dy}{dt} \right|_{t=0} = 2|A|C \left[\sigma_h \cos(\theta_A) - \omega_h \sin(\theta_A) \right] \\ + M_y C \left[\sigma_x \cos\left(\theta_y\right) - \omega_x \sin\left(\theta_y\right) \right] \qquad (3.24)$$

These two equations can be solved for $|A|$ and θ_A. For example, for the numbers given in (3.18) and (3.19), using (3.6) we find that $p_h = -1/4 + j\sqrt{7}/4$, and $A = 0.416 e^{j(0.230\pi)}$. Whatever the specific numbers, *let us assume that $|A|$ and θ_A are now known.* Then, putting all the known numbers back into (3.22) gives a complete solution for $t > 0$.

Scrutinizing the Solution

The first term in the final solution, (3.22), comprises the unforced response and corresponds to the homogeneous solution of the differential equation in conjunction with the information provided by the initial conditions. Notice that this response involves only parameters dependent upon the circuit components, e.g., σ_h and ω_h, and information provided by the initial conditions, $A = |A|e^{j\theta_A}$. The latter term in (3.22) is the forced response. We reemphasize that this part of the response, which corresponds to the nonhomogeneous solution to the differential equation, is of the same form as the forcing function x and that the system has only scaled and time-shifted (as reflected in the phase angle) the input.

It is important to understand that the natural or unforced response is not altogether unrelated to the forcing function. The adjustment that the circuit must make (using its own natural modes) depends on the discrepancy at time zero between the actual initial conditions on the circuit components and the conditions the input would "like" to impose on the components as the forcing begins. Accordingly, we can identify two parts of the natural solution, one due to the initial energy storage, the other to the "shock" of the input at time zero. We can see this in the example above by reconsidering (3.23) and (3.24) and rewriting them as

$$y_0 = 2A_{\text{re}} + M_y \cos\left(\theta_y\right) \tag{3.25}$$

and

$$\frac{i_0}{C} = 2\left[A_{\text{re}}\sigma_h - A_{\text{im}}\omega_h\right] + M_y\left[\sigma_x \cos\left(\theta_y\right) - \omega_x \sin\left(\theta_y\right)\right] \tag{3.26}$$

Solving

$$A_{\text{re}} = \frac{y_0}{2} - \frac{M_y \cos\left(\theta_y\right)}{2} \tag{3.27}$$

$$A_{\text{im}} = \frac{\sigma_h y_0 - (i_0/C)}{2\omega_h} + \frac{M_y \cos\left(\theta_y\right)(\sigma_x - \sigma_h) - M_y \sin\left(\theta_y\right)\omega_x}{2\omega_h} \tag{3.28}$$

We see that each of the real and imaginary parts of the complex number A can be decomposed into a part depending on initial circuit conditions, y_0 and i_0, and a part depending on the system's interaction with the input at the initial instant. Accordingly, we may write

$$A = A_{\text{ic}} + A_{\text{input}} \tag{3.29}$$

where $A_{\text{ic}} = A_{\text{re,ic}} + jA_{\text{im,ic}}$ and $A_{\text{input}} = A_{\text{re,input}} + jA_{\text{im,input}}$. In polar form $A_{\text{ic}} = |A_{\text{ic}}|e^{j\theta_{A_{\text{ic}}}}$ and $A_{\text{input}} = |A_{\text{input}}|e^{j\theta_{A_{\text{input}}}}$. Therefore, the homogeneous solution can be decomposed into two parts

$$\begin{aligned}
\tilde{y}(t) &= \tilde{y}_{\text{ic}}(t) + \tilde{y}_{\text{input}}(t) = 2|A_{\text{ic}}|e^{\sigma_h t}\cos(\omega_h t + \theta_{A_{\text{ic}}}) \\
&\quad + 2|A_{\text{input}}|e^{\sigma_h t}\cos(\omega_h t + \theta_{A_{\text{input}}})
\end{aligned} \tag{3.30}$$

Hence, we observe that a natural response may occur even if the initial conditions on the circuit are zero. The combined natural and forced response in this case, $\tilde{\tilde{y}}(t) + \tilde{y}_{\text{input}}(t)$, is called the **zero-state response** to indicate the state of zero energy storage in the system at time $t = 0$. On the other hand, the response to initial energy storage *only*, $\tilde{y}_{\text{ic}}$, is called the **zero-input response** for the obvious reason.

Generalizing the Phasor Concept: Onward to the Laplace Transform

To begin to understand the meaning of the LT, we reflect on the process of solving the circuit problem above. Although we could examine this solution deeply to understand the LT connections to both the natural and forced responses, it is sufficient for current purposes to examine only the forced solution.

Because the input to the system is an eigensignal in the above, we could assume that the form of $\tilde{y}$ would be identical to that of x, with modifications only to the magnitude and phase. In noting that both x and $\tilde{y}$ would be of the form $Me^{j\theta}$ it seems reasonable that the somewhat tedious nonhomogeneous solution would eventually reduce to an algebraic solution to find M_y and θ_y from M_x and θ_x. All information needed and sought is found in the phasor quantities $\bar{X}$ and $\bar{Y}$ in conjunction with the system information. The critical step which converted the differential equation solution to an algebraic one comes in (3.16) in which the superfluous term $e^{p_x t}$ is divided out of the equation.

Also observe that the ratio $H(p_x) = M_y e^{j\theta_y}/M_x e^{j\theta_x}$ depends only on system parameters and the complex frequency of the input, $p_x = \sigma_x + j\omega_x$. In fact, this ratio, when considered a function, e.g., H, of a general complex frequency, say, $s = \sigma + j\omega$, is called the **system function** for the circuit. In the present example, we see that

$$H(s) = \frac{1}{LCs^2 + RCs + 1} \tag{3.31}$$

The complex number $H(s)$, $s = \sigma + j\omega$, contains the scaling and delay (phase) information induced by the system on a signal with damping σ and frequency ω.

Let us now consider a slightly more general class of driving signals. Suppose we had begun the analysis above with a more complicated input of the form[5]

$$x(t) = \sum_{i=1}^{N} M_i e^{\sigma_i t} \cos(\omega_i t + \theta_i) \tag{3.32}$$

which, for convenience, would have been replaced by

$$x(t) = \sum_{i=1}^{N} M_i e^{j\theta_i} e^{(\sigma_i + j\omega_i)t} = \sum_{i=1}^{N} M_i e^{j\theta_i} e^{p_i t} \tag{3.33}$$

in the nonhomogeneous solution. It follows immediately from linearity that the solution could be obtained by entering each of the components in the input individually, and then combining the N solutions at the output. In each case we would clearly need to rid the analysis of the superfluous term of the form $e^{p_i t}$ by division. This information is equivalent to the form $e^{\sigma_i t} \cos(\omega_i t)$ which is known to automatically carry through to the output.

Now, recalling that $M_i e^{j\theta_i}$ is the generalized phasor for the ith component in (3.33), let us rewrite this expression as

$$x(t) = \sum_{i=1}^{N} \bar{X}(p_i) e^{p_i t} \tag{3.34}$$

where $\bar{X}(p_i) \overset{\text{def}}{=} M_i e^{j\theta_i}$. Expression (3.34) is similar to a Fourier series (see Chapter 4), except that here (unless all $\sigma_i = 0$) the signal is only "pseudo-periodic" in that all of its sinusoidal components may be decaying or expanding in amplitude. The generalized phasors $\bar{X}(p_i)$ are similar to Fourier

---

[5]We omit the unit step u which appears in the input above because we are concerned only with the forced response for $t > 0$.

series coefficients and contain all the information (amplitude and phase) necessary to obtain steady-state solutions. These phasors comprise **frequency domain** information as they contain packets of amplitude and phase information for particular complex frequencies.

With the concepts gleaned from this example, we are now prepared to introduce the LT in earnest.

3.3   Formal Developments

Definitions of the Unilateral and Bilateral Laplace Transforms

Most signals used in engineering analysis and design of circuits and filters can be modeled as a sort of limiting case of (3.34). Such a representation includes not just several complex frequency exponentials as in (3.34), but an **uncountably infinite number** of such exponentials, one for every possible value of frequency ω. Each of these exponentials is weighted by a "generalized phasor" of **infinitesimal magnitude**. The exponential at complex frequency $s = \sigma + j\omega$, for example, is weighted by phasor $\bar{X}(\sigma + j\omega)d\omega/2\pi$, where the differential $d\omega$ assures the infinitesimal magnitude and the scale factor 2π is included by convention. The uncountably infinite number of terms is "summed" by integration as follows

$$x(t) = \int_{-\infty}^{\infty} \bar{X}(\sigma + j\omega)\frac{d\omega}{2\pi}e^{(\sigma+j\omega)t} \tag{3.35}$$

The number σ in this representation is arbitrary as long as the integral exists. In fact, if the integral converges for any σ, then the integral exists for an uncountably infinite number of σ.

The complex function $\bar{X}(\sigma + j\omega)$ in (3.35) is the **Laplace transform** (LT) for the signal $x(t)$. Based on the foregoing discussion, we can interpret the LT as a complex-frequency-dependent, uncountably infinite set of "phasor densities" containing all the magnitude and phase information necessary to find forced solutions for LTI systems. We use the word "density" here to indicate that the LT at complex frequency $\sigma + j\omega$ must be multiplied by the differential $d\omega/2\pi$ to become properly analogous to a phasor. The LT, therefore has, for example, units volts per Hertz. However, we find that the LT is much more than just a phasor-like representation, providing a rich set of analysis tools with which to design and analyze systems, including unforced responses, transients, and stability.

As in the simpler examples above, the solution of differential equations will be made easier by ridding the signals of superfluous complex exponentials of form $e^{(\sigma+j\omega)t}$, that is, by working directly with the LTs. Before doing so we change variables, to put (3.35) into a more conventional form. Let s denote the general complex frequency $s \stackrel{\text{def}}{=} \sigma + j\omega$. Then

$$\boxed{x(t) = \frac{1}{j2\pi} \int_{\sigma-j\infty}^{\sigma+j\infty} X(s)e^{st}\,ds} \tag{3.36}$$

where we have dropped the bar over the LT, X. This integral, which we have interpreted as an "expansion" of the signal x in terms of an uncountably infinite set of infinitesimal generalized phasors and complex exponentials, offers a means for obtaining the signal x from the LT, X. Accordingly, (3.36) is known as the **inverse Laplace transform** (*inverse LT*). The inverse LT operation is often denoted

$$x(t) = \mathcal{L}^{-1}\{X(s)\} \tag{3.37}$$

How one would evaluate such an integral and for what values of s it would exist are issues we shall address later.

In order to rid the signal x of the superfluous factors e^{st}, we can simply compute the LT. Without any rigorous attempt to derive the transform from (3.36), it is believable that

$$X(s) = \int_{-\infty}^{\infty} \frac{x(t)}{e^{st}} \, ds = \int_{-\infty}^{\infty} x(t)e^{-st} \, ds \qquad (3.38)$$

This is the **bilateral**, or **two-sided, Laplace transform** (BLT). The descriptor "bilateral" or "two-sided" is a reference to the fact that the signal may be nonzero in both positive and negative time. In contrast, the **unilateral**, or **one-sided, Laplace transform** (ULT) is defined as

$$X(s) = \int_{0-}^{\infty} \frac{x(t)}{e^{st}} \, ds = \int_{0-}^{\infty} x(t)e^{-st} \, ds \qquad (3.39)$$

When a signal is zero for all $t < 0$, the ULT and BLT are identical. The same inverse LT, (3.36), is applied in either case, with the understanding that the resulting time signal is zero by assumption in the ULT case. While the BLT can be used to treat a more general class of signals, we find that the ULT has the advantage of allowing us to find the component of the natural response due to nonzero initial conditions. In other words, ULT is used to analyze signals that "start" somewhere, a time we conventionally call[6] $t = 0$.

These transformations are reminiscent of the process of dividing through by the complex exponential which was first encountered in the forced solution in the motivating circuit problem [see (3.16)].

Existence of the Laplace Transform

The variable $s = \sigma + j\omega$ is a complex variable over which the LT is calculated. The complex plane with σ along the abscissa and $j\omega$ on the ordinate, is called the s**-plane**. We find some powerful tools centered on the s-plane below. Note that the s-plane is *not* the LT, nor can the LT be "plotted" in the s-plane. The LT is a complex function of the complex variable s, and a plot of the LT would require another two dimensions "over" the s-plane. For this reason, we need to place some constraints on either s or $X(s)$ or both to create a plot. For example, we could use the LT to plot $|X(j\omega)|$ as a function of ω, by evaluating the magnitude of $X(s)$ along the $j\omega$ axis in the s-plane.[7] An illustration of these points is found in Fig. 3.2.

We now address the question: For what values of s (i.e., "where" in the s-plane) does the LT exist? Consider first a two-sided (in time) signal, x. x is assumed piecewise continuous in every finite interval of the real line. We assert that the BLT X ordinarily exists in the s-plane in a strip of the form

$$\sigma_+ < \text{Re}\{s\} = \sigma < \sigma_- \qquad (3.40)$$

as illustrated in Fig. 3.3. In special cases the BLT may converge in the half-plane $\sigma_+ < \sigma$, or the half-plane $\sigma < \sigma_-$, or even in the entire s-plane.

The boundary values σ_+ and σ_- are associated with the positive-time and negative-time portions of the signal, respectively. The minimum possible value of σ_+, and maximum possible value of σ_-, are called the **abscissas of absolute convergence**. We henceforth use the notations σ_+ and σ_- to explicitly mean these extreme values. The vertical strip between, but exclusive of, these abscissas is

---

[6]Note that if we apply the ULT to a signal $x(t)$ that "starts" in negative time, the result will be identical to that for the signal $x(t)u(t)$.

[7]This particular plot is equivalent to the magnitude spectrum of the signal that could be obtained using Fourier techniques discussed in Chapter 4.

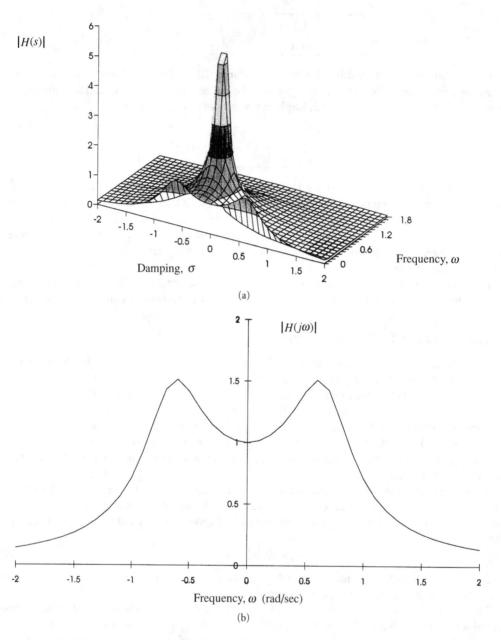

(a)

(b)

3.2   The LT is called "H" rather than "X" in this figure for a reason to be discovered later. (a) A plot of $|H(s)|$ vs. s for the LT $H(s) = (0.5)/(s^2 + 0.5s + 0.5)$. Only the upper-half s-plane is shown ($\omega \geq 0$). Note that the peak occurs near the value $p_h = -1/4 + j\sqrt{7}/4$, a root of the denominator of $H(s)$ which we shall later call a **pole** of the LT. The LT is theoretically infinite at $s = p_h$. (b) Evaluation of $|H(s)|$ along the $j\omega$ axis (corresponding to $\sigma = 0$) with the magnitude plotted as a function of ω.

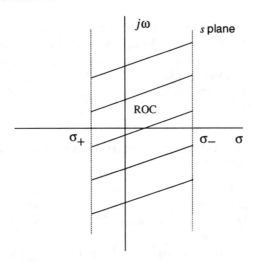

3.3 Except in special cases, the region of convergence for the BLT will be a vertical strip in the s-plane. This strip need **not** contain the $j\omega$ axis as is the case in this illustration.

called the **region of convergence** (ROC) for the LT. In the special cases the ROC extends indefinitely from a single abscissa to the right ($\sigma_+ < \sigma$) or left ($\sigma < \sigma_-$), or covers the entire s-plane.

To verify that (3.40) is a correct description of the ROC, consider the positive-time part of x first. We maintain that $X(s)$ will exist on any s such that $\text{Re}\{s\} = \sigma > \sigma_+$, if and only if (iff) a positive M exists such that x is bounded by $Me^{\sigma_+ t}$ on $t \geq 0$,

$$|x(t)| < Me^{\sigma_+ t} \quad t \geq 0 \tag{3.41}$$

Under this condition (letting X_+ denote the LT of the non-negative-time part of x),

$$
\begin{aligned}
|X_+(s)| &= \left| \int_0^\infty x(t)e^{-st}\, dt \right| \\
&\leq \int_0^\infty \left| x(t)e^{-st} \right|\, dt \\
&= \int_0^\infty |x(t)|\, e^{-\sigma t}\, dt < \int_0^\infty Me^{(\sigma_+ - \sigma)t}\, dt \\
&= \begin{cases} \dfrac{M}{\sigma - \sigma_+}, & \sigma > \sigma_+ \\ \infty, & \text{otherwise} \end{cases}
\end{aligned}
\tag{3.42}
$$

If there is no finite, positive M such that (3.39) holds, then the signal grows faster than $e^{-\sigma_+ t}$ and the LT integral (area under $x(t)e^{-\sigma_+ t}e^{j\omega t}$) will not converge for at least some values of s in the neighborhood of the vertical line $s = \sigma_+$. In this case σ_+ is not a proper abscissa.

By similar means, we can argue that the negative-time part of x must be bounded as

$$|x(t)| < Me^{\sigma_- t} \quad t < 0 \tag{3.43}$$

for some positive M, if the corresponding part of the LT, say, X_-, is to converge at s such that $\text{Re}\{s\} = \sigma < \sigma_-$. Similarly to (3.42) it is shown that

$$
\begin{aligned}
|X_-(s)| &= \left| \int_{-\infty}^0 x(t)e^{-st}\, dt \right| < \int_{-\infty}^0 Me^{(\sigma_- - \sigma)t}\, dt \\
&= \begin{cases} \dfrac{M}{\sigma_- - \sigma}, & \sigma < \sigma_- \\ \infty, & \text{otherwise} \end{cases}
\end{aligned}
\tag{3.44}
$$

and X_- will not converge in some left neighborhood of $s = \sigma_-$ if the condition (3.43) is not met.

It should be clear from the discussion of the BLT that a ULT will ordinarily converge in the half-plane $\mathrm{Re}\{s\} = \sigma > \sigma_+$ iff a positive M exists such (3.41) is met. This follows immediately from the fact that the ULT of $x(t)$ is equivalent to the BLT of $x(t)u(t)$. The "negative-time" part of $x(t)$ yields $X_-(s) = 0$ which converges everywhere in the s-plane. The ULT may also converge everywhere in the s-plane in special cases.

Example of Laplace Transform Computations and Table of Unilateral Laplace Transforms

A listing of commonly used ULTs with ROCs is given in Table 3.1. Each entry in the table can be verified by direct computation of the appropriate LT integral, or, in many cases, properties of the LT can be exploited to make the computation easier. These properties are developed in the section "Properties of the Laplace Transform."

It is rare to find a table of BLTs in engineering material because most of our work is done with the ULT. However, BLTs can often be found by summing the results of two ULTs in the following way. Suppose x is written as

$$x(t) = x_+(t) + x_-(t) \tag{3.45}$$

where x_+ and x_- are the causal and anticausal parts of the signal, respectively. To obtain X_+, we can use a ULT table. To obtain X_-, note the following easily demonstrated property of the BLT: If $\mathcal{L}\{y(t)\} = Y(s)$ with ROC $\{s: \mathrm{Re}\{s\} > a\}$, then $\mathcal{L}\{y(-t)\} = Y(-s)$ with ROC $\{s : \mathrm{Re}\{s\} < -a\}$. Therefore, we can find $\mathcal{L}\{x_-(-t)\}$ in a ULT table, and replace the argument s by $-s$ to obtain $X_-(s)$. Then $X(s) = X_+(s) + X_-(s)$. This strategy is illustrated in Example 1. The ROC of the sum will ordinarily be the intersection of the individual LTs X_+ and X_-, but the total ROC may be larger than this if a pole-zero cancellation occurs (see section following).

TABLE 3.1   Table of Unilateral Laplace Transform Pairs

Signal, $x(t)$	ULT, $X(s)$	ROC
$\delta(t)$	1	Entire s-plane
$\frac{d^k \delta}{dt^k}$	s^k	Entire s-plane
$u(t)$	$\frac{1}{s}$	$\{s : \sigma > 0\}$
$t^{n-1}u(t)$	$\frac{(n-1)!}{s^n}$	$\{s : \sigma > 0\}$
$e^{p_x t}u(t)$	$\frac{1}{(s-p_x)}$	$\{s : \sigma > \sigma_x\}$
$t^n e^{p_x t}u(t)$	$\frac{n!}{(s-p_x)^{n+1}}$	$\{s : \sigma > \sigma_x\}$
$\cos(\omega_x t + \theta_x)u(t)$	$\frac{s\cos(\theta_x)-\omega_x \sin(\theta_x)}{s^2+\omega_x^2}$	$\{s : \sigma > 0\}$
$e^{\sigma_x t}\cos(\omega_x t + \theta_x)u(t)$	$\frac{(s-\sigma_x)\cos(\theta_x)-\omega_x \sin(\theta_x)}{(s-\sigma_x)^2+\omega_x^2}$	$\{s : \sigma > \sigma_x\}$
$M_x t^{n-1} e^{\sigma_x t}\cos(\omega_x t + \theta_x)u(t)$	$\frac{E}{(s-p_x)^n} + \frac{E^*}{(s-p_x^*)^n}$;	
	M_x real and positive, $E = \frac{M_x}{2}(n-1)!\, e^{j\theta_x}$	$\{s : \sigma > \sigma_x\}$

Note: In each entry, s is the general complex number $\sigma + j\omega$, and, where relevant, p_x is the specific complex number $\sigma_x + j\omega_x$.

Let us consider some examples which illustrate the direct forward computation of the LT and the process discussed in the preceding paragraph.

EXAMPLE 3.1:

Find the BLT and ULT for the signal

$$x(t) = Ae^{at}u(-t) + Be^{bt}u(t) \qquad |A|, |B| < \infty \qquad (3.46)$$

Note that A, B, a, and b may be complex.

SOLUTION 3.1   In the bilateral case, we have

$$
\begin{aligned}
X(s) &= \int_{-\infty}^{0^-} Ae^{at}e^{-st}\, dt + \int_{0^-}^{\infty} Be^{bt}e^{-st}\, dt \\[2mm]
&= \frac{Ae^{(a-s)t}}{a-s}\bigg|_{-\infty}^{0^-} + \frac{Be^{(b-s)t}}{a-s}\bigg|_{0^-}^{\infty} \\[2mm]
&= \left[\begin{array}{ll} -\dfrac{A}{s-a}, & \mathrm{Re}\{s\} < \mathrm{Re}\{a\} \\[2mm] \infty, & \text{otherwise} \end{array} \right] \\[2mm]
&\quad + \left[\begin{array}{ll} \dfrac{B}{s-b}, & \mathrm{Re}\{s\} > \mathrm{Re}\{b\} \\[2mm] \infty, & \text{otherwise} \end{array} \right]
\end{aligned}
\qquad (3.47)
$$

The ROC for this LT is $\{s : \mathrm{Re}\{b\} < \mathrm{Re}\{s\} < \mathrm{Re}\{a\}\}$. The LT does not exist for any s for which $\mathrm{Re}\{s\} \geq \mathrm{Re}\{a\}$ or $\mathrm{Re}\{s\} \leq \mathrm{Re}\{b\}$. Note also that when $\mathrm{Re}\{b\} \geq \mathrm{Re}\{a\}$, then no ROC can be found, meaning that the BLT does not exist anywhere for the signal.

The ULT follows immediately from the work above. We have

$$X(s) = \int_{0^-}^{\infty} Be^{bt}e^{-st}\, dt = \begin{cases} \dfrac{B}{s-b}, & \mathrm{Re}\{s\} > \mathrm{Re}\{b\} \\[2mm] \infty, & \text{otherwise} \end{cases} \qquad (3.48)$$

The ROC in this case is $\{s : \mathrm{Re}\{s\} > \mathrm{Re}\{b\}\}$. We need not be concerned about the negative-time part of the signal (and the associated ROC) because the LT effectively zeroes the signal on $t < 0$.

    Note. The result of this example is worth committing to memory as it will reappear frequently.

Note that if we had found the ULT, (3.48), for the causal part of x in a table [call it $X_+(s)$], then we could employ the trick suggested above to find the LT for the anticausal part. Let x_- denote the negative-time part: $x_-(t) = Ae^{at}u(-t)$. We know that the LT of $x_-(-t) = Ae^{-at}u(t)$ (a causal signal) is

$$X_-(-s) = \frac{A}{s+a}, \qquad \text{with ROC } \{s : \mathrm{Re}\{s\} > -a\} \qquad (3.49)$$

Therefore,

$$X_-(s) = \frac{-A}{s-a}, \qquad \text{with ROC } \{s : \mathrm{Re}\{s\} < a\} \qquad (3.50)$$

The overall BLT result is then $X(s) = X_+(s) + X_-(s)$ with ROC equal to the intersection of the individual results. This is consistent with the BLT found by direct integration. $\qquad \Box$

    The simple example above suggests that the BLT can treat a broader class of signals at the expense of greater required care in locating its ROC. A further and related complication of the BLT is the nonuniqueness of the transform with respect to the time signals. Consider the following example.

EXAMPLE 3.2:

Find the BLT for the following signals:

$$x_1(t) = e^{bt} u(t) \quad \text{and} \quad x_2(t) = -e^{bt} u(-t) \tag{3.51}$$

SOLUTION 3.2   From our work in Example 3.1, we find immediately that

$$\begin{aligned} X_1(s) &= \frac{1}{s-b}, \quad \text{Re}\{s\} > \text{Re}\{b\} \quad \text{and} \\ X_2(s) &= \frac{1}{s-b}, \quad \text{Re}\{s\} < \text{Re}\{b\} \end{aligned} \tag{3.52}$$

Neither X_1 nor X_2 can be unambiguously associated with a time signal without knowledge of its ROC. $\qquad\square$

Another drawback of the BLT is its inability to handle initial conditions in problems like the one that motivated our discussion. For this reason, and also because signals tend to be **causal** (occurring only in positive time) in engineering problems, the ULT is more widely used and we shall focus on it exclusively after treating one more important topic in the following section. Before moving to the next section, let us tackle a few more example computations.

EXAMPLE 3.3:

Find the ULT of the impulse function, $\delta(t)$ (see Section 3.6, Appendix A).

SOLUTION 3.3

$$\Delta(s) = \int_{0^-}^{\infty} \delta(t) e^{-st} \, dt = 1 \quad \text{for all } s \tag{3.53}$$

The LT converges everywhere in the s-plane. We note that the lower limit 0^- is important here to yield the answer 1 (which will provide consistency of the theory) instead of $1/2$. $\qquad\square$

EXAMPLE 3.4:

Find the ULT of the unit step function, $u(t)$ [see (3.3)].

SOLUTION 3.4

$$U(s) = \int_{0^-}^{\infty} 1 e^{-st} \, dt = \left. \frac{-e^{-st}}{s} \right|_{0^-}^{\infty} = \frac{1}{s} \quad \text{for Re}\{s\} > 0 \tag{3.54}$$

The ROC for this transform consists of the entire right-half s-plane exclusive of the $j\omega$ axis. $\qquad\square$

EXAMPLE 3.5:

Find the ULT of the damped ($\sigma_x < 0$), undamped ($\sigma_x = 0$), or expanding ($\sigma_x > 0$) sinusoid, $x(t) = M_x e^{\sigma_x t} \cos(\omega_x t + \theta_x) u(t)$.

SOLUTION 3.5   Using Euler's relation, write x as

$$x(t) = \frac{M_x}{2}e^{\sigma_x t}\left[e^{j(\omega_x t + \theta_x)} + e^{-j(\omega_x t + \theta_x)}\right] = \frac{M_x}{2}\left[e^{j\theta_x}e^{p_x t} + e^{-j\theta_x}e^{p_x^* t}\right] \tag{3.55}$$

with $p_x \overset{\text{def}}{=} \sigma_x + j\omega_x$. Taking the LT,

$$\begin{aligned} X(s) &= \frac{M_x}{2}\int_{0^-}^{\infty}\left[e^{j\theta_x}e^{p_x t} + e^{-j\theta_x}e^{p_x^* t}\right]e^{-st}\,dt \\[2mm] &= \int_{0^-}^{\infty}\frac{M_x}{2}e^{j\theta_x}e^{p_x t}e^{-st}\,dt + \int_{0^-}^{\infty}\frac{M_x}{2}e^{-j\theta_x}e^{p_x^* t}e^{-st}\,dt \end{aligned} \tag{3.56}$$

Now using (3.48) on each of the integrals

$$X(s) = \frac{(M_x/2)e^{j\theta_x}}{(s - p_x)} + \frac{(M_x/2)e^{-j\theta_x}}{(s - p_x^*)} \tag{3.57}$$

with the ROC associated with each of the terms being $\text{Re}\{s\} > \text{Re}\{p_x\} = \sigma_x$. Putting the fractions over a common denominator yields

$$\begin{aligned} X(s) &= \frac{M_x}{2}\left[\frac{(s - p_x^*)e^{j\theta_x} + (s - p_x)e^{-j\theta_x}}{(s - p_x)(s - p_x^*)}\right] \\[2mm] &= \frac{M_x}{2}\left[\frac{(se^{j\theta_x} + se^{-j\theta_x} - p_x^*e^{j\theta_x} - p_x e^{-j\theta_x})}{s^2 - 2\,\text{Re}\{p_x\}s + |p_x|^2}\right] \\[2mm] &= M_x\left[\frac{(s - \sigma_x)\cos(\theta_x) - \omega_x\sin(\theta_x)}{(s - \sigma_x)^2 + \omega_x^2}\right] \end{aligned} \tag{3.58}$$

The ROC of X is $\{s : \text{Re}\{s\} > \sigma_x\}$.

Note.   The chain of denominators in (3.58) is worth noting because these relations occur frequently in LT work:

$$\boxed{(s - p_x)(s - p_x^*) = s^2 - 2\,\text{Re}\{p_x\}s + |p_x|^2 = s^2 - 2\sigma_x s + |p_x|^2 = (s - \sigma_x)^2 + \omega_x^2}$$

$$\tag{3.59}$$

$\square$

Poles and Zeroes — Part I

"Pole-zero" analysis is among the most important uses of the LT in circuit and system design and analysis. We need to take a brief look at some elementary theory of functions of complex variables in order to carefully describe the meaning of a pole or zero. When we study methods of inverting the LT in a future section, this side trip will prove to have been especially useful.

Let us begin with a general function, F, of a complex variable s. We stress that $F(s)$ may or may not be a LT. F is said to be **analytic** at $s = a$ if it is differentiable at a and in a neighborhood of a. For example, $F(s) = s - 1$ is analytic everywhere (or **entire**), but $G(s) = |s|$ is nowhere analytic because its derivative exists only at $s = 0$. On the other hand, a point p is an **isolated singular point** of F if the derivative of F does not exist at p, but F is analytic in a neighborhood of p. The function $F(s) = e^{-s}/(s - 1)$ has a singular point at $s = 1$. There is a circular analytic domain around any singular point, p, of F, say $\{s : |s - p| < \rho\}$, in which the function F can be represented by a **Laurent series** [3],

$$F(s) = \sum_{i=0}^{\infty} q_{i,p}(s - p)^i + \sum_{i=1}^{\infty} \frac{r_{i,p}}{(s - p)^i} \tag{3.60}$$

The second sum in (3.60) is called the **principle part** of the function F at p. When the principle part of F at p contains terms up to order n, the isolated singular point p is called an **nth-order pole** of F. Evidently from (3.60), F tends to infinity at a pole and the order of infinity is n. For future reference, we note that the complex number $r_p \overset{\text{def}}{=} r_{1,p}$ is called the **residue** of F at $s = p$.

A **zero** of F is more simply defined as a value of s, say z, at which F is analytic and for which $F(z) = 0$. If all the derivatives up to the $(m - 1)$st are also zero at z, but the mth is nonzero, then z is called an mth-order zero of F. It can be shown that the zeroes of an analytic function F are **isolated**, except in the trivial case $F(s) = 0$ for all s [3].

Most LTs encountered in signal and system problems are quotients of polynomials in s, say

$$X(s) = \frac{N(s)}{D(s)} \tag{3.61}$$

because of the signals employed in engineering work, and because (as we shall see later) rational LTs are naturally associated with LTI systems. N and D connote "numerator" and "denominator." In this case both N and D are analytic everywhere in the s-plane, and the poles and zeroes of X are easily found by factoring the polynomials N and D, to express X in the form

$$X(s) = C \left(\prod_{i=1}^{n_N}(s - z_i) \Big/ \prod_{i=1}^{n_D}(s - p_i) \right) \tag{3.62}$$

where n_N is the number of simple factors in $N(s)$ (order of N in s), n_D the number of simple factors in $D(s)$ (order of D in s), and C is a constant. X is called a **proper rational LT** if $n_D > n_N$. After canceling all factors common to the numerator and denominator, if m terms $(s - z)$ are left in the numerator, then X has an mth order zero at $s = z$. Similarly, if n terms $(s - p)$ are left in the denominator, then X has an nth order pole at $s = p$.

Although the LT does not exist outside the ROC, all of the poles will occur at values of s outside the ROC. None, some, or all of the zeroes may also occur outside the ROC. This does not mean that the LT is valid outside the ROC, but that its poles and zeroes may occur there. A pole is ordinarily indicated in the s-plane by the symbol $\times$; whereas a zero is marked with a small circle $\bigcirc$.

EXAMPLE 3.6:

Find the poles and zeroes of the LT

$$X(s) = \frac{3s^2 + 9s + 9}{(s + 2)(s^2 + 2s + 2)} \tag{3.63}$$

SOLUTION 3.6   Factoring the top and bottom polynomials to put X in form (3.62), we have

$$X(s) = 3 \frac{\left(s + (3 - j\sqrt{3})/2\right)\left(s + (3 + j\sqrt{3})/2\right)}{(s + 2)(s + 1 + j)(s + 1 - j)} \tag{3.64}$$

There are first-order zeroes at $s = (-3 + j\sqrt{3})/2$ and $s = (-3 - j\sqrt{3})/2$, and first-order poles at $s = -2, s = -1 + j$, and $s = -1 - j$. The pole-zero diagram appears in Fig. 3.4.

Two points are worth noting. First, complex poles and zeroes will always occur in conjugate pairs, as they have here, if the LT corresponds to a *real signal*. Second, the denominator of (3.64) can also be expressed as $(s + 2)[(s + 1)^2 + 1]$ [recall (3.59)]. Comparing this form with the LT obtained in Example 3.5 suggests that the latter form might prove useful. □

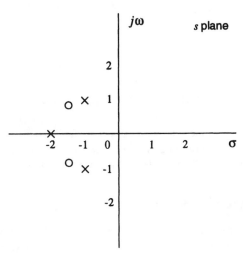

3.4 Pole-zero diagram for Example 3.6.

The purpose of introducing poles and zeroes at this point in our discussion is to note the relationship of these singularities to the ROC. The preceding examples illustrate the following facts:

1. For a "right-sided" (non-negative-time only) signal, x, the ROC of LT X (either ULT or BLT) is $\{s : \text{Re}\{s\} > \text{Re}\{p_+\} = \sigma_+\}$, where p_+ is the pole of X with maximum real part, namely, σ_+. If X has no poles, then the ROC is the entire s-plane.

2. For a "left-sided" (negative-time only) signal, x, the ROC of BLT X is $\{s : \text{Re}\{s\} < \text{Re}\{p_-\} = \sigma_-\}$, where p_- is the pole of X with minimum real part, namely, σ_-. If X has no poles, then the ROC is the entire s-plane.

3. For a "two-sided" signal x, the ROC of the BLT X is $\{s : \text{Re}\{p_+\} = \sigma_+ < \text{Re}\{s\} < \text{Re}\{p_-\} = \sigma_-\}$ where p_+ is the pole of maximum real part associated with the right-sided part of x, and p_- is the pole of minimum real part associated with the left-sided part of x. If the right-sided signal has no pole, then the ROC extends indefinitely to the right in the s-plane. If the left-sided signal has no pole, then the ROC extends indefinitely to the left in the s-plane. Therefore, if neither part of the signal has a pole, then the ROC is the entire s-plane.

Let us revisit three of the examples above to verify these claims. In Example 3.1 we found the ROC for the BLT to be $\{s : \text{Re}\{b\} < \text{Re}\{s\} < \text{Re}\{a\}\}$. The only pole associated with the left-sided sequence is at $s = a$. The only pole associated with the right-sided signal occurs at $s = b$. Following rule 3 in the list above yields exactly the ROC determined by analytical means. The poles of X as well as the ROC are shown in Fig. 3.5(a).

In Example 3.4 we found the ROC to be the entire right-half s-plane, exclusive of the $j\omega$ axis. The single pole of $U(s) = 1/s$ occurs at $s = 0$. Figure 3.5(b) is consistent with rule 1 above.

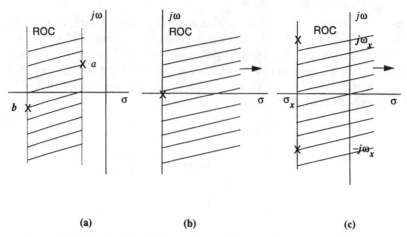

3.5   Pole-zero plots and ROCs for the LTs of (a) Example 3.1, (b) Example 3.4, (c) Example 3.5 above.

The ULT of Example 3.5 has poles at $s = \sigma_x \pm j\omega_x$ and a zero at $s = \sigma_x + \omega_x \tan(\theta_x)$. Rule 1 therefore specifies that the ROC should be $\{s : \text{Re}\{s\} > \text{Re}\{\sigma_x \pm j\omega_x\} = \sigma_x\}$, which is consistent with the solution to Example 3.5. The pole-zero plot and ROC are illustrated in Fig. 3.5(c).

Properties of the Laplace Transform[8]

This section considers some properties of the LT which are useful in computing forward and inverse LTs, and in other manipulations occurring in signal and system design and analysis. A list of properties appears in Table 3.2.

In most cases the verification of these properties follows in a straightforward manner from the definition of the transform. Consider the following examples. For convenience, we define the notation

$$x(t) \longleftrightarrow X(s) \tag{3.65}$$

to mean that x and X are a LT pair, $X(s) = \mathcal{L}\{x(t)\}$.

EXAMPLE 3.7:

Verify the *modulation* property of the LT which states that if $x(t) \longleftrightarrow X(s)$, then $e^{s_0 t}x(t) \longleftrightarrow X(s - s_0)$.

SOLUTION 3.7   By definition,

$$\mathcal{L}\{e^{s_0 t}x(t)\} = \int_{0^-}^{\infty} e^{s_0 t}x(t)e^{-st}\,dt$$

$$= \int_{0^-}^{\infty} x(t)e^{-(s-s_0)t}\,dt = X(s - s_0) \tag{3.66}$$

$\square$

---

[8]Henceforth this study restricts attention to the ULT and uses the acronym "LT" only.

TABLE 3.2   Operational Properties of the Unilateral Laplace Transform

Description of Operation	Formal Operation	Corresponding LT
Linearity	$\alpha x(t) + \beta y(t)$	$\alpha X(s) + \beta Y(s)$
Time delay ($t_0 > 0$)	$x(t - t_0)u(t - t_0)$	$e^{-st_0} X(s)$
Exponential modulation in time (or complex frequency ("s") shift)	$e^{s_0 t} x(t)$	$X(s - s_0)$
Multiplication by t^k, $k = 1, 2, \ldots$	$t^k x(t)$	$(-1)^k \dfrac{d^k X}{ds^k}$
Time differentiation	$\dfrac{d^k x}{dt^k}$	$s^k X(s) - \displaystyle\sum_{i=0}^{k-1} s^i x^{(k-1-i)}(0^-)$ $x^{(i)}(0^-) \stackrel{\text{def}}{=} \dfrac{d^i x}{dt^i}\Big|_{t=0^-}$
Time integration	$\int_{-\infty}^{t} x(\lambda) \, d\lambda$	$\dfrac{X(s)}{s} + \dfrac{x^{(-1)}(0^-)}{s}$ $x^{(-1)}(0^-) \stackrel{\text{def}}{=} \displaystyle\int_{-\infty}^{t} x(\lambda) \, d\lambda\Big|_{t=0^-}$
Convolution	$\int_0^{\infty} x(\lambda) y(t - \lambda) \, d\lambda$	$X(s)Y(s)$
Correlation	$\int_0^{\infty} x(t) y(t - \tau) \, dt$	$X(s)Y(-s)$
Product (s-domain convolution)	$x(t)y(t)$	$\dfrac{1}{j2\pi} \int_{\sigma - j\infty}^{\sigma + j\infty} X(\lambda) Y(s - \lambda) \, d\lambda$
Initial signal value (if time limit exists)	$\lim_{t \to 0^+} x(t)$	$\lim_{s \to \infty} s X(s)$
Final signal value (if time limit exists)	$\lim_{t \to \infty} x(t)$	$\lim_{s \to 0} s X(s)$
Time scaling	$x(\alpha t)$, $\alpha > 0$	$\dfrac{1}{\alpha} X\left(\dfrac{s}{\alpha}\right)$
Periodicity (period T)	$\sum_{i=0}^{\infty} x(t - iT)$ $x(t) = 0$, $t \notin [0, T)$	$\dfrac{X(s)}{(1 - e^{-sT})}$

Note: Throughout, X and Y are LTs of signals x and y, respectively. x and y are causal signals.

EXAMPLE 3.8:

Verify the periodicity property and find the LT for a square wave of period $T = 2$ and duty cycle 1/2.

SOLUTION 3.8   Using the linearity and time-delay properties of the LT

$$\mathcal{L}\left\{\sum_{i=0}^{\infty} x(t - iT)\right\} = \sum_{i=0}^{\infty} X(s)e^{-siT} = X(s) \sum_{i=0}^{\infty} e^{-siT} = \frac{X(s)}{(1 - e^{-sT})} \qquad (3.67)$$

Let us call the square wave $z(t)$ and its LT $Z(s)$. Now one period of z can be written as $x(t) = u(t) - u(t-1)$, $0 \leq t < 2$ (see Fig. 3.6). Using the delay property, therefore, $X(s) = (1/s) - (e^{-s}/s)$. Using (3.67) with $T = 2$, we have

$$Z(s) = \frac{(1/s) - \left(e^{-s}/s\right)}{\left(1 - e^{-2s}\right)} = \frac{\left(1 - e^{-s}\right)}{s\left(1 - e^{-2s}\right)} \qquad (3.68)$$

□

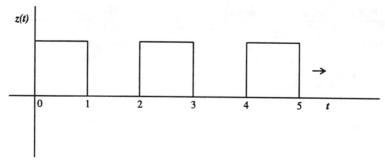

3.6   Square wave of Example 3.8.

EXAMPLE 3.9:

Verify the time-differentiation property of the LT which states that if $x(t) \longleftrightarrow X(s)$, then $dx/dt \longleftrightarrow sX(s) - x(0^-)$.

SOLUTION 3.9   By definition, $\mathcal{L}\{dx/dt\} = \int_{0^-}^{\infty} (dx/dt)e^{-st} \, dt$. Integrating by parts yields

$$\mathcal{L}\left\{\frac{dx}{dt}\right\} = x(t)e^{-st}\Big|_{0^-}^{\infty} + s \int_{0^-}^{\infty} x(t)e^{-st}dt = sX(s) - x(0^-) \tag{3.69}$$

$\square$

EXAMPLE 3.10:

Verify the initial value theorem of the LT which states that if $x(0^+) = \lim_{t \downarrow 0} x(t) < \infty$, then $\lim_{s \to \infty} sX(s) = x(0^+)$.

SOLUTION 3.10   In case a discontinuity exists in x at $t = 0$, define the signal

$$y(t) = x(t) - cu(t) \tag{3.70}$$

where c is the amplitude shift at the discontinuity, $c = x(0^+) - x(0^-)$. Then y will be continuous at $t = 0$ (see Fig. 3.7). Further,

$$\frac{dx}{dt} = \frac{dy}{dt} + c\delta(t) \tag{3.71}$$

so that using the time-differentiation property and the fact that $\mathcal{L}\{c\delta(t)\} = c$, we have

$$sX(s) - x(0^-) = \int_{0^-}^{\infty} \frac{dy}{dt}e^{-st}dt + c \tag{3.72}$$

Because $c = x(0^+) - x(0^-)$,

$$sX(s) = \int_{0^-}^{\infty} \frac{dy}{dt}e^{-st}dt + x(0^+) \tag{3.73}$$

Assuming that the LT of the signal y has a ROC (y is of exponential order), the integral in (3.73) vanishes as $s \to \infty$. Finally, therefore, we obtain that $\lim_{s \to \infty} sX(s) = x(0^+)$. $\square$

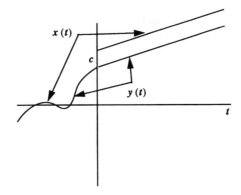

3.7 The signals x and y used in Example 3.10.

EXAMPLE 3.11:

Verify the convolution property of the LT which states that if x and h are causal signals with $x(t) \longleftrightarrow X(s)$, and $h(t) \longleftrightarrow H(s)$, then

$$x(t) * h(t) = \int_{-\infty}^{\infty} x(\xi)h(t - \xi)d\xi \longleftrightarrow X(s)H(s) \qquad (3.74)$$

SOLUTION 3.11   Because $x(t) = 0$ for $t < 0$, we can write

$$\int_{-\infty}^{\infty} x(\xi)h(t - \xi)d\xi = \int_{0^-}^{\infty} x(\xi)h(t - \xi)d\xi \qquad (3.75)$$

Now

$$
\begin{aligned}
\mathcal{L}\left\{ \int_{0^-}^{\infty} x(\xi)h(t - \xi)d\xi \right\} &= \int_{0^-}^{\infty} \int_{0^-}^{\infty} x(\xi)h(t - \xi)e^{-st} d\xi\, dt \\
&= \int_{0^-}^{\infty} \int_{0^-}^{\infty} x(\xi)h(\beta)e^{-s\beta} e^{-s\xi} d\beta\, d\xi \qquad (3.76) \\
&= \int_{0^-}^{\infty} x(\xi)e^{-s\xi} d\xi \int_{0^-}^{\infty} h(\beta)e^{-s\beta} d\beta \\
&= X(s)H(s)
\end{aligned}
$$

The causality of h is used in line (3.76) in setting the lower limit of integration over β to 0^-.   □

The operational properties are used to simplify forward and inverse transform computations and other manipulations involving transforms. To briefly illustrate, three examples follow.

EXAMPLE 3.12:

Using operational properties, rederive the LT for $x(t) = M_x \cos(\omega_x t + \theta_x)u(t)$ which was first considered in Example 3.5.

SOLUTION 3.12   Write x as

$$x(t) = \frac{M_x}{2} e^{j\theta_x} e^{(\sigma_x + j\omega_x)t} u(t) + \frac{M_x}{2} e^{-j\theta_x} e^{(\sigma_x - j\omega_x)t} u(t) \qquad (3.77)$$

The linearity property allows us to ignore the factors $M_x e^{\pm j\theta_x}$ in the process of taking the transform, and returning them afterward. Recalling the previous result, (3.48), we can write immediately

$$\mathcal{L}\{e^{(\sigma_x + j\omega_x)t} u(t)\} = \frac{1}{s - (\sigma_x + j\omega_x)}, \quad \text{Re}\{s\} > \sigma_x \tag{3.78}$$

$$\mathcal{L}\{e^{(\sigma_x - j\omega_x)t} u(t)\} = \frac{1}{s - (\sigma_x - j\omega_x)}, \quad \text{Re}\{s\} > \sigma_x \tag{3.79}$$

so

$$\begin{aligned} X(s) = \;& \frac{M_x}{2} e^{j\theta_x} \frac{1}{s - (\sigma_x + j\omega_x)} \\ & + \frac{M_x}{2} e^{-j\theta_x} \frac{1}{s - (\sigma_x - j\omega_x)} \quad \text{Re}\{s\} > \sigma_x \end{aligned} \tag{3.80}$$

Placing the fractions over a common denominator yields the same result as that found using direct integration in Example 3.5.   □

EXAMPLE 3.13:

Find the time signals corresponding to the following LTs:

$$\begin{aligned} X(s) &= e^{-\pi s} \quad \text{for all } s \\ Y(s) &= \log(7) \frac{e^{-32s}}{s}, \quad \text{Re}\{s\} > 0 \\ Z(s) &= \frac{e^{\sqrt{2}s}}{s + 5} + \frac{\sqrt{3}}{s - 5}, \quad \text{Re}\{s\} > 5 \end{aligned} \tag{3.81}$$

SOLUTION 3.13   Recognize that $X(s) = e^{-\pi s} \Delta(s)$, where $\Delta(s) = 1$ is the LT for the impulse function $\delta(t)$. Using the time-shift property, therefore, we have $x(t) = \delta(t - \pi)$.

Recognize that $Y(s) = \log(7)e^{-32s} U(s)$, where $U(s) = 1/s$ is the LT for the step function $u(t)$. Using linearity and the time-shift property, therefore, $y(t) = \log(7)u(t - 32)$.

In finding z, linearity allows us to treat the two terms separately. Further, from (3.48), we know that $\mathcal{L}^{-1}\{1/(s + 5)\} = e^{-5t}u(t)$ and $\mathcal{L}^{-1}\{1/(s - 5)\} = e^{5t}u(t)$. Therefore, $z(t) = e^{-5(t+\sqrt{2})}u(t + \sqrt{2}) + \sqrt{3}e^{5t}u(t)$. Note that the first term has a ROC $\{s : \text{Re}\{s\} > -5\}$, while the second has ROC $\{s : \text{Re}\{s\} < 5\}$. The overall ROC is therefore consistent with these two components.   □

Inverse Laplace Transform

In principle, finding a time function corresponding to a given LT requires that we compute the integral in (3.36):

$$x(t) = \frac{1}{j2\pi} \int_{\sigma - j\infty}^{\sigma + j\infty} X(s)e^{st} \, ds \tag{3.82}$$

Recall that σ is constant and taken to be in the ROC of X. Direct computation of this line integral requires a knowledge of the theory of complex variables. However, several convenient computational procedures are available that circumvent the need for a detailed understanding of the complex calculus. These measures are the focus of this section. The reader interested in more detailed information on complex variable theory is referred to [3]. "Engineering" treatments of this subject are also found in [10].

We first study the most challenging of the inversion methods, and the one which most directly solves the inversion integral above. The reader interested in quick working knowledge of LT inversion might wish to proceed immediately to the section on partial fraction expansion.

Residue Theory

It is important to be able to compute residues of a function of a complex variable. Recall the Laurent series expansion of a complex function, say, F, which was introduced in "Poles and Zeroes—Part I," equation (3.60). Also, recall that the coefficient $r_{1,p}$ is called the **residue** of F at p, and that we defined the simplified notation $r_p \stackrel{\text{def}}{=} r_{1,p}$ to indicate the residue because the subscript "1" is not useful outside the Laurent series. In the analytic neighborhood of singular point $s = p$ (an nth-order pole) we define the function

$$\varphi_p(s) = (s - p)^n F(s) = r_{1,p}(s - p)^{n-1} + r_{2,p}(s - p)^{n-2}$$
$$+ \cdots + r_{n,p} + \sum_{i=0}^{\infty} q_{i,p}(s - p)^{n+i} \qquad (3.83)$$

in which it is important to note that $r_{n,p} \neq 0$. Because F is not analytic at $s = p$, φ_p is not defined, and is therefore not analytic, at $s = p$. We can, however, make φ_p analytic at p by simply *defining* $\varphi_p(p) \stackrel{\text{def}}{=} r_{n,p}$. In this case φ_p is said to have a **removable singular point** (at p). Note that (3.83) can be interpreted as the Taylor series expansion of φ_p about the point $s = p$. Therefore, the residue is apparently given by

$$r_p \stackrel{\text{def}}{=} r_{1,p} = \frac{\varphi_p^{(n-1)}(p)}{(n-1)!} \qquad (3.84)$$

where $\varphi_p^{(i)}$ indicates the ith derivative of φ_p. When $n = 1$ (first-order pole), which is frequently the case in practice, this expression reduces to

$$r_p = \varphi_p(p) = \lim_{s \to p}(s - p)F(s) \qquad (3.85)$$

The significance of the residues appears in the following key result (e.g., see [3]):

THEOREM 3.1 (*Residue Theorem*): *Let C be a simple closed contour within and on which a function F is analytic except for a finite number of singularity points, $p_1, p_2, \ldots, p_k$ interior to C. If the respective residues at the singularities are $r_{p_1}, r_{p_2}, \ldots, r_{p_k}$, then*

$$\oint_C F(s)\,ds = j2\pi\left(r_{p_1} + r_{p_2} + \cdots + r_{p_k}\right) \qquad (3.86)$$

where the contour C is traversed in the counterclockwise direction.

The relevance of this theorem in our work is as follows: In principle, according to (3.82), we want to integrate the complex function $F(s) = X(s)e^{st}$ on some vertical line in the ROC, for example, $c - j\infty$ to $c + j\infty$, where $c > \sigma_+$. Instead, suppose we integrate over the contour shown in Fig. 3.8. By the residue theorem, we have

$$\oint_C X(e)e^{st}\,ds = j2\pi\left(r_{p_1} + r_{p_2} + \cdots + r_{p_k}\right) \qquad (3.87)$$

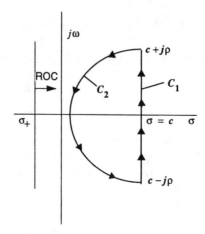

3.8   Contour in the s-plane for evaluating the inverse LT.

where $r_{p_1}, r_{p_2}, \ldots, r_{p_k}$ are the k residues of the function $X(s)e^{st}$. The integral can be decomposed as

$$\oint_C X(s)e^{st}ds = \oint_{C_1} X(s)e^{st}ds + \oint_{C_2} X(s)e^{st}\,ds \tag{3.88}$$

where, as $\rho \to \infty$, C_1 approaches the line over which we wish to integrate according to (3.82). It can be shown[9] that the integral over C_2 contributes nothing to the answer for $t > 0$, provided that X approaches zero uniformly[10] on C_2. Therefore,

$$\lim_{\rho \to \infty} \oint_C X(s)e^{st}\,ds = \lim_{\rho \to \infty} \oint_{C_1} X(s)e^{st}\,ds$$

$$= \lim_{\omega \to \infty} \frac{1}{j2\pi} \int_{c-j\omega}^{c+j\omega} X(s)e^{st}\,ds \tag{3.89}$$

From (3.87), we have

$$\frac{1}{j2\pi} \int_{\sigma-j\infty}^{\sigma+j\infty} X(s)e^{st}ds = r_{p_1} + r_{p_2} + \cdots + r_{p_k}, \quad t > 0 \tag{3.90}$$

Thus, recalling that the left side is the original inversion integral, we have

$$x(t) = \sum_{i=1}^{k} r_{p_i}, \quad t > 0 \tag{3.91}$$

where the r_{p_i} are the residues of $X(s)e^{st}$ at its k singular points.

Note that the residue method returns a time function only over the *positive* time range. We might expect a result beginning at $t = 0$ or $t = 0^-$ because we have defined the forward LT as an integral beginning at $t = 0^-$. The reason for this lower limit is so that an impulse function at the time origin will be transformed "properly." Another important place at which the initial condition "$x(0^-)$" appears is in the LT of a differentiated time function (see Table 3.2 and Example 3.9). Again, the

---

[9]A rigorous mathematical discussion appears in [3], while a particularly clear "engineering" discussion appears in Appendix B of [6].

[10]At the same rate regardless of the angle considered along C_2.

condition is included to properly handle the fact that if x has a discontinuity at $t = 0$, its derivative should include an impulse and the LT of the derivative should include a corresponding constant. The residue cannot properly invert LTs of impulse functions (constants over the s-plane) because such LTs do not converge uniformly to zero over the semicircular part of the contour C_2. Such constants in the LT must be inverted in a more *ad hoc* way. If no apparent impulses occur and the residue method has provided x for $t > 0$, it is, in fact, possible to assign an arbitrary finite value to $x(0)$. This is because one point in the time signal will not affect the LT, so the proper correspondence between x and X remains.[11] If it is necessary to assign a value to $x(0)$, the most natural procedure is to let $x(0) \stackrel{\text{def}}{=} x(0^+)$. Then when we write the final answer as $x(t) = $ [time signal determined by residues]$u(t)$, the signal takes an implied value $x(0^+)/2$ at $t = 0$.

The discussion above emphasizes that in general, successful application of the residue method depends on the uniform convergence of the LT to zero on the contour segment C_2. In principle each LT to be inverted must be checked against this criterion. However, this seemingly foreboding process is usually not necessary in linear signal and system analysis. The reason is that a practical LT ordinarily will take the form of a ratio of polynomials in s like (3.61). The check for proper convergence is a simple matter of assuring that the order of the denominator in s exceeds that of the numerator. When this is not the case, a simple remedy exists which we illustrate in Example 3.15. Another frequent problem is the occurrence of a LT of form

$$X(s) = \frac{N(s)e^{st_0}}{D(s)}, \quad t_0 < 0 \tag{3.92}$$

where N and D are polynomials. The trick here is to recognize the factor e^{st_0} as corresponding to a time shift which can be taken care of at the end of the problem, once the rational part of the transform is inverted.

Finally, we remark that similar results apply to the bilateral LT. When the signal is two sided in time, the causal part is obtained using the procedure above, but including only residues of poles known to be associated with the non-negative-time part of the signal. The noncausal part of the signal is found by summing residues belonging to "noncausal" poles. Note that the association of poles with the causal and noncausal parts of the signal follows from a specification of the ROC. Poles belonging to the causal signal are to the left of the ROC, while noncausal poles are to the right.

Let us now illustrate the procedure with two examples.

EXAMPLE 3.14:

In Example 3.5 we showed that

$$\mathcal{L}\{M_x e^{\sigma_x t} \cos(\omega_x t + \theta_x) u(t)\} = M_x \left[\frac{(s - \sigma_x)\cos(\theta_x) - \omega_x \sin(\theta_x)}{(s - \sigma_x)^2 + \omega_x^2} \right] \tag{3.93}$$

with ROC $\{s : \text{Re}\{s\} > \sigma_x\}$. Verify that this is correct by finding the inverse LT using residues. Call the signal x and the LT X.

SOLUTION 3.14   We can ignore the scalar M_x until the end due to linearity. Two poles are in the transform: $p_x = \sigma_x + j\omega_x$ and $p_x^* = \sigma_x - j\omega_x$ ($k = 2$ in the discussion above). These can be obtained by expanding the denominator and using the quadratic equation, but it is useful to

---

[11] In fact, any two signals that differ only on a set of measure zero (e.g., see [7]) will have the same LT.

remember the relationship between a quadratic polynomial written in this form and the conjugate roots [recall (3.59)]. The residue for the pole at p_x is given by

$$
\begin{aligned}
\varphi_{p_x}(p_x) = X(s)e^{st}(s - p_x)\big|_{s=p_x} &= \frac{[(s - \sigma_x)\cos(\theta_x) - \omega_x \sin(\theta_x)]\,e^{st}}{(s - p_x^*)}\bigg|_{s=p_x} \\
&= \frac{[(p_x - \sigma_x)\cos(\theta_x) - \omega_x \sin(\theta_x)]\,e^{p_x t}}{(p_x - p_x^*)} \qquad (3.94) \\
&= \frac{[(j\omega_x)\cos(\theta_x) - \omega_x \sin(\theta_x)]\,e^{(\sigma_x + j\omega_x)t}}{(2j\omega_x)}
\end{aligned}
$$

Similarly, we have for the pole at p_x^*, $\varphi_{p_x^*}(p_x^*) = [(-j\omega_x)\cos(\theta_x) - \omega_x \sin(\theta_x)]\,e^{(\sigma_x - j\omega_x)t}/(-2j\omega_x)$. For $t > 0$, therefore,

$$
\begin{aligned}
x(t) &= \varphi_{p_x}(p_x) + \varphi_{p_x^*}(p_x^*) \\
&= \frac{e^{\sigma_x t}\left[j\omega_x \cos(\theta_x)\left(e^{j\omega_x t} + e^{-j\omega_x t}\right) - \omega_x t \sin(\theta_x)\left(e^{j\omega_x t} - e^{-j\omega_x t}\right)\right]}{2j\omega_x} \qquad (3.95) \\
&= e^{\sigma_x t}[\cos(\theta_x)\cos(\omega_x t) - \sin(\theta_x)\sin(\omega_x t)] = e^{\sigma_x t}\cos(\omega_x t + \theta_x) \qquad (3.96)
\end{aligned}
$$

and the transform is verified. $\square$

The following example illustrates the technique for handling polynomial quotients in which the order of the denominator does not exceed the numerator (X is not a proper rational LT).

EXAMPLE 3.15:

Find the causal time signal corresponding to LT

$$
X(s) = \frac{N(s)}{D(s)} = \frac{Gs^2}{(s - p)^2}, \qquad G,\, p \text{ are real} \qquad (3.97)
$$

SOLUTION 3.15   To use the residue method, we must reduce the order of the numerator to at most unity. By dividing polynomial D into N using long division, we can express X as

$$
X(s) = G + \frac{2Gps - Gp^2}{(s - p)^2} \overset{\text{def}}{=} X_1(s) + X_2(s) \qquad (3.98)
$$

First note that we can use linearity and invert the two terms separately. The first is simple because (see Table 3.1, $\mathcal{L}^{-1}\{X_1(s)\} = \mathcal{L}^{-1}\{G\} = G\delta(t)$. In the second term we find a pole of order $n = 2$ at $s = p$. To use (3.84) to compute the residue, we require $\varphi_p(s) = X_2(s)e^{st}(s - p)^2 = \left[2Gps - Gp^2\right]e^{st}$. Then, for $t > 0$, the residue is

$$
r_p = \frac{1}{1!}\frac{d\varphi_p}{ds}\bigg|_{s=p} = G\left[2pe^{pt} + 2p^2 t e^{pt} - p^2 t e^{pt}\right] = Gpe^{pt}\,[2 + pt] \qquad (3.99)
$$

Because r_p is the only residue, we have $x(t) = G\delta(t) + Gpe^{pt}\,[2 + pt]\,u(t)$.

REMARK 3.1   If the order of the numerator *exceeds* the denominator by $k > 0$, the long division will result in a polynomial of the form $A_k s^k + A_{k-1}s^{k-1} + \cdots + A_0$. Consequently, the time domain signal will contain derivatives of the impulse function. In particular, if $k = 1$, a "doublet" will be present in the time signal. (See Table 3.1 and Appendix A for more information.) $\qquad \square$

We can usually find several ways to solve inverse LT problems, residues often being among the most challenging. In this example, for instance, we could use (3.98) to write

$$X(s) = G + \frac{2Gps}{(s-p)^2} - \frac{Gp^2}{(s-p)^2} = G + X_3(s) - X_4(s) \qquad (3.100)$$

then use Table 3.1 to find $\mathcal{L}^{-1}\{G\} = G\delta(t)$ and $\mathcal{L}^{-1}\{X_4\} = x_4(t)$ [or use residues to find $x_4(t)$]. Noting that $X_3(s) = 2ps\,X_4(s)$, we could then use the s-differentiation property (Table 3.2) to find x_3 from x_4. This alternative solution illustrates a general method that ordinarily is used regardless of the fundamental inversion technique. Linearity allows us to divide the problem into a sum of smaller, easier problems to invert, and then combine solutions at the end. The method below is probably the most popular, and clearly follows this paradigm.

Partial Fraction Expansion

The **partial fraction expansion (PFE)** method can be used to invert only rational LTs, with the exception that factors of the form e^{st_0} can be handled as discussed near (3.92). As noted above, this is not practically restricting for most engineering analyses. Partial fraction expansion is closely related to the residue method, a relationship evident via our examples.

As in the case of residues, a rational LT to be inverted must be proper, having a numerator polynomial whose order is strictly less than that of the denominator. If this is not the case, long division should be employed in the same manner as in the residue method (see Example 3.15).

Consider the LT $X(s) = N(s)/D(s)$. Suppose X has k poles, $p_1, p_2, \ldots, p_k$ and D is factored as

$$D(s) = (s - p_1)^{n_1}(s - p_2)^{n_2} \cdots (s - p_k)^{n_k} \qquad (3.101)$$

where n_i is the order of the ith pole. (Note that $\sum_{i=1}^{k} n_i = n_D$.) Now if x is a real signal, the complex poles will appear in conjugate pairs. It will *sometimes* be convenient to combine the corresponding factors into a quadratic following (3.59),

$$(s - p)(s - p^*) = s^2 + \beta s + \gamma = s^2 - 2\,\mathrm{Re}\{p\}s + |p|^2 = s^2 - 2\sigma_p s + |p|^2 \qquad (3.102)$$

for each conjugate pair. Assuming that the poles are ordered in (3.101) so that the first k' are real and the last $2k''$ are complex ($k = k' + 2k''$), D can be written

$$
\begin{aligned}
D(s) \;=\; & (s - p_1)^{n_{\mathrm{re},1}}(s - p_2)^{n_{\mathrm{re},2}} \cdots (s - p_{k'})^{n_{\mathrm{re},k'}} (s^2 + \beta_1 s + \gamma_1)^{n_{\mathrm{c},1}} \\
& \times (s^2 + \beta_2 s + \gamma_2)^{n_{\mathrm{c},2}} \cdots (s^2 + \beta_{k''} s + \gamma_{k''})^{n_{\mathrm{c},k''}}
\end{aligned}
\qquad (3.103)
$$

Terms of the form $(s - p)$ are called **simple linear factors**, while those of form $(s^2 + \beta s + \gamma)$ are **simple quadratic factors**. When a factor is raised to a power greater than one, we say that there are **repeated linear (or quadratic) factors**. We emphasize that linear factors need not always be combined into quadratics when they represent complex poles. In other words, in the term $(s - p_i)^{n_i}$, the pole p_i may be complex. Whether quadratics are used depends on the approach taken to solution.

The idea behind PFE is to decompose the larger problem into the sum of smaller ones. We expand the LT as

$$
\begin{aligned}
X(s) = \frac{N(s)}{D(s)} \;&=\; \frac{N(s)}{\mathrm{factor}_1\,\mathrm{factor}_2 \cdots \mathrm{factor}_{k'}\,\mathrm{factor}_{k'+1} \cdots \mathrm{factor}_{k'+k''}} \qquad (3.104) \\[2mm]
&=\; \frac{N_1(s)}{\mathrm{factor}_1} + \frac{N_2(s)}{\mathrm{factor}_2} + \cdots + \frac{N_{k'}(s)}{\mathrm{factor}_{k'}} + \frac{N_{k'+1}(s)}{\mathrm{factor}_{k'+1}} \\[2mm]
&\quad + \frac{N_{k'+k''}(s)}{\mathrm{factor}_{k'+k''}} \qquad (3.105)
\end{aligned}
$$

Now because of linearity, we can invert each of the **partial fractions** in the sum individually, then add the results. Each of these "small" problems is easy and ordinarily can be looked up in a table or solved by memory.

We now consider a series of cases and examples.

CASE 3.1   Simple Linear Factors. Let $X(s) = N(s)/D(s)$. Assume that the order of N is strictly less than the order of D Without loss of generality (for the case under consideration), assume that the first factor in D is a simple linear factor so that D can be written $D(s) = (s - p_1)D_{\text{other}}(s)$ where D_{other} is the product of all remaining factors. Then the PFE will take the form

$$X(s) \quad = \quad \frac{N(s)}{D(s)} = \frac{N(s)}{(s - p_1)D_{\text{other}}(s)} = \frac{N_1(s)}{(s - p_1)}$$
$$+ \Big[\text{other PFs corresponding to factors in } D_{\text{other}}(s) \Big] \qquad (3.106)$$

Now note that

$$(s - p_1)X(s) = \frac{N(s)}{D_{\text{other}}(s)} = N_1(s)$$
$$+ (s - p_1)\Big[\text{other PFs corresponding to factors in } D_{\text{other}}(s) \Big] \qquad (3.107)$$

Letting $s = p_1$ reveals that

$$N_1(p_1) = A_1 = \frac{N(p_1)}{D_{\text{other}}(p_1)} = [(s - p_1)X(s)]|_{s=p_1} \qquad (3.108)$$

Note that the number A_1 is the residue of the pole at $s = p_1$. In terms of our residue notation

$$\varphi_{p_1}(s) = (s - p_1) X(s) = \frac{N(s)}{D_{\text{other}}(s)} \quad \text{and} \quad r_{p_1} = A_1 = \varphi_{p_1}(p_1) \qquad (3.109)$$

Carefully note that we are computing residues of poles of $X(s)$, not $X(s)e^{st}$, in this case.

EXAMPLE 3.16:

Given LT

$$X(s) = \frac{s^2 + 3s}{(s + 1)(s + 2)(s + 4)} \qquad (3.110)$$

with ROC $\{s : \text{Re}\{s\} > -1\}$, find the corresponding time signal, x.

SOLUTION 3.16   Check that the order of the denominator exceeds that of the numerator. Because it does, we can proceed by writing

$$X(s) = \frac{A_1}{(s + 1)} + \frac{A_2}{(s + 2)} + \frac{A_3}{(s + 4)} \qquad (3.111)$$

Using the method above, we find that

$$A_1 = \frac{s^2 + 3s}{(s + 2)(s + 4)}\bigg|_{s=-1} = -\frac{2}{3} \qquad (3.112)$$

In a similar manner we find that $A_2 = 1$ and $A_3 = 2/3$. Therefore

$$X(s) = -\frac{2/3}{(s+1)} + \frac{1}{(s+2)} + \frac{2/3}{(s+4)} \tag{3.113}$$

Now using linearity and Table 3.1 [or recalling (3.48)], we can immediately write

$$x(t) = \left[-\frac{2}{3}e^{-t} + e^{-2t} + \frac{2}{3}e^{-4t} \right] u(t) \tag{3.114}$$

$\square$

CASE 3.2   **Simple Quadratic Factors.** When $D(s)$ contains a simple quadratic factor, the LT can be expanded as

$$X(s) = \frac{N(s)}{D(s)} = \frac{N(s)}{(s^2 - \beta s + \gamma)D_{\text{other}}(s)} \tag{3.115}$$

$$= \frac{Bs + C}{(s^2 - \beta s + \gamma)} + \left[\text{other PFs corresponding to factors in } D_{\text{other}}(s) \right]$$

The usefulness of this form is illustrated below.

EXAMPLE 3.17:

Find the time signal x corresponding to LT

$$X(s) = \frac{(s+4)}{(s+2)(s^2 + 6s + 34)}, \quad \text{ROC}\{s : \text{Re}\{s\} > -3\} \tag{3.116}$$

SOLUTION 3.17   The order of D exceeds the order of N, so we may proceed. The roots of the quadratic term are $p, p^* = (-6 \pm \sqrt{36 - 136})/2 = -3 \pm j5$, so we leave it as a quadratic. (If the roots were real, we would use the simple linear factor approach.) Expand the LT into PFs

$$X(s) = \frac{A}{(s+2)} + \frac{Bs + C}{(s^2 + 6s + 34)} \tag{3.117}$$

Using the method for simple linear factors, we find that $A = 1/13$. Now multiply both sides of (3.117) by $D(s) = (s+2)(s^2 + 6s + 34)$ to obtain

$$(s+4) = \frac{1}{13}(s^2 + 6s + 34) + (Bs + C)(s+2) \tag{3.118}$$

Equating like powers of s on the two sides of the equation yields $B = -1/13$ and $C = 9/13$, so

$$X(s) = \frac{1/13}{(s+2)} + \frac{[(-1/13)s + (9/13)]}{(s^2 + 6s + 34)} \tag{3.119}$$

The first fraction has become familiar by now and corresponds to time function $(1/13)e^{-2t}u(t)$. Let us focus on the second fraction. Note that [recall (3.59)]

$$s^2 + 6s + 34 = (s - p)(s - p^*) = (s + 3 - j5)(s + 3 + j5) = (s + 3)^2 + 5^2 \tag{3.120}$$

The second fraction, therefore, can be written as

$$-\frac{1}{13}\left[\frac{(s+3)}{(s+3)^2+5^2} - \frac{12}{(s+3)^2+5^2}\right]$$

$$= -\frac{1}{13}\left[\frac{(s+3)}{(s+3)^2+5^2} - \frac{(12/5)5}{(s+3)^2+5^2}\right] \tag{3.121}$$

The terms in brackets correspond to a cosine and sine, respectively, according to Table 3.1. Therefore,

$$x(t) = \left[\frac{1}{13}e^{-2t} - \frac{1}{13}e^{-3t}\cos(5t) + \frac{12}{5(13)}e^{-3t}\sin(5t)\right]u(t) \tag{3.122}$$

□

Simple quadratic factors need not be handled by the above procedure, but can be treated as simple linear factors, as illustrated by example.

EXAMPLE 3.18:

Repeat the previous problem using simple linear factors.

SOLUTION 3.18　　Expand X into simple linear PFs

$$X(s) = \frac{A}{(s+2)} + \frac{E}{(s-p)} + \frac{E^*}{(s-p^*)}, \quad p = -3 + j5 \tag{3.123}$$

Note that the PFE coefficients corresponding to complex-conjugate pole pairs will themselves be complex conjugates. In order to derive a useful general relationship, we ignore the specific numbers for the moment. Using the familiar inverse transform, we can write

$$x(t) = \left[Ae^{-2t} + Ee^{pt} + E^*e^{p^*t}\right]u(t) \tag{3.124}$$

Letting $E = |E|e^{j\theta_E}$ and $p = \sigma_p + j\omega_p$, we have

$$\boxed{Ee^{pt} + E^*e^{p^*t} = |E|e^{j\theta_E}e^{(\sigma_p+j\omega_p)t} + |E|e^{-j\theta_E}e^{(\sigma_p-j\omega_p)t} = 2|E|e^{\sigma_p t}\cos(\omega_p t + \theta_E)}$$

$$\tag{3.125}$$

This form should be noted for use with complex-conjugate pairs.

Using the method for finding simple linear factor coefficients, we find that $A = 1/13$. Also,

$$E = X(s)(s-p)\big|_{s=p} = \frac{(s+4)}{(s+2)(s-p^*)}\bigg|_{s=p} = \frac{(p+4)}{(p+2)(j2\,\mathrm{Im}\{p\})}$$

$$= \frac{(1+j5)}{(-1+j5)(j10)} = 0.1\,e^{-j0.626\pi} \tag{3.126}$$

Therefore, the answer can be written:

$$x(t) = \left[\frac{1}{13}e^{-2t} + 0.2e^{-3t}\cos(5t - 0.626\pi)\right]u(t) \tag{3.127}$$

This solution is shown to be consistent with that of the previous example using the trigonometric identity $\cos(\alpha + \beta) = \cos(\alpha)\cos(\beta) - \sin(\alpha)\sin(\beta)$. ☐

CASE 3.3 Repeated Linear Factors. When $D(s)$ contains a repeated linear factor, for example, $(s - p)^n$, the LT must be expanded as

$$X(s) = \frac{N(s)}{D(s)} = \frac{N(s)}{(s - p)^n D_{\text{other}}(s)} = \frac{A_n}{(s - p)^n} + \frac{A_{n-1}}{(s - p)^{n-1}} + \cdots + \frac{A_1}{(s - p)}$$
$$+ \left[\text{other PFs corresponding to factors in } D_{\text{other}}(s)\right] \tag{3.128}$$

The PFE coefficients of the n fractions are found as follows: Define

$$\varphi_p(s) = (s - p)^n X(s) \tag{3.129}$$

Then

$$A_{n-i} = \frac{1}{i!} \frac{d^i}{ds^i} \varphi_p \bigg|_{s=p} \tag{3.130}$$

The similarity of these computations to residues is apparent, but only A_1 can be interpreted as the residue of pole p. We illustrate this method by example.

EXAMPLE 3.19:

Find the time signal x corresponding to LT $X(s) = (s + 4)/(s + 1)^3$.

SOLUTION 3.19   The order of D exceeds the order of N, so we may proceed. X has a third-order pole at $s = -1$, so we expand X as

$$X(s) = \frac{(s + 4)}{(s + 1)^3} = \frac{A_3}{(s + 1)^3} + \frac{A_2}{(s + 1)^2} + \frac{A_1}{(s + 1)} \tag{3.131}$$

Let $\varphi_{-1}(s) = (s + 1)^3 X(s) = (s + 4)$. Then

$$A_3 = \varphi_{-1}(-1) = 3, \quad A_2 = \frac{d\varphi_{-1}}{ds}\bigg|_{s=-1} = 1, \quad \text{and} \quad A_1 = \frac{1}{2}\frac{d^2\varphi_{-1}}{ds^2}\bigg|_{s=-1} = 0 \tag{3.132}$$

So

$$X(s) = \frac{3}{(s + 1)^3} + \frac{1}{(s + 1)^2} \tag{3.133}$$

Using Table 3.1, we have

$$x(t) = \left[\frac{3}{2}t^2 e^{-t} + t e^{-t}\right] u(t) \tag{3.134}$$

☐

CASE 3.4 Repeated Quadratic Factors. When $D(s)$ contains a repeated quadratic factor, e.g., $(s^2 + \beta s + \gamma)^n$, the LT may be either inverted by separating the quadratic into repeated linear factors [one nth-order factor for each of the complex roots; see (3.155)], then treated using the method of Case

3, or it can be expanded as

$$X(s) = \frac{N(s)}{D(s)} = \frac{N(s)}{(s^2 + \beta s + \gamma)^n D_{other}(s)} \tag{3.135}$$

$$= \sum_{i=0}^{n-1} \frac{(B_{n-i}s + C_{n-i})}{(s^2 + \beta s + \gamma)^n} + \left[\text{other PFs corresponding to factors in } D_{other}(s) \right]$$

The PFE coefficients of the n fractions are found algebraically as we illustrate by example.

EXAMPLE 3.20:

Find the time signal x corresponding to LT $X(s) = 4s^2/(s^2 + 1)^2(s + 1)$ [10].

SOLUTION 3.20   Recognize the factor $(s^2 + 1)^2$ as an $n = 2$-order quadratic factor with $\beta = 0$ and $\gamma = 1$. We know from our previous work that $\beta = -2 \, \text{Re}\{p\}$, where p and p^* are the poles associated with the quadratic factor, in this case $\pm j$. In turn, the real part of p provides the damping term in front of the sinusoid represented by the quadratic factor. In this case, therefore, we should expect one of the terms in the time signal to be a pure sinusoid.

Write X as

$$X(s) = \frac{(B_2 s + C_2)}{(s^2 + 1)^2} + \frac{(B_1 s + C_1)}{(s^2 + 1)} + \frac{A}{(s + 1)} \tag{3.136}$$

Using the familiar technique for simple linear factors, we find that $A = 1$. Now multiplying both sides of (3.136) by $(s^2 + 1)^2(s + 1)$, we obtain $4s^2 = (B_2 s + C_2)(s + 1) + (B_1 s + C_1)(s^2 + 1)(s + 1) + (s^2 + 1)^2$. Equating like powers of s, we obtain $B_2 = -1$, $C_2 = 1$, $B_1 = 2$, and $C_1 = -2$. Hence,

$$X(s) = \frac{1}{(s + 1)} + \frac{(2s - 2)}{(s^2 + 1)^2} - \frac{(s - 1)}{(s^2 + 1)} \tag{3.137}$$

We can now use Tables 3.1 and 3.2 to invert the three fractions. Note that the third term will yield an undamped sinusoid as predicted. Also note that the middle term is related to the third by differentiation. This fact can be used in obtaining the inverse. $\square$

3.4   Laplace Transform Analysis of Linear Systems

Let us return to the example circuit problem which originally motivated our discussion and discover ways in which the LT can be used in system analysis. Three fundamental means are available for using the LT in such problems. The most basic is the use of LT theory to solve the differential equation governing the circuit dynamics. The differential equation solution methods are quite general and apply to linear, constant coefficient differential equations arising in any context. The second method involves the use of the "system function," a LT-based representation of a LTI system, which embodies all the relevant information about the system dynamics. Finally, we preview "LT equivalent circuits," a LT method which is primarily used for circuit analysis.

Solution of the System Differential Equation

Consider a system is governed by a linear, constant coefficient differential equation of the form

$$\sum_{\ell=0}^{n_D} a_\ell \frac{d^\ell}{dt^\ell} y(t) = \sum_{\ell=0}^{n_N} b_\ell \frac{d^\ell}{dt^\ell} x(t) \tag{3.138}$$

with appropriate initial conditions given. (The numbers n_N and n_D should be considered fixed integers for now, but are later seen to be consistent with similar notation used in the discussion of rational LTs.) Using the linearity and time-differentiation properties of the LT, we can transform both sides of the equation to obtain

$$\sum_{\ell=0}^{n_D} a_\ell \left[s^\ell Y(s) - \sum_{i=0}^{\ell-1} y^{(i)}(0^-)s^{\ell-i-1} \right]$$
$$= \sum_{\ell=0}^{n_N} b_\ell \left[s^\ell X(s) - \sum_{i=0}^{\ell-1} x^{(i)}(0^-)s^{\ell-i-1} \right] \quad (3.139)$$

where $y^{(i)}$ and $x^{(i)}$ are the ith derivatives of y and x. Rearranging, we have

$$Y(s) = \frac{X(s) \sum_{\ell=0}^{n_N} b_\ell s^\ell - \sum_{\ell=0}^{n_N} b_\ell \sum_{i=0}^{\ell-1} x^{(i)}(0^-)s^{\ell-i-1} + \sum_{\ell=0}^{n_D} a_\ell \sum_{i=0}^{\ell-1} y^{(i)}(0^-)s^{\ell-i-1}}{\sum_{\ell=0}^{n_D} a_\ell s^\ell}$$

$$(3.140)$$

Given the input signal x and all necessary initial conditions on x and y, all quantities on the right side of (3.140) are known and can be combined to yield Y. Our knowledge of LT inversion will then, in principle, allow us to deduce y. This process often turns an unwieldy differential equation solution into simple algebraic operations. The price paid for this simplification, however, is that the process of inverting Y to obtain y is sometimes challenging.

Recall that in the motivating example, a similar conversion of the (nonhomogeneous) differential equation solution to algebraic operations occurred [recall (3.16) and surrounding discussion], as the superfluous term $e^{p_x t}$ was divided out of the equation. Except for the lack of attention paid to initial conditions, what remains in (3.16) is tantamount to a LT equation of form (3.139), as shown below. The homogeneous solution and related initial conditions were not included in this discussion to avoid obfuscating the main issue. The reader was encouraged to think of the LT as a process of "dividing" the "e^{st}" term out of the signals before starting the solution. We can now clearly see this fundamental connection between the differential equation and LT solutions.

EXAMPLE 3.21:

Return to the motivating example in Section 3.2 and solve the problem using LT analysis.

SOLUTION 3.21   Recall the differential equation governing the circuit, (3.1), $x(t) = LC(d^2y/dt^2) + RC(dy/dt) + y(t)$. The initial conditions are $y(0^-) = y_0$ and $i(0^-) = i_0$. Recall also that for convenience we seek the solution for $x(t) = M_x e^{j\theta_x} e^{(\sigma_x + j\omega_x)t} u(t) = M_x e^{j\theta_x} e^{p_x t} u(t)$, recognizing that the "correct" solution will be the real part of that obtained.

Taking the LT of each side of the differential equation, we have

$$X(s) = LC \left[s^2 Y(s) - sy(0^-) - y^{(1)}(0^-) \right] + RC \left[sY(s) - y(0^-) \right] + Y(s) \quad (3.141)$$

or

$$Y(s) = \frac{X(s)}{(LCs^2 + RCs + 1)} + \frac{(sLC - RC)y(0^-) + LCy^{(1)}(0^-)}{(LCs^2 + RCs + 1)} \quad (3.142)$$

Dividing both numerator and denominator of each fraction by LC, and inserting the LT $X(s) = M_x e^{j\theta_x}/(s - p_x)$ and the initial conditions [recall that $Cy^{(1)}(t) = i(t) \Rightarrow Cy^{(1)}(0^-) = i_0$], we have

$$Y(s) = \frac{M_x e^{j\theta_x}/LC}{(s - p_x)[s^2 + (R/L)s + 1]} + \frac{[s - (R/L)]y_0 + i_0/C}{[s^2 + (R/L)s + 1]} \quad (3.143)$$

Using PFE this can be written

$$
\begin{aligned}
Y(s) &= \frac{\dfrac{M_x e^{j\theta_x}/LC}{[p_x^2 + (R/L)p_x + 1]}}{(s - p_x)} + \frac{\dfrac{M_x e^{j\theta_x}/LC}{(p_h - p_x)}}{(s - p_h)} \\
&\quad + \frac{\left(\dfrac{M_x e^{j\theta_x}/LC}{(p_h - p_x)}\right)^*}{(s - p_h^*)} + \frac{[s - (R/L)]y_0 + i_0/C}{(s - p_h)(s - p_h^*)}
\end{aligned}
\tag{3.144}
$$

where p_h and p_h^* are the system poles. We have expanded the first fraction in (3.143) using simple linear factors involving the poles p_x, p_h, and p_h^*. The latter two poles correspond to the system [roots of $(s^2 + (R/L)s + 1)$], and the resulting terms in (3.144) are part of the natural response. The first pole, p_x, is attributable to the forcing function, and the resulting term in (3.144) will yield the forced response in the time domain. Finally, the last term in (3.144), which arises from the last term in (3.143), is also part of the natural response. The separation of the natural response into these three LT terms harkens back to the time-domain discussion about the distinct contributions of the input and initial conditions to the natural response. The last term is clearly related to initial conditions and the circuit's natural means of dissipating that energy. The former terms, which will also yield a damped sinusoid of identical complex frequency to that of the third term, are clearly "caused" by the input.   □

REMARK 3.2   The curious reader may wonder why the first fraction in (3.144) and (3.17), both of which represent the forced response, are not identical. After all, we have been encouraged to view the LT as a kind of generalized phasor representation. Recall, however, that the LT at a particular s must be thought of as a "phasor density." If an eigensignal of the system represents one complex frequency, e.g., $s = p_x$, the LT is infinitely dense at that point in the s-plane, corresponding to the existence of a pole there. If the signal does have a conventional phasor representation such as $\bar{Y} = M_y e^{j\theta_y}$, then this phasor will be related to the LT as

$$
\bar{Y} = Y(s)\frac{ds}{2\pi}\bigg|_{s=p_x} = \lim_{s \to p_x} Y(s)(s - p_x)
\tag{3.145}
$$

which we recognize as the residue of the LT Y at the pole p_x. The reader can easily verify this assertion using (3.143). This discussion is closely related to the interconnection between the Fourier series coefficients (which are similar to conventional sinusoidal phasors) and the Fourier transform. These signal representations are discussed in Chapter 4.

The System Function

DEFINITION 3.1   In our motivating example an input sinusoid with generalized phasor $M_x e^{j\theta_x}$ produced an output sinusoid with generalized phasor $M_y e^{j\theta_y}$. We discovered that the ratio of phasors was dependent only upon system parameters and the complex frequency, p_x. We noted that when considered as a function of the general complex frequency, s, this ratio is called the **system function**. That is, the system function is a complex function, H, of complex frequency, s, such that if a damped sinusoid of complex frequency $s = p_x$ and with generalized phasor $\bar{X} = M_x e^{j\theta_x}$ is used as input to the system, the forced response will be a sinusoid of complex frequency p_x and with generalized phasor

$$
M_y e^{j\theta_y} = H(p_x)M_x e^{j\theta_x}
\tag{3.146}
$$

Preview of Magnitude and Phase Responses

It is often important to know how a system will respond to a pure sinusoid of radian frequency, e.g., ω_x. In particular, we would like to know the amplitude and phase changes imposed on the sinusoid by the system. In terms of the definition just given, we see that this information is contained in the system function evaluated at frequency $p_x = j\omega_x$. In particular, $|H(j\omega_x)|$ represents the magnitude factor and $\arg\{H(j\omega_x)\}$ the phase change at this frequency. When plotted as functions of general frequency ω, the real functions $|H(j\omega)|$ and $\arg\{H(j\omega)\}$ are called the **magnitude** (or sometimes **frequency**) **response**, and **phase response** of the system, respectively. The complex function $H(j\omega)$ will be seen to be the Fourier transform of the impulse response (see Appendix B and Chapter 4) and is sometimes called the **transfer function** for the system. Returning to Fig. 3.2(b), the reader will discover that we have plotted the magnitude response for the series RLC circuit of Section 3.2 with numerical values given in (3.18).

The magnitude and phase responses of the system can be obtained graphically from the pole-zero diagram. Writing $H(s)$ similarly to (3.62), $H(s) = C \prod_{i=1}^{n_N}(s - z_i) / \prod_{i=1}^{n_D}(s - p_i)$. Therefore,

$$|H(j\omega)| = C \frac{\prod_{i=1}^{n_N} |j\omega - z_i|}{\prod_{i=1}^{n_D} |j\omega - p_i|} \quad \text{and}$$

$$\arg\{H(j\omega)\} = \sum_{i=1}^{n_N} \arg\{(j\omega - z_i)\} - \sum_{i=1}^{n_D} \arg\{(j\omega - p_i)\} \quad (3.147)$$

where we have assumed $C > 0$ (if not, add π radians to $\arg\{H(j\omega)\}$). By varying ω the desired plots are obtained as illustrated in Fig. 3.9.

DEFINITION 3.2 More generally, the **system function**, $H(s)$, for a LTI system can be defined as the ratio of the LT, for example, $Y(s)$, of the output, resulting from any input with LT, for example, $X(s)$, when the system is initially at rest (zero initial conditions). In other words, H is the ratio of the LT of the **zero-state response** to the LT of the input,

$$H(s) \overset{\text{def}}{=} \frac{Y(s)}{X(s)} \quad \text{when all initial conditions are zero} \quad (3.148)$$

While the latter definition is more general, the two definitions are consistent when the input is an eigensignal, as we show by example.

EXAMPLE 3.22:

Show that the two definitions of the system function above are consistent for the RLC circuit of Section 3.2.

SOLUTION 3.22 Replacing the specific complex frequency p_x by a general complex frequency s in the initial example using phasor analysis, we found that [recall (3.31)] $H(s) = 1/(LCs^2 + RCs + 1)$. On the other hand, using more formal LT analysis on the same problem, we derived (3.142). Forming the ratio $Y(s)/X(s)$ where both initial conditions are set to zero, yields an identical result. $\quad\square$

Finally, another very useful definition of the system function is as follows.

DEFINITION 3.3 The output of a LTI system to an impulse excitation, $x(t) = \delta(t)$ (see Section 3.6, Appendix A), when all initial conditions are zero, is called the **impulse response** of the system. The

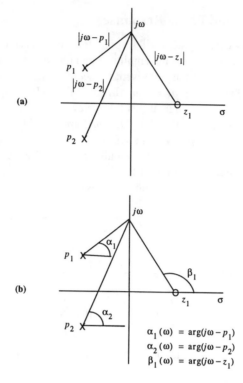

3.9   Pole-zero plot for an example system function, $H(s) = C(s - z_1)/(s - p_1)(s - p_2)$. (a) To obtain the magnitude response $|H(j\omega)|$ at frequency ω, the product of lengths from all zeroes to $s = j\omega$ is divided by the product of lengths from all poles to $s = j\omega$. The result must be multiplied by the gain term $|C|$. (b) To obtain the phase response $\arg\{H(j\omega)\}$ at frequency ω, the sum of angles from all poles to $s = j\omega$ is subtracted from the sum of angles from all zeroes to $s = j\omega$. An additional π radians is added if $C < 0$.

impulse response is usually denoted $h(t)$. The system function can be defined as the LT of the impulse response $H(s) = \mathcal{L}\{h(t)\}$.

The consistency of this definition with the first two is easy to demonstrate. Let H denote the system function for the system, regardless of how it might be related to $h(t)$. From Definition 2, we have $H(s) = Y(s)/X(s)$ for any valid LTs X and Y. Let $x(t) = \delta(t)$, in which case $X(s) = 1$. By definition, $y(t) = h(t)$, so $\mathcal{L}\{h(t)\} = \mathcal{L}\{y(t)\} = H(s)X(s) = H(s)$.

This interpretation of H enables us to find the impulse response of the system, a task which is not always easy in the time domain because of the pitfalls of working with impulse functions.

EXAMPLE 3.23:

Find the impulse response, h, for the circuit of Section 3.2.

SOLUTION 3.23   Using various means we showed that the system function is $H(s) = 1/(LCs^2 + RCs + 1)$. Let us find h by computing $h(t) = \mathcal{L}^{-1}\{H(s)\}$. Using the quadratic equation, we find the roots of the denominator (poles of the system) to be [c.f., (3.6)]

$$p_1, p_2 = \frac{-RC \pm \sqrt{R^2C^2 - 4LC}}{2LC} = -\frac{R}{2L} \pm \sqrt{\frac{R^2}{4L^2} - \frac{1}{LC}} \tag{3.149}$$

Assume that these poles are complex conjugates and call them p_h, p_h^*, where

$$p_h = -\frac{R}{2L} + j\sqrt{\frac{1}{LC} - \frac{R^2}{4L^2}} \overset{\text{def}}{=} \sigma_h + j\omega_h \tag{3.150}$$

Comparing to our initial work on finding the homogeneous solution of the differential equation, we see that these system poles are the roots of the characteristic equation. The reason for using subscript "h" in our early work should now be clear. In terms of (3.150) we can rewrite H as

$$H(s) = \frac{1/LC}{(s - \sigma_h)^2 + \omega_h^2} \tag{3.151}$$

[recall (3.59)]. Now using Table 3.1: $\mathcal{L}\{e^{\sigma_h t}\cos(\omega_h t + \theta_h)\} = [(s - \sigma_h)\cos(\theta_h) - \omega_h \sin(\theta_h)]/[(s - \sigma_h)^2 + \omega_h^2]$. Letting $\theta_h = \pi/2$ and using linearity, we have

$$h(t) = -\frac{1}{LC\omega_h}e^{\sigma_h t}\cos\left(\omega_h t + \frac{\pi}{2}\right)u(t) = \frac{1}{LC\omega_h}e^{\sigma_h t}\sin(\omega_h t)u(t) \tag{3.152}$$

We could also note that $|p_h| = \sqrt{1/LC}$ and write the initial scale factor as $|p_h|^2/\omega_h$. □

We see clearly that the impulse response is closely related to the natural responses of a system, which in turn are tied to the homogeneous solution of the differential equation. These transient responses depend only on properties of the system, and not on properties of the input signal (beyond the initial instant of excitation). The form of the homogeneous differential equation solution is specified by the number and values of the roots of the characteristic equation. These roots are, thus, the poles of the system function H. The system function offers an extremely valuable tool for the design and analysis of systems. We now turn to this important topic.

Poles and Zeroes—Part II: Stability Analysis of Systems

We return to the issue of poles and zeroes, this time with attention restricted to rational LTs. In particular, we focus on the poles and zeroes of a system function and their effects on the performance of the system.

Natural Modes

The individual time responses corresponding to the poles of H are often called **natural modes** of the system. These modes are indicators of the physical properties of the system (e.g., circuit values), which in turn determine the natural way in which the system will dissipate, store, amplify, or respond to energy of various frequencies. Consider two general cases. Suppose H has a real pole of order n at $s = p$, so that it can be written

$$H(s) = \sum_{i=1}^{n}\frac{A_i}{(s - p)^i} + \text{[other terms]} \tag{3.153}$$

We know from previous work, therefore, that (see Table 3.1) h has corresponding modal components,

$$h(t) = \sum_{i=1}^{n}\frac{A_i}{(i - 1)!}t^{i-1}e^{pt}u(t) + \text{[other terms]} \tag{3.154}$$

When $|p| < 0$, the modal components due to pole p will decay exponentially with time (modulated by the terms t^{i-1}). When $|p| > 0$, these modal components will increase exponentially with time

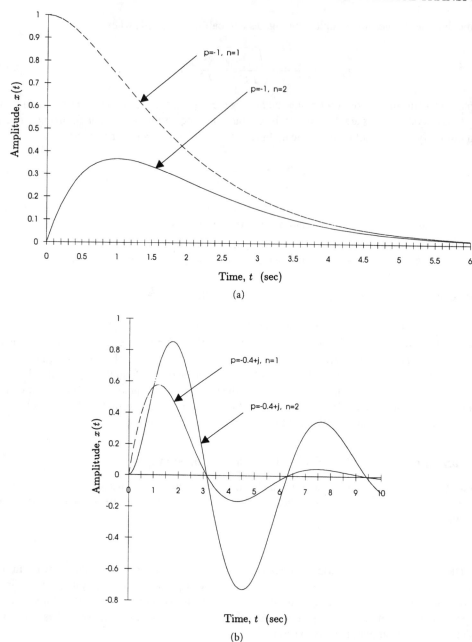

3.10   Modal components in the impulse response corresponding to (a) a real pole of orders $n = 1$ and $n = 2$ at $s = p$, and (b) a complex pole pair of orders $n = 1$ and $n = 2$ at $s = p, p^*$.

(modulated by the terms t^{i-1}). When $p = 0$, the term will either remain bounded if $n = 1$, or the terms will increase with power n as t increases. These cases are illustrated in Fig. 3.10(a).

Next, let H have a complex pole pair of order n at $s = p, p^*$, so that it can be written

$$H(s) = \sum_{i=1}^{n} \left[\frac{E_i}{(s-p)^i} + \frac{E_i^*}{(s-p^*)^i} \right] + [\text{other terms}] \qquad (3.155)$$

Using Table 3.1 we see that h has a corresponding modal component,

$$h(t) = \sum_{i=1}^{n} \frac{2|E_i|}{(i-1)!} t^{i-1} e^{\sigma_p t} \cos(\omega_p t + \theta_{E_i}) u(t) + [\text{other terms}] \tag{3.156}$$

where $p = \sigma_p + j\omega_p$, and $\theta_{E_i} = \arg\{E_i\}$. In this case, when $|p| < 0$, the sinusoidal modal components will decay exponentially with time (modulated by the terms t^{i-1}); when $|p| > 0$, these terms will increase exponentially with time (modulated by the terms t^{i-1}); and when $p = 0$, the terms due to p will either represent a constant or increasing sinusoid, depending on the value of n. These cases are illustrated in Fig. 3.10(b).

BIBO stability

We digress momentarily to discuss the concept of stability. To avoid some unnecessary complications, we restrict the discussion to *causal* systems, and, as usual, to the unilateral LT. There are various ways to define stability, but the most frequent and useful definition for a LTI system is that of BIBO stability. A system is said to be **bounded-input–bounded-output** (BIBO) **stable** iff every bounded input produces a bounded output. Formally, any bounded input, for example, x such that $|x(t)| \le B_x < \infty$ for all t, must result in an output, y, for which $B_y < \infty$ exists so that $|y(t)| \le B_y$ for all t.

A necessary and sufficient condition for BIBO stability of a LTI system is that its impulse response be absolutely integrable,

$$\int_0^\infty |h(t)| \, dt < \infty \tag{3.157}$$

This is easy to show. First, assume $|x(t)| < B_x$ for all t. Then, from the convolution integral

$$|y(t)| = \left| \int_0^\infty x(t-\lambda) h(\lambda) \, d\lambda \right| \tag{3.158}$$

Now, using the Schwarz inequality (see [7]), we have

$$|y(t)| \le \int_0^\infty |x(t-\lambda)| \, |h(\lambda)| \, d\lambda \le B_x \int_0^\infty |h(\lambda)| \, d\lambda \tag{3.159}$$

Consequently, it is sufficient that (3.157) be true for $B_y < \infty$ to exist. On the other hand, suppose that condition (3.157) were not true, but the system were BIBO stable. For a fixed t, consider the input

$$x(t-\lambda) = \begin{cases} 1, & h(\lambda) > 0 \\ -1, & h(\lambda) < 0 \\ 0, & h(\lambda) = 0 \end{cases} \tag{3.160}$$

For this input, the output at time t is

$$y(t) = \int_0^\infty |h(\lambda)| \, d\lambda \tag{3.161}$$

which is not bounded according to assumption. Therefore, we encounter a contradiction showing that the condition (3.157) is also necessary for BIBO stability.

Stability and Natural Modes

Now we tie the stability analysis to the discussion of modal components above. We assert that *a causal LTI system with proper rational system function H will be BIBO stable iff all of its poles are in the left-half s-plane.* That this is true is easily seen. If any pole is in the right-half s-plane, we know that h will contain at least one mode that will increase without bound. Therefore, (3.157) cannot hold and the system is not BIBO stable. Conversely, if all poles are in the left-half s-plane, (3.157) will hold. The case in which one or more *simple* poles fall exactly on the $s = j\omega$ axis (and none in the right-half s-plane) is called **marginal BIBO stability**. In this case (3.157) does not hold, so the system is not strictly BIBO stable. However, the system does theoretically produce bounded outputs for some inputs. Finally, note that we show by example below Example 3.24(a) that an improper rational system function cannot represent a BIBO stable system.

Based on our earlier discussion of LT ROCs, an equivalent way to state the BIBO stability condition is as follows: *A causal, LTI system is BIBO stable iff the ROC of its system function H includes the $s = j\omega$ axis in the s-plane.* We recall that the failure to include the $j\omega$ axis would imply at least one pole of H in the right-half s-plane.

Let us conclude this discussion by considering some examples.

EXAMPLE 3.24:

Comment on the BIBO stability of the following systems:

(a) $H(s) = \dfrac{N(s)}{D(s)} = \dfrac{(1/2)s^3}{s^2 + 2s + 1} = \dfrac{(1/2)s^3}{(s+1)^2}$

(b) $H(s) = \dfrac{(s-3)}{s^2 + 5s + 4}$

(c) $H(s) = \dfrac{(s-3)}{(s^2 + 5s + 4)(s^2 - 2s - 3)}$

(d) $H(s) = \dfrac{6}{[(s+3)^2 + 25](s+8)}$

(e) $H(s) = \dfrac{s}{s^2 + 4}$

SOLUTION 3.24

(a) H is not a proper rational fraction because the order of $N(s) >$ order of $D(s)$. We illustrate that such a system is not BIBO stable. Dividing D into N, we can write

$$H(s) = \frac{1}{2}s - 1 + \frac{3/2}{s(s^2 + 2s + 1)} \qquad (3.162)$$

Suppose we enter $x(t) = u(t)$ as a (bounded) input. Then

$$Y(s) = H(s)X(s) = \frac{H(s)}{s} = \frac{1}{2} - \frac{1}{s} + \frac{3/2}{s(s^2 + 2s + 1)} \qquad (3.163)$$

Therefore,

$$y(t) = \frac{1}{2}\delta(t) - u(t) + \left[\text{terms resulting from } \frac{3/2}{s(s^2 + 2s + 1)} \right] \qquad (3.164)$$

The output is unbounded in response to a bounded input and is therefore not BIBO stable. *A similar result will occur whenever H is not proper.*

(b) H has poles at $s = -1, -4$, and a zero at $s = 3$. The system is BIBO stable. Note that the right-half plane zero has no adverse effect on stability.

(c) $H(s) = N(s)/D(s)$ has a second-order pole at $s = -1$, and a simple pole at $s = -4$. Both N and D have roots at $s = 3$, thus, neither a pole nor zero is found there (they cancel). The system is therefore BIBO stable.

REMARK 3.3 The response to *nonzero initial conditions* of a system which has one or more "canceled poles" in the right-half s-plane will increase without bound, and therefore could be considered "unstable" in some sense, even though it is BIBO stable. The reader is invited to show that the present system will respond in this undesirable manner to nonzero initial conditions. A system is said to be **asymptotically stable** if it is BIBO stable and its response to initial conditions approaches zero as $t \to \infty$. We see that asymptotic stability implies BIBO stability but the converse is not true.

(d) Recalling (3.59) we find that H has poles at $s = -3 \pm j5$ and $s = -8$. No finite zeroes are included. The system is BIBO stable and will have both an oscillatory mode and a damped exponential mode.

(e) H has poles at $s = \pm j2$ which are on the $j\omega$ axis. The system is *marginally* BIBO stable. □

EXAMPLE 3.25:

Return to the series RLC circuit of Section 3.2 and discuss the system's stability as R varies for a fixed L and C. Assume that $R < 2\sqrt{L/C}$.

SOLUTION 3.25 In light of (3.149) the upper bound on R means that the poles of the system will always be complex conjugates, e.g., p_h, p_h^* [h is oscillatory]. From (3.150) we see that the poles of the system can be written in polar form as

$$p_h, p_h^* = \sqrt{\sigma_h^2 + \omega_h^2}\, e^{\pm j\,\tan^{-1}(\omega_h/\sigma_h)} = \frac{1}{\sqrt{LC}} e^{\pm j\,\tan^{-1}(\omega_h/\sigma_h)} \tag{3.165}$$

For fixed L and C, the poles remain on a circle of radius $1/\sqrt{LC}$ in the s-plane. Having established this fact, recall the Cartesian form of p_h given in (3.149) and note that as $R \to 0$, $p_h \to j\sqrt{1/LC}$. On the other hand, as $R \to 2\sqrt{L/C}$, $p_h \to -\sqrt{1/LC}$ (see Fig. 3.11). Therefore, the poles remain in the left-half s-plane over the specified range of R, except when $R = 0$, in which case the poles are exactly on the $j\omega$-axis at $s = \pm j(1/\sqrt{LC})$. Therefore, the system is BIBO stable, except when $R = 0$, in which case it is marginally stable. Only if R were to take negative values would the circuit go unstable. This is not realistic unless active circuit elements are present which effectively present negative resistance. The reader is encouraged to explore what happens as R continues to increase beyond the given upper bound. □

Laplace-Domain Phasor Circuits

After completing a sufficient number of circuit problems involving differential equations like Example 3.21, certain patterns would become apparent in a Laplace-transformed differential equation like (3.141). These patterns occur precisely because of the invariant relationships between currents and voltages across lumped parameter components. For example, because the current and voltage through and inductor are related as $v_L(t) = L(di_L/dt)$, whenever this relationship is encountered and transformed in a circuit problem, it becomes $V_L(s) = sLI_L(s) - Li_L(0^-)$. This relationship

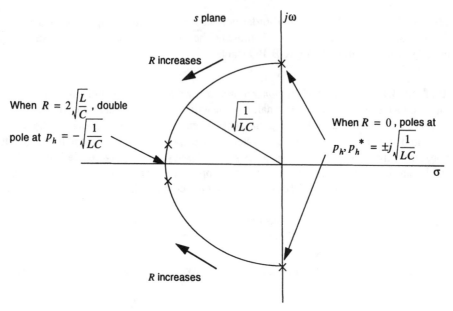

3.11   Locus of the poles as R varies in Example 3.25.

may be written without recourse to the time domain by treating $V_L(s)$ and $I_L(s)$ as a "DC voltage" and a "DC current," by replacing the inductor with a "resistor" with value sL, by adding a voltage source in series with value $Li_L(0^-)$ having proper polarity, a small DC circuit problem.

In general, we can make "Laplace" substitutions for each component in a circuit, then use "DC" analysis to write the Laplace-transformed differential equation directly. The appropriate substitutions and variations for both mesh and nodal analysis are discussed in detail in Chapter 36, along with more advanced frequency-domain uses of such LT replacements. After studying that material, the reader may wish to provide closure to the present discussion by solving the problem below.

EXAMPLE 3.26:

Derive the Laplace-transformed differential equation, (3.141), using Laplace circuit equivalents.

3.5   Conclusions and Further Reading

The Laplace transform (LT) is a powerful tool in the analysis and design of linear, time-invariant circuits and systems. In this exposition we have developed the LT by appealing to the manner in which it turns differential equation solutions into simple algebraic problems. We have focused not only on the technique for exploiting this advantage, but the reasons that this happens. Along the way we have discovered some useful properties of the LT, as well as the meaning of poles and zeroes. We also discussed numerous techniques for recovering the time signal from the LT, noting that this sometimes difficult task is the price paid for alleviating the difficulties of time-domain solution. In the latter part of the chapter we turned our attention to the analysis of systems, in particular the meaning and derivation of the system function. This study led to the understanding of the relationship between the system function and impulse response, and between the pole locations of the system function and stability of the system.

Most of our work has focused on the unilateral LT (ULT), although the bilateral LT (BLT) was carefully discussed at the outset. The advantage of the ULT is that it provides a way to incorporate initial condition information, a very important property in many design and analysis problems, particularly because the initial conditions play an important role in the transient response of the system. On the other hand, the BLT can handle a much broader class of signals, a feature that is often advantageous in theoretical and formal developments. The region of convergence (ROC) of the LT in the s-plane becomes much more important for the BLT as the LT itself is not unique without it. Accordingly, we spent a significant amount of effort studying the ROC. We later saw that the ROC is closely related to stability analysis, even for ULT analysis.

The LT is closely related to the Fourier and z-transforms which are the subjects of subsequent chapters. Section 3.7, Appendix B previews these ideas.

Finally, we note that several LT topics have not been treated here. These topics, such as the analysis of feedback systems and state-space models, deal principally with control theory and applications and are therefore outside of the scope of this book. Many fine books on signals and systems are available to which the reader can turn to explore these subjects. Some of these texts are referenced below. Many excellent books on circuit analysis and filter design, too numerous to cite here, are also available.

3.6 Appendix A: The Dirac Delta (Impulse) Function

The **Dirac delta, or impulse, function**, $\delta(t)$, is defined as the function with the following properties: If signal x is continuous at t_0 ("where the impulse is located in time"), then,

$$\int_a^b x(t)\delta(t - t_0)\, dt = \begin{cases} 0, & t_0 \notin [a, b] \\ x(t_0), & t_0 \in (a, b) \end{cases} \tag{3.166}$$

Two special cases are noted:

1. Note that (3.166) does not cover the case in which the impulse is located *exactly* at one of the limits of integration. In such cases whether x is continuous at t_0,

   The integral takes the value $\frac{1}{2}x(t_0^+)$ if t_0 is the lower limit, $t_0 = a$.

   The integral takes the value $\frac{1}{2}x(t_0^-)$ if t_0 is the upper limit, $t_0 = b$.

2. The only case not explicitly covered is one in which x is *discontinuous* at t_0 and t_0 is not a limit of integration. In this case the integral takes the value $\frac{1}{2}[x(t_0^-) + x(t_0^+)]$ if $t_0 \in (a, b)$, and 0 otherwise. Note that this answer is also valid if x is continuous at t_0, but it is unnecessarily complicated.

Note what happens in the special case in which $x(t) = 1$, $a = -\infty$, $b = t$, and $t_0 = 0$. From the definition, we can write

$$\int_{-\infty}^t \delta(\lambda)\, d\lambda = \begin{cases} 0, & t < 0 \\ 1, & t > 0 \\ \frac{1}{2}, & t = 0 \end{cases} \tag{3.167}$$

We see that

$$\int_{-\infty}^t \delta(\lambda)\, d\lambda = u(t) \tag{3.168}$$

Therefore, except at $t = 0$,

$$\frac{du}{dt} = \delta(t) \tag{3.169}$$

What emerges here is a very strange function. We see from (3.168) that δ must be zero *everywhere except at $t = 0$* because, apparently we accumulate area at only that point. The area under that one point is unity because the integral takes a jump from 0 to 1 as t crosses zero. Because δ has zero width and unity total area, it must have *infinite amplitude* (at that one point!). To indicate the delta function, therefore, we draw "arrows" as shown in Fig. 3.12. It is sometimes mathematically useful to indicate a delta function with area other than unity. In this case we simply label the arrow with a number called the "weight" of the impulse. Note that it makes no sense to draw taller and shorter arrows for different weights, although this is sometimes done in textbooks, because the weight does not indicate the "height" of the function (∞!), rather, its *area*!

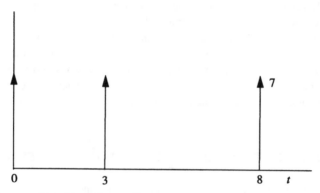

3.12   The impulse functions $\delta(t)$, $\delta(t-3)$, and $7\delta(t-8)$.

Finally, note that computing integrals with delta functions in them is very easy because one need only follow the rules of the definition. For example,

$$\int_{-18}^{\infty} e^{\pi(\lambda-4)}\delta(\lambda-6)\,d\lambda = e^{2\pi} \tag{3.170}$$

and

$$\int_{0}^{2} 3u(t)\delta(t)\,dt = 3\frac{u(0^{+})}{2} = \frac{3}{2} \tag{3.171}$$

REMARK 3.4   The kth derivative of the impulse (usually $k \leq 2$) occasionally appears in LT work. The notation

$$\delta^{(k)}(t) \stackrel{\text{def}}{=} \frac{d^{k}\delta}{dt^{k}} \tag{3.172}$$

is often used to denote this signal. The signal $\delta^{(1)}(t)$ is called a **doublet** and is plotted as shown in Fig. 3.13.

3.7   Appendix B: Relationships among the Laplace, Fourier, and z-Transforms

The transforms previewed here are most frequently defined and discussed for two-sided signals. Therefore, it is most natural to base this discussion on the BLT, as defined in (3.38).

The **Fourier transform** (FT), X_F, of a signal x is defined by the integral

$$X_F(\omega) = \int_{-\infty}^{\infty} x(t)e^{j\omega t}\,dt \tag{3.173}$$

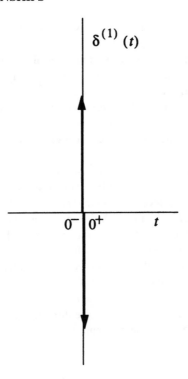

3.13   The doublet.

Upon comparison to (3.38), it is apparent that the FT evaluated at radian frequency ω is equivalent to the BLT of x evaluated at $s = j\omega$ in the s-plane. The FT can, therefore, be obtained over all ω by evaluating the BLT, e.g., X_L, along the $j\omega$ axis:

$$X_F(\omega) = X_L(s)|_{s=j\omega} \tag{3.174}$$

Evidently, the FT will only exist for a signal x if its BLT has a ROC that includes the $j\omega$ axis.

One very important class of signals whose BLT ROCs include the entire left-half s-plane, but not the $j\omega$ axis, is the periodic signals. For this purpose, the **Fourier series** (FS) can be used to expand the signal on a set of discrete, harmonically related, basis functions (either complex exponentials or sinusoids). The complex exponential version is

$$x(t) = \sum_{\ell=-\infty}^{\infty} c_\ell e^{j\ell\omega_0 t} \tag{3.175}$$

where $\omega_0 = 2\pi/T_0$ is the fundamental radian frequency, with T_0 the period of the waveform. The complex numbers c_ℓ, $\ell = \ldots, -1, 0, 1, 2, \ldots$ are the **FS coefficients** computed as

$$c_\ell = \frac{1}{T_0} \int_{T_0} x(t) e^{-j\ell\omega_0 t} \, dt \tag{3.176}$$

where the integral is taken over any period of the waveform.

Comparing (3.176) and (3.173), we see that the FS coefficients are equivalent to (scaled) samples of the FT *of one period* of x, where the samples are taken at frequencies $\ell\omega_0$. If we have a periodic waveform, therefore, we can always represent it by samples of the FT of one period. Similarly, if we have a "short" signal and want to represent it using only frequency samples, we can let it be periodic and represent it using the FS coefficients. In this case we simply need to recall that the signal is not truly periodic and work with only one period.

Conversely, the FT may be represented using only **samples** of the time waveform by artificially letting the FT become periodic, then letting the time samples play the role of the FS coefficients. In this case we, in effect, let the BLT become periodic along the $j\omega$ axis. This "backward FS" is what is known as the **discrete-time Fourier transform** (DTFT). The DTFT is discussed in Chapter 4 along with the **discrete Fourier transform**, a Fourier-type transform which is discrete and periodic in both time and frequency. For the latter, the connections to the BLT are too obtuse to describe in brief terms here.

Finally, if we let the BLT become periodic in ω with some fixed period *along each σ line*, the BLT can also be represented by discrete-time samples. This is similar to writing a "FS" which changes for each σ. This **discrete-time Laplace transform** (DTLT) could, in principle, be used in the design and analysis of discrete-time systems in much the same way the BLT is used in continuous-time work. For historical reasons and for mathematical convenience, however, the z-**transform** (ZT) is almost universally used. The ZT is obtained from the DTLT using the mapping $e^{sT} \rightarrow z$, where T is the sample period on the time signal. As a consequence of this mapping, "strips" in the s-plane map into annuli in the z-plane. Therefore, the ROC of a ZT takes the form of an annulus and the unit circle in the z-plane plays the role of the $j\omega$ axis in the s-plane.

These ideas will become clearer and more precise through the study of successive chapters. The reader is also encouraged to see [2] for an elementary approach to discrete FTs.

References

[1] Carlson, G.E., *Signal and Linear System Analysis,* Boston, Houghton-Mifflin, 1992.

[2] Deller, J.R., "Tom, Dick, and Mary discover the DFT," *IEEE Signal Processing Mag.,* 11, 36–50, April 1994.

[3] Churchill, R.V., Brown, J.W., and Verhey, R.F., *Complex Variables and Applications,* 3rd ed., New York, McGraw-Hill, 1976.

[4] Doetsch, G., *Guide to the Applications of Laplace Transforms,* New York, Van Nostrand, 1961.

[5] Hayt, W.H. and Kemmerly, J.E., *Engineering Circuit Analysis,* New York, McGraw-Hill, 1971.

[6] Houts, R.C., *Signal Analysis in Linear Systems,* Philadelphia, Saunders, 1989.

[7] Kolmogorov, A.N. and Fomin, S.V., *Introductory Real Analysis,* New York, Dover, 1975. Translated and edited by R. A. Silverman.

[8] Kraniauskas, P., *Transforms in Signals and Systems,* Reading, MA, Addison-Wesley, 1992.

[9] LePage, W., *Complex Variables and the Laplace Transform for Engineers,* New York, McGraw-Hill, 1961.

[10] MacGillem, C.D. and Cooper, G.R., *Continuous and Discrete Signal and System Analysis,* 3rd ed., Philadelphia, Saunders, 1991.

4

Fourier Series, Fourier Transforms and the DFT

W. Kenneth Jenkins
University of Illinois,
Urbana-Champaign

4.1   Introduction

Fourier methods are commonly used for signal analysis and system design in modern telecommunications, radar, and image processing systems. Classical Fourier methods such as the Fourier series and the Fourier integral are used for continuous time (CT) signals and systems, i.e., systems in which a characteristic signal, $s(t)$, is defined at all values of t on the continuum $-\infty < t < \infty$. A more recently developed set of Fourier methods, including the discrete time Fourier transform (DTFT) and the discrete Fourier transform (DFT), are extensions of basic Fourier concepts that apply to discrete time (DT) signals. A characteristic DT signal, $s[n]$, is defined only for values of n where n is an integer in the range $-\infty < n < \infty$. The following discussion presents basic concepts and outlines important properties for both the CT and DT classes of Fourier methods, with a particular emphasis on the relationships between these two classes. The class of DT Fourier methods is particularly useful as a basis for digital signal processing (DSP) because it extends the theory of classical Fourier analysis to DT signals and leads to many effective algorithms that can be directly implemented on general computers or special purpose DSP devices.

The relationship between the CT and the DT domains is characterized by the operations of sampling and reconstruction. If $s_a(t)$ denotes a signal $s(t)$ that has been uniformly sampled every T seconds, then the mathematical representation of $s_a(t)$ is given by

$$s_a(t) = \sum_{n=-\infty}^{\infty} s(t)\delta(t - nT) \tag{4.1}$$

where $\delta(t)$ is a CT impulse function defined to be zero for all $t \neq 0$, undefined at $t = 0$, and has unit area when integrated from $t = -\infty$ to $t = +\infty$. Because the only places at which the product $s(t)\delta(t - nT)$ is not identically equal to zero are at the sampling instants, $s(t)$ in (4.1) can be replaced with $s(nT)$ without changing the overall meaning of the expression. Hence, an alternate expression for $s_a(t)$ that is often useful in Fourier analysis is given by

$$s_a(t) = \sum_{n=-\infty}^{\infty} s(nT)\delta(t - nT) \tag{4.2}$$

The CT sampling model $s_a(t)$ consists of a sequence of CT impulse functions uniformly spaced at intervals of T seconds and weighted by the values of the signal $s(t)$ at the sampling instants, as depicted in Fig. 4.1. Note that $s_a(t)$ is not defined at the sampling instants because the CT impulse function itself is not defined at $t = 0$. However, the values of $s(t)$ at the sampling instants are imbedded as "area under the curve" of $s_a(t)$, and as such represent a useful mathematical model of the sampling process. In the DT domain the sampling model is simply the sequence defined by taking the values of $s(t)$ at the sampling instants, i.e.,

$$s[n] = s(t)|_{t=nT} \tag{4.3}$$

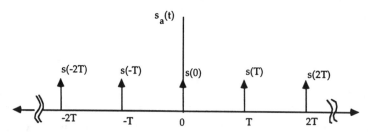

4.1   CT model of a sampled CT signal.

In contrast to $s_a(t)$, which is not defined at the sampling instants, $s[n]$ is well defined at the sampling instants, as illustrated in Fig. 4.2. Thus, it is now clear that $s_a(t)$ and $s[n]$ are different but equivalent models of the sampling process in the CT and DT domains, respectively. They are both useful for signal analysis in their corresponding domains. Their equivalence is established by the fact that they have equal spectra in the Fourier domain, and that the underlying CT signal from which $s_a(t)$ and $s[n]$ are derived can be recovered from either sampling representation, provided a sufficiently large sampling rate is used in the sampling operation (see below).

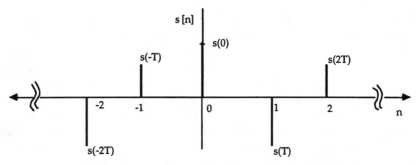

4.2   DT model of a sampled CT signal.

4.2 Fourier Series Representation of Continuous Time Periodic Signals

It is convenient to begin this discussion with the classical Fourier series representation of a periodic time domain signal, and then derive the Fourier integral from this representation by finding the limit of the Fourier coefficient representation as the period goes to infinity. The conditions under which a periodic signal $s(t)$ can be expanded in a Fourier series are known as the Dirichet conditions. They require that in each period $s(t)$ has a finite number of discontinuities, a finite number of maxima and minima, and that $s(t)$ satisfies the following absolute convergence criterion [1]:

$$\int_{-T/2}^{T/2} |s(t)|\, dt < \infty \tag{4.4}$$

It is assumed in the following discussion that these basic conditions are satisfied by all functions that will be represented by a Fourier series.

Exponential Fourier Series

If a CT signal $s(t)$ is periodic with a period T, then the classical complex Fourier series representation of $s(t)$ is given by

$$s(t) = \sum_{n=-\infty}^{\infty} a_n e^{jn\omega_0 t} \tag{4.5a}$$

where $\omega_0 = 2\pi/T$, and where the a_n are the complex Fourier coefficients given by

$$a_n = (1/T) \int_{-T/2}^{T/2} s(t) e^{-jn\omega_0 t}\, dt \tag{4.5b}$$

It is well known that for every value of t where $s(t)$ is continuous, the right-hand side of (4.5a) converges to $s(t)$. At values of t where $s(t)$ has a finite jump discontinuity, the right-hand side of (4.5a) converges to the average of $s(t^-)$ and $s(t^+)$, where $s(t^-) \equiv \lim_{\epsilon \to 0} s(t - \epsilon)$ and $s(t^+) \equiv \lim_{\epsilon \to 0} s(t + \epsilon)$.

For example, the Fourier series expansion of the sawtooth waveform illustrated in Fig. 4.3 is characterized by $T = 2\pi$, $\omega_0 = 1$, $a_0 = 0$, and $a_n = a_{-n} = A\cos(n\pi)/(jn\pi)$ for $n = 1, 2, \ldots,$. The coefficients of the exponential Fourier series represented by (4.5b) can be interpreted as the spectral representation of $s(t)$, because the a_n-th coefficient represents the contribution of the $(n\omega_0)$-th frequency to the total signal $s(t)$. Because the a_n are complex valued, the Fourier domain representation has both a magnitude and a phase spectrum. For example, the magnitude of the a_n is plotted in Fig. 4.4 for the sawtooth waveform of Fig. 4.3. The fact that the a_n constitute a discrete set is consistent with the fact that a periodic signal has a "line spectrum," i.e., the spectrum contains only integer multiples of the fundamental frequency ω_0. Therefore, the equation pair given by (4.5a) and (4.5b) can be interpreted as a transform pair that is similar to the CT Fourier transform for periodic signals. This leads to the observation that the classical Fourier series can be interpreted as a special transform that provides a one-to-one invertible mapping between the discrete-spectral domain and the CT domain. The next section shows how the periodicity constraint can be removed to produce the more general classical CT Fourier transform which applies equally well to periodic and aperiodic time domain waveforms.

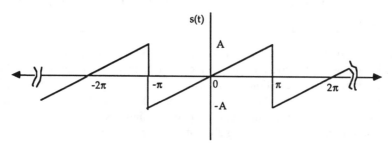

4.3   Periodic CT signal used in Fourier series example.

The Trigonometric Fourier Series

Although Fourier series expansions exist for complex periodic signals, and Fourier theory can be generalized to the case of complex signals, the theory and results are more easily expressed for real-valued signals. The following discussion assumes that the signal $s(t)$ is real-valued for the sake of simplifying the discussion. However, all results are valid for complex signals, although the details of the theory will become somewhat more complicated.

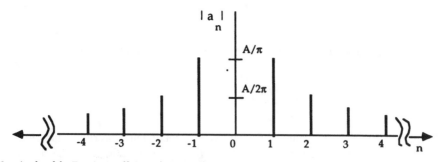

4.4   Magnitude of the Fourier coefficients for example of Figure 4.3.

For real-valued signals $s(t)$, it is possible to manipulate the complex exponential form of the Fourier series into a trigonometric form that contains $\sin(\omega_0 t)$ and $\cos(\omega_0 t)$ terms with corresponding real-valued coefficients [1]. The trigonometric form of the Fourier series for a real-valued signal $s(t)$ is given by

$$s(t) = \sum_{n=0}^{\infty} b_n \cos(n\omega_0 t) + \sum_{n=1}^{\infty} c_n \sin(n\omega_0 t) \qquad (4.6a)$$

where $\omega_0 = 2\pi/T$. The b_n and c_n are real-valued Fourier coefficients determined by

$$
\begin{aligned}
b_0 &= (1/T) \int_{-T/2}^{T/2} s(t)\, dt \\
b_n &= (2/T) \int_{-T/2}^{T/2} s(t) \cos(n\omega_0 t)\, dt, \qquad n = 1, 2, \ldots, \\
c_n &= (2/T) \int_{-T/2}^{T/2} s(t) \sin(n\omega_0 t)\, dt, \qquad n = 1, 2, \ldots,
\end{aligned}
\qquad (4.6b)
$$

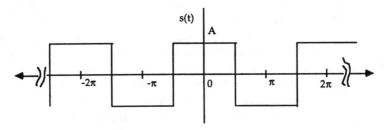

4.5   Periodic CT signal used in Fourier series example 2.

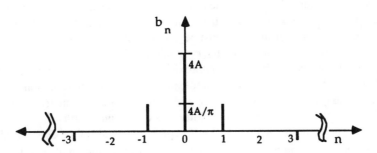

4.6   Fourier coefficients for example of Figure 4.5.

An arbitrary real-valued signal $s(t)$ can be expressed as a sum of even and odd components, $s(t) = s_{\text{even}}(t) + s_{\text{odd}}(t)$, where $s_{\text{even}}(t) = s_{\text{even}}(-t)$ and $s_{\text{odd}}(t) = -s_{\text{odd}}(-t)$, and where $s_{\text{even}}(t) = [s(t) + s(-t)]/2$ and $s_{\text{odd}}(t) = [s(t) - s(-t)]/2$. For the trigonometric Fourier series, it can be shown that $s_{\text{even}}(t)$ is represented by the (even) cosine terms in the infinite series, $s_{\text{odd}}(t)$ is represented by the (odd) sine terms, and b_0 is the DC level of the signal. Therefore, if it can be determined by inspection that a signal has a DC level, or if it is even or odd, then the correct form of the trigonometric series can be chosen to simplify the analysis. For example, it is easily seen that the signal shown in Fig. 4.5 is an even signal with a zero DC level. Therefore it can be accurately represented by the cosine series with $b_n = 2A \sin(\pi n/2)/(\pi n/2), n = 1, 2, \ldots$, as illustrated in Fig. 4.6. In contrast, note that the sawtooth waveform used in the previous example is an odd signal with zero DC level; thus, it can be completely specified by the sine terms of the trigonometric series. This result can be demonstrated by pairing each positive frequency component from the exponential series with its conjugate partner, i.e., $c_n = \sin(n\omega_0 t) = a_n e^{jn\omega_0 t} + a_{-n} e^{-jn\omega_0 t}$, whereby it is found that $c_n = 2A\cos(n\pi)/(n\pi)$ for this example. In general it is found that $a_n = (b_n - jc_n)/2$ for $n = 1, 2, \ldots, a_0 = b_0$, and $a_{-n} = a_n^*$. The trigonometric Fourier series is common in the signal processing literature because it replaces complex coefficients with real ones and often results in a simpler and more intuitive interpretation of the results.

Convergence of the Fourier Series

The Fourier series representation of a periodic signal is an approximation that exhibits mean squared convergence to the true signal. If $s(t)$ is a periodic signal of period T, and $s'(t)$ denotes the Fourier series approximation of $s(t)$, then $s(t)$ and $s'(t)$ are equal in the mean square sense if

$$\text{MSE} = \int_{-T/2}^{T/2} |s(t) - s(t)'|^2 \, dt = 0 \tag{4.7}$$

Even with (4.7) satisfied, mean square error (MSE) convergence does not mean that $s(t) = s'(t)$ at every value of t. In particular, it is known that at values of t, where $s(t)$ is discontinuous, the Fourier series converges to the average of the limiting values to the left and right of the discontinuity. For example, if t_0 is a point of discontinuity, then $s'(t_0) = [s(t_0^-) + s(t_0^+)]/2$, where $s(t_0^-)$ and $s(t_0^+)$ were defined previously. (Note that at points of continuity, this condition is also satisfied by the definition of continuity.) Because the Dirichet conditions require that $s(t)$ have at most a finite number of points of discontinuity in one period, the set S_t, defined as all values of t within one period where $s(t) \neq s'(t)$, contains a finite number of points, and S_t is a set of measure zero in the formal mathematical sense. Therefore, $s(t)$ and its Fourier series expansion $s'(t)$ are *equal almost everywhere,* and $s(t)$ can be considered identical to $s'(t)$ for the analysis of most practical engineering problems.

Convergence almost everywhere is satisfied only in the limit as an infinite number of terms are included in the Fourier series expansion. If the infinite series expansion of the Fourier series is truncated to a finite number of terms, as it must be in practical applications, then the approximation will exhibit an oscillatory behavior around the discontinuity, known as the Gibbs phenomenon [1]. Let $s'_N(t)$ denote a truncated Fourier series approximation of $s(t)$, where only the terms in (4.5a) from $n = -N$ to $n = N$ are included if the complex Fourier series representation is used, or where only the terms in (4.6a) from $n = 0$ to $n = N$ are included if the trigonometric form of the Fourier series is used. It is well known that in the vicinity of a discontinuity at t_0 the Gibbs phenomenon causes $s'_N(t)$ to be a poor approximation to $s(t)$. The peak magnitude of the Gibbs oscillation is 13% of the size of the jump discontinuity $s(t_0^-) - s(t_0^+)$ regardless of the number of terms used in the approximation. As N increases, the region that contains the oscillation becomes more concentrated in the neighborhood of the discontinuity, until, in the limit as N approaches infinity, the Gibbs oscillation is squeezed into a single point of mismatch at t_0.

If $s'(t)$ is replaced by $s'_N(t)$ from (4.7), it is important to understand the behavior of the error MSE_N as a function of N, where

$$\text{MSE}_N = \int_{-T/2}^{T/2} |s(t) - s'_N(t)|^2 \, dt \tag{4.8}$$

An important property of the Fourier series is that the exponential basis functions $e^{jn\omega_0 t}$ (or $\sin(n\omega_0 t)$ and $\cos(n\omega_0 t)$ for the trigonometric form) for $n = 0, \pm 1, \pm 2, \ldots$ (or $n = 0, 1, 2, \ldots$ for the trigonometric form) constitute an orthonormal set, i.e., $t_{nk} = 1$ for $n = k$, and $t_{nk} = 0$ for $n \neq k$, where

$$t_{nk} = (1/T) \int_{-T/2}^{T/2} (e^{-jn\omega_0 t})(e^{jk\omega_0 t}) \, dt \qquad (4.9)$$

As terms are added to the Fourier series expansion, the orthogonality of the basis functions guarantees that the error decreases in the mean square sense, i.e., that MSE_N monotonically decreases as N is increased. Therefore, a practitioner can proceed with the confidence that when applying Fourier series analysis more terms are always better than fewer in terms of the accuracy of the signal representations.

4.3   The Classical Fourier Transform for Continuous Time Signals

The periodicity constraint imposed on the Fourier series representation can be removed by taking the limits of (4.5a) and (4.5b) as the period T is increased to infinity. Some mathematical preliminaries are required so that the results will be well defined after the limit is taken. It is convenient to remove the $(1/T)$ factor in front of the integral by multiplying (4.5b) through by T, and then replacing $T a_n$ by a'_n in both (4.5a) and (4.5b). Because $\omega_0 = 2\pi/T$, as T increases to infinity, ω_0 becomes infinitesimally small, a condition that is denoted by replacing ω_0 with $\Delta\omega$. The factor $(1/T)$ in (4.5a) becomes $(\Delta\omega/2\pi)$. With these algebraic manipulations and changes in notation (4.5a) and (4.5b) take on the following form prior to taking the limit:

$$s(t) = (1/2\pi) \sum_{n=-\infty}^{\infty} a'_n e^{jn\Delta\omega t} \Delta\omega \qquad (4.10a)$$

$$a'_n = \int_{-T/2}^{T/2} s(t) e^{-jn\Delta\omega t} \, dt \qquad (4.10b)$$

The final step in obtaining the CT Fourier transform is to take the limit of
both (4.10a) and (4.10b) as $T \to \infty$. In the limit the infinite summation in (4.10a) becomes an integral, $\Delta\omega$ becomes $d\omega$, $n\Delta\omega$ becomes ω, and a'_n becomes the CT Fourier transform of $s(t)$, denoted by $S(j\omega)$. The result is summarized by the following transform pair, which is known throughout most of the engineering literature as the classical CT Fourier transform (CTFT):

$$s(t) = (1/2\pi) \int_{-\infty}^{\infty} S(j\omega) e^{j\omega t} \, d\omega \qquad (4.11a)$$

$$S(j\omega) = \int_{-\infty}^{\infty} s(t) e^{-j\omega t} \, dt \qquad (4.11b)$$

Often (4.11a) is called the Fourier integral and (4.11b) is simply called the Fourier transform. The relationship $S(j\omega) = \mathcal{F}\{s(t)\}$ denotes the Fourier transformation of $s(t)$, where $\mathcal{F}\{\cdot\}$ is a symbolic notation for the Fourier transform operator, and where ω becomes the continuous frequency variable after the periodicity constraint is removed. A transform pair $s(t) \leftrightarrow S(j\omega)$ represents a one-to-one invertible mapping as long as $s(t)$ satisfies conditions which guarantee that the Fourier integral converges. (More mathematical details of the CTFT are presented in Chapter 6.)

From (4.11a) it is easily seen that $\mathcal{F}\{\delta(t - t_0)\} = e^{-j\omega t_0}$, and from (4.11b) that $\mathcal{F}^{-1}\{2\pi\delta(\omega - \omega_0)\} = e^{j\omega_0 t}$, so that $\delta(t - t_0) \leftrightarrow e^{-j\omega t_0}$ and $e^{j\omega_0 t} \leftrightarrow 2\pi\delta(\omega - \omega_0)$ are valid Fourier transform pairs. Using these relationships it is easy to establish the Fourier transforms of $\cos(\omega_0 t)$ and $\sin(\omega_0 t)$, as well as many other useful waveforms that are encountered in common signal analysis problems. A number of such transforms are shown in Table 4.1.

TABLE 4.1   Some Basic CTFT Pairs

Signal	Fourier Transform	Fourier Series Coefficients (if periodic)
$\displaystyle\sum_{k=-\infty}^{+\infty} a_k e^{jk\omega_0 t}$	$\displaystyle 2\pi \sum_{k=-\infty}^{+\infty} a_k \delta(\omega_k \omega_0)$	a_k
$e^{j\omega_0 t}$	$2\pi\delta(\omega + \omega_0)$	$a_1 = 1$ $a_k = 0, \quad$ otherwise
$\cos \omega_0 t$	$\pi[\delta(\omega - \omega_0) + \delta(\omega + \omega_0)]$	$a_1 = a_{-1} = \dfrac{1}{2}$ $a_k = 0, \quad$ otherwise
$\sin \omega_0 t$	$\dfrac{\pi}{j}[\delta(\omega - \omega_0) - \delta(\omega + \omega_0)]$	$a_1 = -a_{-1} = \dfrac{1}{2j}$ $a_k = 0, \quad$ otherwise
$x(t) = 1$	$2\pi\delta(\omega)$	$a_0 = 1, \quad a_k = 0, \quad k \neq 0$ $\left(\begin{array}{l}\text{has this Fourier series representation for any}\\ \text{choice of } T_0 > 0\end{array}\right)$
Periodic square wave $x(t) = \begin{cases} 1, & \lvert t\rvert < T_1 \\ 0, & T_1 < \lvert t\rvert \le \dfrac{T_0}{2}\end{cases}$ and $x(t + T_0) = x(t)$	$\displaystyle\sum_{k=-\infty}^{+\infty} \dfrac{2\sin k\omega_0 T_1}{k}\delta(\omega_k \omega_0)$	$\dfrac{\omega_0 T_1}{\pi}\sin c\left(\dfrac{k\omega_0 T_1}{\pi}\right) = \dfrac{\sin k\omega_0 T_1}{k\pi}$
$\displaystyle\sum_{n=-\infty}^{+\infty} \delta(t - nT)$	$\dfrac{2\pi}{T}\displaystyle\sum_{k=-\infty}^{+\infty}\delta\left(\omega - \dfrac{2\pi k}{T}\right)$	$a_k = \dfrac{1}{T} \quad$ for all k
$x(t) = \begin{cases} 1, & \lvert t\rvert < T_1 \\ 0, & \lvert t\rvert > T_1\end{cases}$	$2T_1 \sin c\left(\dfrac{\omega T_1}{\pi}\right) = \dfrac{2\sin\omega T_1}{\omega}$	—
$\dfrac{W}{\pi}\sin c\left(\dfrac{Wt}{\pi}\right) = \dfrac{\sin Wt}{\pi t}$	$X(\omega) = \begin{cases} 1, & \lvert\omega\rvert < W \\ 0, & \lvert\omega\rvert > W\end{cases}$	—
$\delta(t)$	1	—
$u(t)$	$\dfrac{1}{j\omega} + \pi\delta(\omega)$	—
$\delta(t - t_0)$	$e^{j\omega t_0}$	—
$e^{-at}u(t), \text{Re}\{a\} > 0$	$\dfrac{1}{a + j\omega}$	—
$te^{-at}u(t), \text{Re}\{a\} > 0$	$\dfrac{1}{(a + j\omega)^2}$	—
$\dfrac{t^{n-1}}{(n-1)!}e^{-at}u(t),$ $\text{Re}\{a\} > 0$	$\dfrac{1}{(a + j\omega)^n}$	—

Source:A. V. Oppenheim et al., Signals and Systems, Englewood Cliffs, NJ: Prentice-Hall, 1983.

The CTFT is useful in the analysis and design of CT systems, i.e., systems that process CT signals. Fourier analysis is particularly applicable to the design of CT filters which are characterized by Fourier magnitude and phase spectra, i.e., by $\lvert H(j\omega)\rvert$ and $\arg H(j\omega)$, where $H(j\omega)$ is commonly called the frequency response of the filter. For example, an **ideal transmission channel** is one which passes a signal without distorting it. The signal may be scaled by a real constant A and delayed by a fixed time increment t_0, implying that the impulse response of an ideal channel is $A\delta(t - t_0)$, and its corresponding frequency response is $Ae^{-j\omega t_0}$. Hence, the frequency response of an ideal channel is specified by constant amplitude for all frequencies, and a phase characteristic which is a linear function given by ωt_0.

Properties of the Continuous Time Fourier Transform

The CTFT has many properties that make it useful for the analysis and design of linear CT systems. Some of the more useful properties are stated below. A more complete list of the CTFT properties is given in Table 4.2. Proofs of these properties can be found in [2] and [3]. In the following discussion $\mathcal{F}\{\cdot\}$ denotes the Fourier transform operation, $\mathcal{F}^{-1}\{\cdot\}$ denotes the inverse Fourier transform operation, and $*$ denotes the convolution operation defined as

$$f_1(t) * f_2(t) = \int_{-\infty}^{\infty} f_1(t - \tau) f_2(\tau) \, d\tau$$

1. Linearity (superposition): $\mathcal{F}\{af_1(t) + bf_2(t)\} = a\mathcal{F}\{f_1(t)\} + b\mathcal{F}\{f_2(t)\}$ (a and b, complex constants)
2. Time shifting: $\mathcal{F}\{f(t - t_0)\} = e^{-j\omega t_0}\mathcal{F}\{f(t)\}$
3. Frequency shifting: $e^{j\omega_0 t} f(t) = \mathcal{F}^{-1}\{F(j(\omega - \omega_0))\}$
4. Time domain convolution: $\mathcal{F}\{f_1(t) * f_2(t)\} = \mathcal{F}\{f_1(t)\}\mathcal{F}\{f_2(t)\}$
5. Frequency domain convolution: $\mathcal{F}\{f_1(t)f_2(t)\} = (1/2\pi)\mathcal{F}\{f_1(t)\} * \mathcal{F}\{f_2(t)\}$
6. Time differentiation: $-j\omega F(j\omega) = \mathcal{F}\{d(f(t))/dt\}$
7. Time integration: $\mathcal{F}\{\int_{-\infty}^{t} f(\tau) \, d\tau\} = (1/j\omega)F(j\omega) + \pi F(0)\delta(\omega)$

The above properties are particularly useful in CT system analysis and design, especially when the system characteristics are easily specified in the frequency domain, as in linear filtering. Note that properties 1, 6, and 7 are useful for solving differential or integral equations. Property 4 provides the basis for many signal processing algorithms because many systems can be specified directly by their impulse or frequency response. Property 3 is particularly useful in analyzing communication systems in which different modulation formats are commonly used to shift spectral energy to frequency bands that are appropriate for the application.

Fourier Spectrum of the Continuous Time Sampling Model

Because the CT sampling model $s_a(t)$, given in (4.1), is in its own right a CT signal, it is appropriate to apply the CTFT to obtain an expression for the spectrum of the sampled signal:

$$\mathcal{F}\{s_a(t)\} = \mathcal{F}\left\{\sum_{n=-\infty}^{\infty} s(t)\delta(t - nT)\right\} = \sum_{n=-\infty}^{\infty} s(nT)e^{-j\omega Tn} \tag{4.12}$$

Because the expression on the right-hand side of (4.12) is a function of $e^{j\omega T}$ it is customary to denote the transform as $F(e^{j\omega T}) = \mathcal{F}\{s_a(t)\}$. Later in the chapter this result is compared to the result of operating on the DT sampling model, namely $s[n]$, with the DT Fourier transform to illustrate that the two sampling models have the same spectrum.

Fourier Transform of Periodic Continuous Time Signals

Was saw earlier that a periodic CT signal can be expressed in terms of its Fourier series. The CTFT can then be applied to the Fourier series representation of $s(t)$ to produce a mathematical expression

TABLE 4.2   Properties of the CTFT

Name	If $\mathcal{F}f(t) = F(j\omega)$, then
Definition	$f(j\omega) = \int_{-\infty}^{\infty} f(t)e^{j\omega t}\,dt$
	$f(t) = \dfrac{1}{2\pi}\int_{-\infty}^{\infty} F(j\omega)e^{j\omega t}\,d\omega$
Superposition	$\mathcal{F}[af_1(t) + bf_2(t)] = aF_1(j\omega) + bF_2(j\omega)$
Simplification if:	
(a) $f(t)$ is even	$F(j\omega) = 2\int_0^{\infty} f(t)\cos\omega t\,dt$
(b) $f(t)$ is odd	$F(j\omega) = 2j\int_0^{\infty} f(t)\sin\omega t\,dt$
Negative t	$\mathcal{F}f(-t) = F^*(j\omega)$
Scaling:	
(a) Time	$\mathcal{F}f(at) = \dfrac{1}{|a|}F\left(\dfrac{j\omega}{a}\right)$
(b) Magnitude	$\mathcal{F}af(t) = aF(j\omega)$
Differentiation	$\mathcal{F}\left[\dfrac{d^n}{dt^n}f(t)\right] = (j\omega)^n F(j\omega)$
Integration	$\mathcal{F}\left[\int_{-\infty}^{t} f(x)\,dx\right] = \dfrac{1}{j\omega}F(j\omega) + \pi F(0)\delta(\omega)$
Time shifting	$\mathcal{F}f(t-a) = F(j\omega)e^{j\omega a}$
Modulation	$\mathcal{F}f(t)e^{j\omega_0 t} = F[j(\omega - \omega_0)]$
	$\{\mathcal{F}f(t)\cos\omega_0 t = \tfrac{1}{2}F[j(\omega - \omega_0)] + F[j(\omega + \omega_0)]\}$
	$\{\mathcal{F}f(t)\sin\omega_0 t = \tfrac{1}{2}j\{F[j(\omega - \omega_0)] - F[j(\omega + \omega_0)]\}$
Time convolution	$\mathcal{F}^{-1}[F_1(j\omega)F_2(j\omega)] = \int_{-\infty}^{\infty} f_1(\tau)f_2(\tau)f_2(t_\tau)\,d\tau$
Frequency convolution	$\mathcal{F}[f_1(t)f_2(t)] = \dfrac{1}{2\pi}\int_{-\infty}^{\infty} F_1(j\lambda)F_2[j(\omega_\lambda)]\,d\lambda$

Source:M. E. VanValkenburg, Network Analysis, (3rd edition), Englewood Cliffs, NJ: Prentice-Hall, 1974.

for the "line spectrum" characteristic of periodic signals.

$$\mathcal{F}\{s(t)\} = \mathcal{F}\left\{\sum_{n=-\infty}^{\infty} a_n e^{jn\omega_0 t}\right\} = 2\pi \sum_{n=-\infty}^{\infty} a_n \delta(\omega - n\omega_0) \tag{4.13}$$

The spectrum is shown pictorially in Fig. 4.7. Note the similarity between the spectral representation of Fig. 4.7 and the plot of the Fourier coefficients in Fig. 4.4, which was heuristically interpreted as a "line spectrum." Figures 4.4 and 4.7 are different but equivalent representations of the Fourier spectrum. Note that Fig. 4.4 is a DT representation of the spectrum, while Fig. 4.7 is a CT model of the same spectrum.

The Generalized Complex Fourier Transform

The CTFT characterized by (4.11a) and (4.11b) can be generalized by considering the variable $j\omega$ to be the special case of $u = \sigma + j\omega$ with $\sigma = 0$, writing (4.11a and b) in terms of u, and interpreting

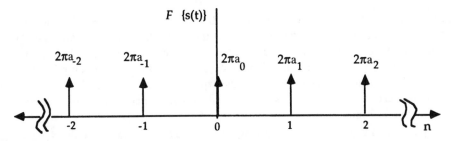

4.7 Spectrum of the Fourier series representation of $s(t)$.

u as a complex frequency variable. The resulting complex Fourier transform pair is given by (4.14a) and (4.14b)

$$s(t) \;=\; (1/2\pi j) \int_{\sigma - j\infty}^{\sigma + j\infty} S(u)e^{jut}\, du \qquad (4.14a)$$

$$S(u) \;=\; \int_{-\infty}^{\infty} s(t)e^{-jut}\, dt \qquad (4.14b)$$

The set of all values of u for which the integral of (4.14b) converges is called the region of convergence (ROC). Because the transform $S(u)$ is defined only for values of u within the ROC, the path of integration in (4.14a) must be defined by σ so that the entire path lies within the ROC. In some literature this transform pair is called the **bilateral Laplace transform** because it is the same result obtained by including both the negative and positive portions of the time axis in the classical Laplace transform integral. [Note that in (4.14a and b) the complex frequency variable was denoted by u rather than by the more common s, in order to avoid confusion with earlier uses of $s(\cdot)$ as signal notation.] The complex Fourier transform (bilateral Laplace transform) is not often used in solving practical problems, but its significance lies in the fact that it is the most general form that represents the point at which Fourier and Laplace transform concepts become the same. Identifying this connection reinforces the notion that Fourier and Laplace transform concepts are similar because they are derived by placing different constraints on the same general form. (For further reading on the Laplace transform see Chapter 3.)

4.4 The Discrete Time Fourier Transform

The discrete time Fourier transform (DTFT) can be obtained by using the DT sampling model and considering the relationship obtained in (4.12) to be the definition of the DTFT. Letting $T = 1$ so that the sampling period is removed from the equations and the frequency variable is replaced with a normalized frequency $\omega' = \omega T$, the DTFT pair is defined in (4.15a and b). Note that in order to simplify notation it is not customary to distinguish between ω and ω', but rather to rely on the context of the discussion to determine whether ω refers to the normalized ($T = 1$) or the unnormalized ($T \neq 1$) frequency variable.

$$S(e^{j\omega'}) \;=\; \sum_{n=-\infty}^{\infty} s[n]e^{-j\omega'n} \qquad (4.15a)$$

$$s[n] \;=\; (1/2\pi) \int_{-\pi}^{\pi} S(e^{j\omega'})e^{jn\omega'}\, d\omega' \qquad (4.15b)$$

The spectrum $S(e^{j\omega'})$ is periodic in ω' with period 2π. The fundamental period in the range $-\pi < \omega' \leq \pi$, sometimes referred to as the baseband, is the useful frequency range of the DT system because frequency components in this range can be represented unambiguously in sampled form (without aliasing error). In much of the signal processing literature the explicit primed notation is omitted from the frequency variable. However, the explicit primed notation will be used throughout this section because the potential exists for confusion when so many related Fourier concepts are discussed within the same framework.

By comparing (4.12) and (4.15a), and noting that $\omega' = \omega T$, it is established that

$$\mathcal{F}\{s_a(t)\} = \text{DTFT}\{s[n]\} \tag{4.16}$$

where $s[n] = s(t)_{t=nT}$. This demonstrates that the spectrum of $s_a(t)$, as calculated by the CT Fourier transform is identical to the spectrum of $s[n]$ as calculated by the DTFT. Therefore, although $s_a(t)$ and $s[n]$ are quite different sampling models, they are equivalent in the sense that they have the same Fourier domain representation.

A list of common DTFT pairs is presented in Table 4.3. Just as the CT Fourier transform is useful in CT signal system analysis and design, the DTFT is equally useful in the same capacity for DT systems. It is indeed fortuitous that Fourier transform theory can be extended in this way to apply to DT systems.

TABLE 4.3   Some Basic DTFT Pairs

Sequence	Fourier Transform
1. $\delta[n]$	1
2. $\delta[n - n_0]$	$e^{-j\omega n_0}$
3. $1 \quad (-\infty < n < \infty)$	$\displaystyle\sum_{k=-\infty}^{\infty} 2\pi\delta(\omega + 2\pi k)$
4. $a^n u[n] \quad (\lvert a\rvert < 1)$	$\dfrac{1}{1 - ae^{-j\omega}}$
5. $u[n]$	$\dfrac{1}{1 - e^{-j\omega}} + \displaystyle\sum_{k=-\infty}^{\infty} \pi\delta(\omega + 2\pi k)$
6. $(n + 1)a^n u[n] \quad (\lvert a\rvert < 1)$	$\dfrac{1}{(1 - ae^{-j\omega})^2}$
7. $\dfrac{r^2 \sin \omega_p(n + 1)}{\sin \omega_p} u[n] \quad (\lvert r\rvert < 1)$	$\dfrac{1}{1 - 2r\cos\omega_p e^{-j\omega} + r^2 e^{j2\omega}}$
8. $\dfrac{\sin \omega_c n}{\pi n}$	$Xe^{j\omega} = \begin{cases} 1, & \lvert\omega\rvert < \omega_c \\ 0, & \omega_c < \lvert\omega\rvert \leq \pi \end{cases}$
9. $x[n] = \begin{cases} 1, & 0 \leq n \leq M \\ 0, & \text{otherwise} \end{cases}$	$\dfrac{\sin[\omega(M + 1)/2]}{\sin(\omega/2)} e^{-j\omega M/2}$
10. $e^{j\omega_0 n}$	$\displaystyle\sum_{k=-\infty}^{\infty} 2\pi\delta(\omega - \omega_0 + 2\pi k)$
11. $\cos(\omega_0 n + \phi)$	$\pi \displaystyle\sum_{k=-\infty}^{\infty} [e^{j\phi}\delta(\omega - \omega_0 + 2\pi k) + e^{-j\phi}\delta(\omega + \omega_0 + 2\pi k)]$

Source: A. V. Oppenheim and R. W. Schafer, Digital Signal Processing, Englewood Cliffs, NJ: Prentice-Hall, 1975.

In the same way that the CT Fourier transform was found to be a special case of the complex Fourier transform (or bilateral Laplace transform), the DTFT is a special case of the bilateral z-transform with $z = e^{j\omega't}$. The more general bilateral z-transform is given by:

$$S(z) = \sum_{n=-\infty}^{\infty} s[n]z^{-n} \tag{4.17a}$$

$$s[n] = (1/2\pi j) \int_C S(z)z^{n-1} dz \tag{4.17b}$$

where C is a counterclockwise contour of integration which is a closed path completely contained within the region of convergence of $S(z)$. Recall that the DTFT was obtained by taking the CT Fourier transform of the CT sampling model represented by $s_a(t)$. Similarly, the bilateral z-transform results by taking the bilateral Laplace transform of $s_a(t)$. If the lower limit on the summation of (4.17a) is taken to be $n = 0$, then (4.17a) and (4.17b) become the one-sided z-transform, which is the DT equivalent of the one-sided LT for CT signals. The hierarchical relationship among these various concepts for DT systems is discussed later in this chapter, where it will be shown that the family structure of the DT family tree is identical to that of the CT family. For every CT transform in the CT world there is an analogous DT transform in the DT world, and vice versa. (For further reading on the z-transform see Chapter 5.)

Properties of the Discrete Time Fourier Transform

Because the DTFT is a close relative of the classical CT Fourier transform it should come as no surprise that many properties of the DTFT are similar to those presented for the CT Fourier transform in the previous section. In fact, for many of the properties presented earlier an analogous property exists for the DTFT. The following list parallels the list that was presented in the previous section for the CT Fourier transform, to the extent that the same property exists. A more complete list of DTFT pairs is given in Table 4.4. (Note that the primed notation on ω' is dropped in the following to simplify the notation, and to be consistent with standard usage.)

1. Linearity (superposition): $\text{DTFT}\{af_1[n] + bf_2[n]\} = a\text{DTFT}\{f_1[n]\} + b\text{DTFT}\{f_2[n]\}$
   (a and b, complex constants)
2. Index shifting: $\text{DTFT}\{f[n - n_0]\} = e^{-j\omega n_0}\text{DTFT}\{f[n]\}$
3. Frequency shifting: $e^{j\omega_0 n} f[n] = \text{DTFT}^{-1}\{F(e^{j(\omega-\omega_0)})\}$
4. Time domain convolution: $\text{DTFT}\{f_1[n] * f_2[n]\} = \text{DTFT}\{f_1[n]\}\text{DTFT}\{f_2[n]\}$
5. Frequency domain convolution: $\text{DTFT}\{f_1[n]f_2[n]\} = (1/2\pi)\text{DTFT}\{f_1[n]\}*\text{DTFT}\{f_2[n]\}$
6. Frequency differentiation: $nf[n] = \text{DTFT}^{-1}\{dF(e^{j\omega})/d\omega\}$

Note that the time differentiation and time-integration properties of the CTFT do not have analogous counterparts in the DTFT because time domain differentiation and integration are not defined for DT signals. When working with DT systems practitioners must often manipulate difference equations in the frequency domain. For this purpose property 1 and property 2 are very important. As with the CTFT, property 4 is very important for DT systems because it allows engineers to work with the frequency response of the system, in order to achieve proper shaping of the input spectrum or to achieve frequency selective filtering for noise reduction or signal detection. Also, property 3 is useful for the analysis of modulation and filtering operations common in both analog and digital communication systems.

The DTFT is defined so that the time domain is discrete and the frequency domain is continuous. This is in contrast to the CTFT that is defined to have continuous time and continuous frequency

TABLE 4.4　　Properties of the DTFT

Sequence	Fourier Transform
$x[n]$	$X(e^{j\omega})$
$y[n]$	$Y(e^{j\omega})$
1. $ax[n] + by[n]$	$aX(e^{j\omega}) + bY(e^{j\omega})$
2. $x[n - n_d]$　　(n_d an integer)	$e^{-j\omega n_d}X(e^{j\omega})$
3. $e^{j\omega_0 n}x[n]$	$X(e^{j(\omega - \omega_0)})$
4. $x[-n]$	$X(e^{-j\omega})$　　if $x[n]$ is real
	$X^*(e^{j\omega})$
5. $nx[n]$	$j\dfrac{dX(e^{j\omega})}{d\omega}$
6. $x[n] * y[n]$	$X(e^{j\omega})Y(e^{j\omega})$
7. $x[n]y[n]$	$\dfrac{1}{2\pi}\displaystyle\int_{-x}^{x} X(e^{j\theta})Y(e^{j(\omega - \theta)})\, d\theta$

Parseval's Theorem

8. $\displaystyle\sum_{n=-\infty}^{\infty} |x[n]|^2 = \frac{1}{2\pi}\int_{-\pi}^{\pi} |X(e^{j\omega})|^2\, d\omega$

9. $\displaystyle\sum_{n=-\infty}^{\infty} x[n]y^*[n] = \frac{1}{2\pi}\inf_{-\pi}^{\pi} X(e^{j\omega})Y^*(e^{j\omega})\, d\omega$

Source: A. V. Oppenheim and R. W. Schafer, Digital Signal Processing, Englewood Cliffs, NJ: Prentice-Hall, 1975.

domains. The mathematical dual of the DTFT also exists, which is a transform pair that has a continuous time domain and a discrete frequency domain. In fact, the dual concept is really the same as the Fourier series for periodic CT signals presented earlier in the chapter, as represented by (4.5a) and (4.5b). However, the classical Fourier series arises from the assumption that the CT signal is inherently periodic, as opposed to the time domain becoming periodic by virtue of sampling the spectrum of a continuous frequency (aperiodic time) function [8]. The dual of the DTFT, the discrete frequency Fourier transform (DFFT), has been formulated and its properties tabulated as an interesting and useful transform in its own right [5]. Although the DFFT is similar in concept to the classical CT Fourier series, the formal properties of the DFFT [5] serve to clarify the effects of frequency domain sampling and time domain aliasing. These effects are obscured in the classical treatment of the CT Fourier series because the emphasis is on the inherent "line spectrum" that results from time domain periodicity. The DFFT is useful for the analysis and design of digital filters that are produced by frequency sampling techniques.

Relationship between the Continuous and Discrete Time Spectra

Because DT signals often originate by sampling CT signals, it is important to develop the relationship between the original spectrum of the CT signal and the spectrum of the DT signal that results. First, the CTFT is applied to the CT sampling model, and the properties listed above are used to produce the following result:

$$\begin{aligned}
\mathcal{F}\{s_a(t)\} &= \mathcal{F}\left\{ s(t) \sum_{n=-\infty}^{\infty} \delta(t - nT) \right\} \\
&= (1/2\pi)S(j\omega) * \mathcal{F}\left\{ \sum_{n=-\infty}^{\infty} \delta(t - nT) \right\}
\end{aligned} \tag{4.18}$$

In this section it is important to distinguish between ω and ω', so the explicit primed notation is used in the following discussion where needed for clarification. Because the sampling function (summation of shifted impulses) on the right-hand side of the above equation is periodic with period T it can be replaced with a CT Fourier series expansion as follows:

$$S(e^{j\omega T}) = \mathcal{F}\{s_a(t)\} = (1/2\pi)S(j\omega) * \mathcal{F}\left\{\sum_{n=-\infty}^{\infty}(1/T)e^{j(2\pi/T)nt}\right\}$$

Applying the frequency domain convolution property of the CTFT yields

$$S(e^{j\omega T}) = (1/2\pi)\sum_{n=-\infty}^{\infty}S(j\omega) * (2\pi/T)\delta(\omega - (2\pi/T)n)$$

The result is

$$S(e^{j\omega T}) = (1/T)\sum_{n=-\infty}^{\infty}S(j[\omega - (2\pi/T)n]) = (1/T)\sum_{n=-\infty}^{\infty}S(j[\omega - n\omega_s]) \qquad (4.19a)$$

where $\omega_s = (2\pi/T)$ is the sampling frequency expressed in radians per second. An alternate form for the expression of (4.19a) is

$$S(e^{j\omega'}) = (1/T)\sum_{n=-\infty}^{\infty}S(j[(\omega' - n2\pi)/T]) \qquad (4.19b)$$

where $\omega' = \omega T$ is the normalized DT frequency axis expressed in radians. Note that $S(e^{j\omega T}) = S(e^{j\omega'})$ consists of an infinite number of replicas of the CT spectrum $S(j\omega)$, positioned at intervals of $(2\pi/T)$ on the ω axis (or at intervals of 2π on the ω' axis), as illustrated in Fig. 4.8. Note that if $S(j\omega)$ is band limited with a bandwidth ω_c, and if T is chosen sufficiently small so that $\omega_s > 2\omega_c$, then the DT spectrum is a copy of $S(j\omega)$ (scaled by $1/T$) in the baseband. The limiting case of $\omega_s = 2\omega_c$ is called the Nyquist sampling frequency. Whenever a CT signal is sampled at or above the Nyquist rate, no aliasing distortion occurs (i.e., the baseband spectrum does not overlap with the higher-order replicas) and the CT signal can be exactly recovered from its samples by extracting the baseband spectrum of $S(e^{j\omega'})$ with an ideal low-pass filter that recovers the original CT spectrum by removing all spectral replicas outside the baseband and scaling the baseband by a factor of T.

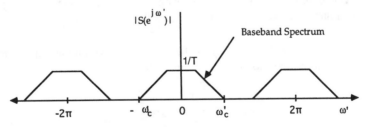

4.8   Illustration of the relationship between the CT and DT spectra.

4.5   The Discrete Fourier Transform

To obtain the discrete Fourier transform (DFT) the continuous frequency domain of the DTFT is sampled at N points uniformly spaced around the unit circle in the z-plane, i.e., at the points

$\omega_k = (2\pi k/N), k = 0, 1, \ldots, N - 1$. The result is the DFT pair defined by (4.20a) and (4.20b). The signal $s[n]$ is either a finite length sequence of length N, or it is a periodic sequence with period N.

$$S[k] \quad = \quad \sum_{n=0}^{N-1} s[n]e^{-j2\pi kn/N} \qquad k = 0, 1, \ldots, N - 1 \tag{4.20a}$$

$$s[n] \quad = \quad (1/N)\sum_{k=0}^{N-1} S[k]e^{j2\pi kn/N} \qquad n = 0, 1, \ldots, N - 1 \tag{4.20b}$$

Regardless of whether $s[n]$ is a finite length or periodic sequence, the DFT treats the N samples of $s[n]$ as though they are one period of a periodic sequence. This is an important feature of the DFT, and one that must be handled properly in signal processing to prevent the introduction of artifacts. Important properties of the DFT are summarized in Table 4.5. The notation $((k))_N$ denotes k modulo N, and $R_N[n]$ is a rectangular window such that $R_N[n] = 1$ for $n = 0, \ldots, N - 1$, and $R_N[n] = 0$ for $n < 0$ and $n \geq N$. The transform relationship given by (4.20a) and (4.20b) is also valid when $s[n]$ and $S[k]$ are periodic sequences, each of period N. In this case n and k are permitted to range over the complete set of real integers, and $S[k]$ is referred to as the discrete Fourier series (DFS). The DFS is developed by some authors as a distinct transform pair in its own right [6]. Whether the DFT and the DFS are considered identical or distinct is not very important in this discussion. The important point to be emphasized here is that the DFT treats $s[n]$ as though it were a single period of a periodic sequence, and all signal processing done with the DFT will inherit the consequences of this assumed periodicity.

Properties of the Discrete Fourier Series

Most of the properties listed in Table 4.5 for the DFT are similar to those of the z-transform and the DTFT, although some important differences exist. For example, property 5 (time-shifting property), holds for *circular* shifts of the finite length sequence $s[n]$, which is consistent with the notion that the DFT treats $s[n]$ as one period of a periodic sequence. Also, the multiplication of two DFTs results in the **circular convolution** of the corresponding DT sequences, as specified by property 7. This latter property is quite different from the **linear convolution** property of the DTFT. Circular convolution is the result of the assumed periodicity discussed in the previous paragraph. Circular convolution is simply a linear convolution of the periodic extensions of the finite sequences being convolved, in which each of the finite sequences of length N defines the structure of one period of the periodic extensions.

For example, suppose one wishes to implement a digital filter with finite impulse response (FIR) $h[n]$. The output $y(n)$ in response to input $s[n]$ is given by

$$y[n] = \sum_{k=0}^{N-1} h[k]s[n - k] \tag{4.21}$$

where $y(n)$ is obtained by transforming $h[n]$ and $s[n]$ into $H[k]$ and $S[k]$ using the DFT, multiplying the transforms point-wise to obtain $Y[k] = H[k]S[k]$, and then using the inverse DFT to obtain $y[n] = \text{DFT}^{-1}\{Y[k]\}$. If $s[n]$ is a finite sequence of length M, then the results of the circular convolution implemented by the DFT will correspond to the desired linear convolution iff the block length of the DFT, N_{DFT}, is chosen sufficiently large so that $N_{\text{DFT}} \geq N + M - 1$ and both $h[n]$ and $s[n]$ are padded with zeroes to form blocks of length N_{DFT}.

TABLE 4.5 Properties of the DFT

Finite-Length Sequence (Length N)	N-Point DFT (Length N)
1. $x[n]$	$X[k]$
2. $x_1[n], x_2[n]$	$X_1[k], X_2[k]$
3. $ax_1[n] + bx_2[n]$	$aX_1[k] + bX_2[k]$
4. $X[n]$	$Nx[((-k))_N]$
5. $x[((n_m))_N]$	$W_N^{km} X[k]$
6. $W_N^{-ln} x[n]$	$X[((k-l))_N]$
7. $\displaystyle\sum_{m=0}^{N-1} x_1(m)x_2[((n_m))_N]$	$X_1[k]X_2[k]$
8. $x_1[n]x_2[n]$	$\displaystyle\frac{1}{N}\sum_{l=0}^{N-1} X_1(l)X_2[((k-l)_N]$
9. $x^*[n]$	$X^*[((-k))_N]$
10. $x^*[((-n))_N]$	$X^*[k]$
11. $\mathrm{Re}\{x[n]\}$	$X_{ep}[k] = \frac{1}{2}\{X[((k))_N] + K^*[((-k))_N]\}$
12. $j\,\mathrm{Im}\{x[n]\}$	$X_{op}[k] = \frac{1}{2}\{X[((k))_N] - X^*[((-k))_N]\}$
13. $x_{ep}[n] = \frac{1}{2}\{x[n] + x^*[((-n))_N]\}$	$\mathrm{Re}\{X[k]\}$
14. $x_{op}[n] = \frac{1}{2}\{x[n] - x^*[((-n))_N]\}$	$j\,\mathrm{Im}\{X[k]\}$
Properties 15–17 apply only when $x[n]$ is real	
15. Symmetry properties	$\begin{cases} X[k] &=& X^*[((-k))_N] \\ \mathrm{Re}\{X[k]\} &=& \mathrm{Re}\{X[((-k))_N]\} \\ \mathrm{Im}\{X[k]\} &=& -\mathrm{Im}\{X[((-k))_N]\} \\ |X[k]| &=& |X[((-k))_N]| \\ \sphericalangle\{X[k]\} &=& -\sphericalangle\{X[((-k))_N]\} \end{cases}$
16. $x_{ep}[n] = \frac{1}{2}\{x[n] + x[((-n))_N]\}$	$\mathrm{Re}\{X[k]\}$
17. $x_{op}[n] = \frac{1}{2}\{x[n] - x[((-n))_N]\}$	$j\,\mathrm{Im}\{X[k]\}$

Source: A. V. Oppenheim and R. W. Schafer, *Discrete-Time Signal Processing*, Englewood Cliffs, NJ: Prentice-Hall, 1989.

Fourier Block Processing in Real-Time Filtering Applications

In some practical applications either the value of M is too large for the memory available, or $s[n]$ may not actually be finite in length, but rather a continual stream of data samples that must be processed by a filter at real-time rates. Two well-known algorithms are available that partition $s[n]$ into smaller blocks and process the individual blocks with a smaller-length DFT: (1) overlap-save partitioning and (2) overlap-add partitioning. Each of these algorithms is summarized below.

Overlap-Save Processing

In this algorithm N_{DFT} is chosen to be some convenient value with $N_{\mathrm{DFT}} > N$. The signal $s[n]$ is partitioned into blocks which are of length N_{DFT} and which overlap by $N - 1$ data points. Hence, the kth block is $s_k[n] = s[n + k(N_{\mathrm{DFT}} - N + 1)], n = 0, \ldots, N_{\mathrm{DFT}} - 1$. The filter is an augmented filter with $N_{\mathrm{DFT}} - N$ zeroes to produce

$$h_{\mathrm{pad}}[n] = \left[\begin{array}{ll} h[n], & n = 0, \ldots, N-1 \\ 0, & n = N, \ldots, N_{\mathrm{DFT}} - 1 \end{array} \right] \tag{4.22}$$

The DFT is then used to obtain $Y_{\mathrm{pad}}[n] = \mathrm{DFT}\{h_{\mathrm{pad}}[n]\} \cdot \mathrm{DFT}\{s_k[n]\}$, and $y_{\mathrm{pad}}[n] = \mathrm{IDFT}\{Y_{\mathrm{pad}}[n]\}$. From the $y_{\mathrm{pad}}[n]$ array the values that correctly correspond to the linear convolution are saved; values that are erroneous due to wraparound error caused by the circular convolution of the DFT are discarded. The kth block of the filtered output is obtained by

$$y_k[n] = \left[\begin{array}{ll} y_{\mathrm{pad}}[n], & n = N-1, \ldots, N_{\mathrm{DFT}} - 1 \\ 0, & n = 0, \ldots, N-2 \end{array} \right] \tag{4.23}$$

For the overlap-save algorithm, each time a block is processed there are $N_{DFT} - N + 1$ points saved and $N - 1$ points discarded. Each block moves forward by $N_{DFT} - N + 1$ data points and overlaps the previous block by $N - 1$ points.

Overlap-Add Processing

This algorithm is similar to the previous one except that the kth input block is defined as

$$s_k[n] = \begin{bmatrix} s[n + kL], & n = 0, \ldots, L - 1 \\ 0, & n = L, \ldots, N_{DFT} - 1 \end{bmatrix} \tag{4.24}$$

where $L = N_{DFT} - N + 1$. The filter function $h_{pad}[n]$ is augmented with zeroes, as before, to create $h_{pad}[n]$, and the DFT processing is executed as before. In each block $y_{pad}[n]$ that is obtained at the output the first $N - 1$ points are erroneous, the last $N - 1$ points are erroneous, and the middle $N_{DFT} - 2(N - 1)$ points correctly correspond to the linear convolution. However, if the last $N - 1$ points from block k are overlapped with the first $N - 1$ points from block $k + 1$ and added pairwise, correct results corresponding to linear convolution are also obtained from these positions. Hence, after this addition the number of correct points produced per block is $N_{DFT} - N + 1$, which is the same as that for the overlap-save algorithm. The overlap-add algorithm requires approximately the same amount of computation as the overlap-save algorithm, although the addition of the overlapping portions of blocks is extra. This feature, together with the added delay of waiting for the next block to be finished before the previous one is complete, has resulted in more popularity for the overlap-save algorithm in practical applications.

Block filtering algorithms make it possible to efficiently filter continual data streams in real time because the fast Fourier transform (FFT) algorithm can be used to implement the DFT, thereby minimizing the total computation time and permitting reasonably high overall data rates. However, block filtering generates data in bursts, i.e., a delay occurs during which no filtered data appear, and then an entire block is suddenly generated. In real-time systems buffering must be used. The block algorithms are particularly effective for filtering very long sequences of data that are prerecorded on magnetic tape or disk.

Fast Fourier Transform Algorithms

The DFT is typically implemented in practice with one of the common forms of the FFT algorithm. The FFT is not a Fourier transform in its own right, but simply a computationally efficient algorithm that reduces the complexity of the computing DFT from order $\{N^2\}$ to order $\{N \log_2 N\}$. When N is large, the computational savings provided by the FFT algorithm is so great that the FFT makes real-time DFT analysis practical in many situations that would be entirely impractical without it. Fast Fourier transform algorithms abound, including decimation-in-time (D-I-T) algorithms, decimation-in-frequency (D-I-F) algorithms, bit-reversed algorithms, normally ordered algorithms, mixed-radix algorithms (for block lengths that are not powers of 2), prime factor algorithms, and Winograd algorithms [7]. The D-I-T and the D-I-F radix-2 FFT algorithms are the most widely used in practice. Detailed discussions of various FFT algorithms can be found in [3, 6, 7], and [10].

The FFT is easily understood by examining the simple example of $N = 8$. The FFT algorithm can be developed in numerous ways, all of which deal with a nested decomposition of the summation operator of (4.20a). The development presented here is called an **algebraic development** of the FFT because it follows straightforward algebraic manipulation. First, the summation indices (k, n) in (4.20a) are expressed as explicit binary integers, $k = k_2 4 + k_1 2 + k_0$ and $n = n_2 4 + n_1 2 + n_0$, where k_i and n_i are bits that take on the values of either 0 or 1. If these expressions are substituted into (4.20a), all terms in the exponent that contain the factor $N = 8$ can be deleted because $e^{-j2\pi l} = 1$ for any integer l. Upon deleting such terms and regrouping the remaining terms, the product nk can be expressed in either of two ways:

$$nk = (4k_0)n_2 + (4k_1 + 2k_0)n_1 + (4k_2 + 2k_1 + k_0)n_0 \qquad (4.25a)$$

$$nk = (4n_0)k_2 + (4n_1 + 2n_0)k_1 + (4n_2 + 2n_1 + n_0)k_0 \qquad (4.25b)$$

Substituting (4.25a) into (4.20a) leads to the D-I-T FFT, whereas substituting (4.25b) leads to the D-I-F FFT. Only the D-I-T FFT is discussed further here. The D-I-F and various related forms are treated in detail in [6].

The D-I-T FFT decomposes into $\log_2 N$ stages of computation, plus a stage of bit reversal,

$$x_1[k_0, n_1, n_0] = \sum_{n_2=0}^{1} s[n_2, n_1, n_0] W_8^{4k_0 n_2} \qquad \text{(stage 1)} \qquad (4.26a)$$

$$x_2[k_0, k_1, n_0] = \sum_{n_1=0}^{1} x[k_0, n_1, n_0] W_8^{(4k_1 + 2k_0)n_2} \qquad \text{(stage 2)} \qquad (4.26b)$$

$$x_3[k_0, k_1, k_2] = \sum_{n_0=0}^{1} x[k_0, k_1, n_0] W_8^{(4k_2 + 2k_1 + k_0)n_0} \qquad \text{(stage 3)} \qquad (4.26c)$$

$$S[k_2, k_1, k_0] = x_3[k_0, k_1, k_2] \qquad \text{(bit reversal)} \qquad (4.26d)$$

In each summation above one of the n_i is summed out of the expression, while at the same time a new k_i is introduced. The notation is chosen to reflect this. For example, in stage 3, n_0 is summed out, k_2 is introduced as a new variable, and n_0 is replaced by k_2 in the result. The last operation, called bit reversal, is necessary to correctly locate the frequency samples $X[k]$ in the memory. It is easy to show that if the samples are paired correctly, an **in-place computation** can be done by a sequence of butterfly operations. The term in-place means that each time a butterfly is to be computed, a pair of data samples is read from memory, and the new data pair produced by the butterfly calculation is written back into the memory locations where the original pair was stored, thereby overwriting the original data. An in-place algorithm is designed so that each data pair is needed for only one butterfly, and thus the new results can be immediately stored on top of the old in order to minimize memory requirements.

For example, in stage 3 the $k = 6$ and $k = 7$ samples should be paired, yielding a "butterfly" computation that requires one complex multiply, one complex add, and one subtract:

$$x_3(1, 1, 0) = x_2(1, 1, 0) + W_8^3 x_2(1, 1, 1) \qquad (4.27a)$$

$$x_3(1, 1, 0) = x_2(1, 1, 0) - W_8^3 x_2(1, 1, 1) \qquad (4.27b)$$

Samples $x_2(6)$ and $x_2(7)$ are read from the memory, the butterfly is executed on the pair, and $x_3(6)$ and $x_3(7)$ are written back to the memory, overwriting the original values of $x_2(6)$ and $x_2(7)$. In general, $N/2$ butterflies are found in each stage and there are $\log_2 N$ stages, so the total number of butterflies is $(N/2) \log_2 N$. Because one complex multiplication per butterfly is the maximum, the total number of multiplications is bounded by $(N/2) \log_2 N$ (some of the multiplies involve factors of unity and should not be counted).

Figure 4.9 shows the signal flow graph of the D-I-T FFT for $N = 8$. This algorithm is referred to as an in-place FFT with normally ordered input samples and bit-reversed outputs. Minor variations that include both bit-reversed inputs and normally ordered outputs and non-in-place algorithms with normally ordered inputs and outputs are possible. Also, when N is not a power of 2, a mixed-radix algorithm can be used to reduce computation. The mixed-radix FFT is most efficient when N is highly composite, i.e., $N = p_1^{r_1} p_2^{r_2} \cdots p_L^{r_L}$, where the p^i are small prime numbers and the r^i are positive integers. It can be shown that the order of complexity of the mixed radix FFT is order$\{N(r_1(p_1 - 1) + r_2(p_2 - 1) + \cdots + r_L(p_L - 1)\}$. Because of the lack of uniformity of structure among stages, this algorithm has not received much attention for hardware implementation. However, the mixed-radix FFT is often used in software applications, especially for processing data recorded in laboratory experiments in which it is not convenient to restrict the block lengths to be powers of 2. Many advanced FFT algorithms, such as higher-radix forms, the mixed-radix form, the prime-factor algorithm, and the Winograd algorithm are described in [9]. Algorithms specialized for real-valued data reduce the computational cost by a factor of two. A radix-2 D-I-T FFT program, written in C language, is listed in Table 4.6.

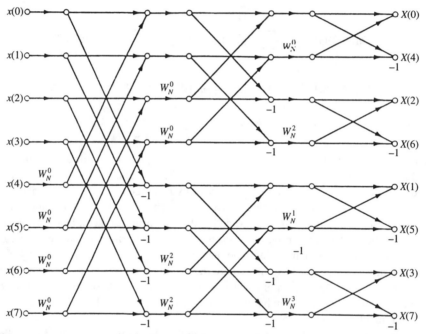

4.9   D-I-T FFT algorithm with normally ordered inputs and bit-reversed outputs.

4.6   Family Tree of Fourier Transforms

It is now possible to illustrate the functional relationships among the various forms of Fourier transforms that have been discussed in the previous sections. The family tree of CT Fourier transform is shown in Fig. 4.10, where the most general, and consequently the most powerful, Fourier transform is the classical complex Fourier transform (or equivalently, the bilateral Laplace transform). Note that the complex Fourier transform is identical to the bilateral Laplace transform, and it is at this level that the classical Laplace transform and Fourier transform techniques become identical. Each special member of the CT Fourier family is obtained by impressing certain constraints on the general form, thereby producing special transforms that are simpler and more useful in practical problems where the constraints are met.

TABLE 4.6   An In-Place D-I-T FFT Program in C Language

```
*********************************************************************/
*      fft: in-place radix-2 DFT of a complex input               */
*                                                                 */
*      input:                                                     */
*         n:      length of FFT: must be a power of two           */
*         m:      n = 2**m                                        */
*      input/output:                                              */
*         x:      float array of length n with real part of data  */
*         y:      float array of lengtn n with image part of data */
*********************************************************************/
fft(n,m,x,y)
tnt       n,m;
float     x[ ], y[ ]:
{
          int    i,j,k,n1,n2:
          float  c,s,e,a,t1,t2;

          j = 0;                          /*BIT-REVERSE  */
          n2 = n/2;
          for (i=1; 1 < n-1; i++)              /*bit-reverse counter */
          {
           n1 = n1/2;
           while ( j >= n1)
             {
              j = j - n1;
              n1 = n1/2;
             }
           j = j + n1;
           if (i < j)                         /*swap data */
             {
              t1 = x[i]; x[i] = x[j]; x[j] = t1;
              t1 = y[i]; y[i] = y[j]; y[j] = t1;
             }
          }
          n1 = 0; n2 = 1;               /* FFT */
          for (i = 0; i < m; i++)             /*state loop */
          {
           n1 = n2;   n2 = n2 + n2;
           e = -6.283185307179586/n2;
           a = 0.0;

           for (j=0; j < n1; j++)             /*flight loop */
             {
              c = cos(a); s=sin (a);
              a = a + e;

              for (k=j; k < n; k=k+n2)         /*butterfly loop */
               {
                t1 = c*x[k+n1] - s*y[k+n1];
                t2 = s*x[k+n1] + c*y[k+n1];
                x[k+n1] = x[k] - t1;
                y[k+n1] = y[k] - t2;
                x[k] = x[k] + t1;
                y[k] = y[k] + t2;
               }
             }
          }
          return;
}
```

The analogous family of DT Fourier techniques is presented in Fig. 4.11, in which the bilateral z-transform is analogous to the complex Fourier transform, the unilateral z-transform is analogous to the classical (one-sided) Laplace transform, the DTFT is analogous to the classical Fourier (CT) transform and the DFT is analogous to the classical (CT) Fourier series.

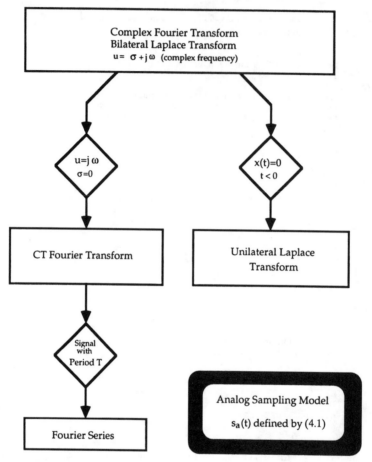

4.10   Relationships among CT Fourier concepts.

4.7   Selected Applications of Fourier Methods

Fast Fourier Transform in Spectral Analysis

An FFT program is often used to perform spectral analysis on signals that are sampled and recorded as part of laboratory experiments, or in certain types of data acquisition systems. Several issues must be addressed when spectral analysis is performed on (sampled) analog waveforms that are observed over a finite interval of time.

Windowing

The FFT treats the block of data as though it were one period of a periodic sequence. If the underlying waveform is not periodic, then harmonic distortion may occur because the periodic waveform created by the FFT may have sharp discontinuities at the boundaries of the blocks. This effect is minimized by removing the mean of the data (it can always be reinserted) and by windowing the data so the ends of the block are smoothly tapered to zero. A good rule of thumb is to taper 10% of the data on each end of the block using either a cosine taper or one of the other common windows shown in Table 4.7. An alternate interpretation of this phenomenon is that the finite length observation has already windowed the true waveform with a rectangular window that has large spectral sidelobes (see Table 4.7). Hence, applying an additional window results in a more desirable window that minimizes frequency domain distortion.

TABLE 4.7   Common Window Functions

Name	Function	Peak Side-Lobe Amplitude (dB)	Mainlobe Width	Minimum Stopband Attenuation (dB)
Rectangular	$\omega(n) = 1. \quad 0 \le n \le N-1$	-13	$4\pi/N$	-21
Bartlett	$\omega(n) = \begin{cases} 2/N, & 0 \le n \le (N-1)/2 \\ 22n/N, & (N-1)/2 \le n \le N-1 \end{cases}$	-25	$8\pi/N$	-25
Hanning	$\omega(n) = (1/2)[1 - \cos(2\pi n/N)]$ $0 \le n \le N-1$	-31	$8\pi/N$	-44
Hamming	$\omega(n) = 0.54 - 0.46\cos(2\pi n/N),$ $0 \le n \le N-1$	-43 -43	$8\pi/N$ $8\pi/N$	-53 -53
Backman	$\omega(n) = 0.42 - 0.5\cos(2\pi n/N)$ $+ 0.08\cos(4\pi n/N), \quad 0 \le n \le N-1$	-57	$12\pi/N$	-74

Zero Padding

An improved spectral analysis is achieved if the block length of the FFT is increased. This can be done by taking more samples within the observation interval, increasing the length of the observation interval, or augmenting the original data set with zeroes. First, it must be understood that the finite observation interval results in a fundamental limit on the spectral resolution, even before the signals are sampled. The CT rectangular window has a $(\sin x)/x$ spectrum, which is convolved with the true spectrum of the analog signal. Therefore, the frequency resolution is limited by the width of the mainlobe in the $(\sin x)/x$ spectrum, which is inversely proportional to the length of the observation interval. Sampling causes a certain degree of aliasing, although this effect can be minimized by sampling at a high enough rate. Therefore, lengthening the observation interval increases the fundamental resolution limit, while taking more samples within the observation interval minimizes aliasing distortion and provides a better definition (more sample points) on the underlying spectrum.

Padding the data with zeroes and computing a longer FFT does give more frequency domain points (improved spectral resolution), but it does not improve the fundamental limit, nor does it alter the effects of aliasing error. The resolution limits are established by the observation interval and the sampling rate. No amount of zero padding can improve these basic limits. However, zero padding is a useful tool for providing more spectral definition, i.e., it allows a better view of the (distorted) spectrum that results once the observation and sampling effects have occurred.

Leakage and the Picket Fence Effect

An FFT with block length N can accurately resolve only frequencies $\omega_k = (2\pi/N)k, k = 0, \ldots, N-1$ that are integer multiples of the fundamental $\omega_1 = (2\pi/N)$. An analog waveform that is sampled and subjected to spectral analysis may have frequency components between the harmonics. For example, a component at frequency $\omega_{k+1/2} = (2\pi/N)(k+1/2)$ will appear scattered throughout the spectrum. The effect is illustrated in Fig. 4.12 for a sinusoid that is observed through a rectangular window and then sampled at N points. The **picket fence effect** means that not all frequencies can be seen by the FFT. Harmonic components are seen accurately, but other components "slip through the picket fence" while their energy is "leaked" into the harmonics. These effects produce artifacts in the spectral domain that must be carefully monitored to assure that an accurate spectrum is obtained from FFT processing.

Finite Impulse Response Digital Filter Design

A common method for designing FIR digital filters is by use of windowing and FFT analysis. In general, window designs can be carried out with the aid of a hand calculator and a table of well-known window functions. Let $h[n]$ be the impulse response that corresponds to some desired frequency response, $H(e^{j\omega})$. If $H(e^{j\omega})$ has sharp discontinuities, such as the low-pass example shown in Fig. 4.13, then $h[n]$ will represent an infinite impulse response (IIR) function. The objective is to time limit $h[n]$ in such a way as to not distort $H(e^{j\omega})$ any more than necessary. If $h[n]$ is simply truncated, a ripple (Gibbs phenomenon) occurs around the discontinuities in the spectrum, resulting in a distorted filter (Fig. 4.13).

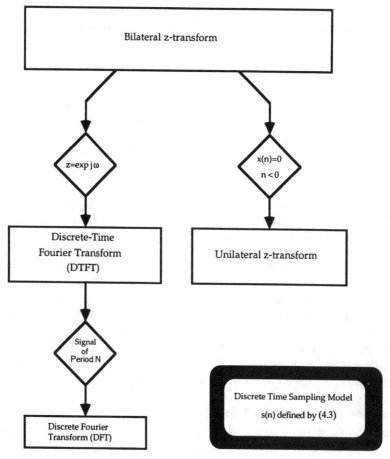

4.11   Relationships among DT concepts.

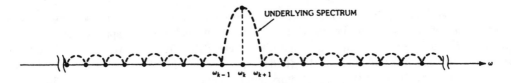

(A) FFT of a windowed sinusoid with frequency $\omega_1 = 2\pi k/N$.

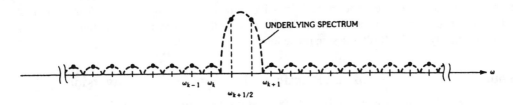

(B) Leakage for a nonharmonic sinusoidal component.

4.12 Illustration of leakage and the picket-fence effects.

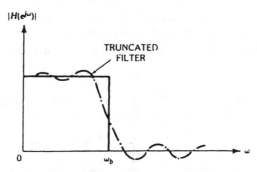

4.13 Gibbs effect in a low-pass filter caused by truncating the impulse response.

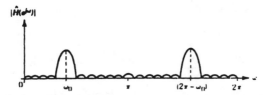

4.14 Design of a simple bandpass FIR filter by windowing.

Suppose that $w[n]$ is a window function that time limits $h[n]$ to create an FIR approximation, $h'[n]$; i.e., $h'[n] = w[n]h[n]$. Then if $W(e^{j\omega})$ is the DTFT of $w[n]$, $h'[n]$ will have a Fourier transform given by $H'(e^{j\omega}) = W(e^{j\omega}) * H(e^{j\omega})$, where $*$ denotes convolution. Thus, the ripples in $H'(e^{j\omega})$ result from the sidelobes of $W(e^{j\omega})$. Ideally, $W(e^{j\omega})$ should be similar to an impulse so that $H'(e^{j\omega})$ is approximately equal to $H(e^{j\omega})$.

   Special Case.   Let $h[n] = \cos n\omega_0$, for all n. Then $h[n] = w[n]\cos n\omega_0$, and

$$H'(e^{j\omega}) = (1/2)W(e^{j(\omega+\omega_0)}) + (1/2)W(e^{j(\omega-\omega_0)}) \tag{4.28}$$

as illustrated in Fig. 4.14. For this simple class, the center frequency of the bandpass is controlled by ω_0, and both the shape of the bandpass and the sidelobe structure are strictly determined by the choice of the window. While this simple class of FIRs does not allow for very flexible designs, it is a simple technique for determining quite useful low-pass, bandpass, and high-pass FIRs.

General Case. Specify an ideal frequency response, $H(e^{j\omega})$, and choose samples at selected values of ω. Use a long inverse FFT of length N' to find $h'[n]$, an approximation to $h[n]$, where if N is the desired length of the final filter, then $N' \gg N$. Then use a carefully selected window to truncate $h'[n]$ to obtain $h[n]$ by letting $h[n] = \omega[n]h'[n]$. Finally, use an FFT of length N' to find $H'(e^{j\omega})$. If $H'(e^{j\omega})$ is a satisfactory approximation to $H(e^{j\omega})$, the design is finished. If not, choose a new $H(e^{j\omega})$ or a new $w[n]$ and repeat. Throughout the design procedure it is important to choose $N' = kN$, with k an integer that is typically in the range of 4 to 10. Because this design technique is a trial and error procedure, the quality of the result depends to some degree on the skill and experience of the designer. Table 4.7 lists several well-known window functions that are often useful for this type of FIR filter design procedure.

Fourier Analysis of Ideal and Practical Digital-to-Analog Conversion

From the relationship characterized by (4.19b) and illustrated in Fig. 4.8, a CT signal $s(t)$ can be recovered from its samples by passing $s_a(t)$ through an ideal lowpass filter that extracts only the baseband spectrum. The ideal lowpass filter, shown in Fig. 4.15, is a zero-phase CT filter whose magnitude response is a constant of value T in the range $-\pi < \omega' \le \pi$, and zero elsewhere. The impulse response of this "reconstruction filter" is given by $h(t) = T \operatorname{sinc}((\pi/T)t)$, where $\operatorname{sinc} x = (\sin x)/x$. The reconstruction can be expressed as $s(t) = h(t) * s_a(t)$ which, after some mathematical manipulation, yields the following classical reconstruction formula:

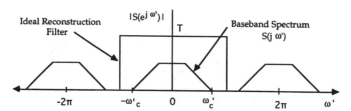

4.15   Illustration of ideal reconstruction.

$$s(t) = \sum_{n=-\infty}^{\infty} s(nT) \operatorname{sinc}((\pi/T)(t - nT)) \tag{4.29}$$

Note that the signal $s(t)$ is exactly recovered from its samples only if an infinite number of terms is included in the summation of (4.29). However, a good approximation of $s(t)$ can be obtained with only a finite number of terms if the lowpass reconstruction filter $h(t)$ is modified to have a finite interval of support, i.e., if $h(t)$ is nonzero only over a finite time interval. The reconstruction formula of (4.29) is an important result in that it represents the inverse of the sampling operation. By this means Fourier transform theory establishes that as long as CT signals are sampled at a sufficiently high rate, the information content contained in $s(t)$ can be represented and processed in either a CT or DT format. Fourier sampling and reconstruction theory provides the theoretical mechanism for translation between one format or the other without loss of information.

A CT signal $s(t)$ can be perfectly recovered from its samples using (4.29) as long as the original sampling rate was high enough to satisfy the Nyquist sampling criterion, because the sampling frequency exceeds the Nyquist rate, i.e., $\omega_s > 2\omega_B$. If the sampling rate does not satisfy the Nyquist criterion the adjacent periods of the analog spectrum will overlap, causing a distorted spectrum. This effect, called **aliasing distortion**, is rather serious because it cannot be corrected easily once it has occurred. In general, an analog signal should always be prefiltered with an CT low-pass filter prior to sampling so that aliasing distortion does not occur.

Figure 4.16 shows the frequency response of a fifth-order elliptic analog low-pass filter that meets industry standards for prefiltering speech signals. These signals are subsequently sampled at an 8-kHz sampling rate and transmitted digitally across telephone channels. The band-pass ripple is less than ±0.01 dB from DC up to the frequency 3.4 kHz (too small to be seen in Fig. 4.16), and the stopband rejection reaches at least −32.0 dB at 4.6 kHz and remains below this level throughout the stopband.

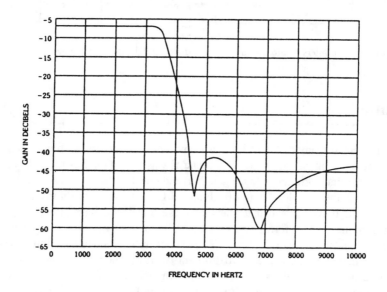

4.16 A fifth-order elliptic analog anti-aliasing filter used in the telecommunications industry with an 8-kHz sampling rate.

Most practical systems use digital-to-analog converters for reconstruction, which results in a staircase approximation to the true analog signal, i.e.,

$$\hat{s}(t) = \sum_{n=-\infty}^{\infty} s(nT)\{u(t - nT) - u[t - (n + 1)]\}, \tag{4.30}$$

where $\hat{s}(t)$ denotes the reconstructed approximation to $s(t)$, and $u(t)$ denotes a CT unit step function. The approximation $\hat{s}(t)$ is equivalent to a result obtained by using an approximate reconstruction filter of the form

$$H_a(j\omega) = 2T e^{-j\omega T/2} \sin c(\omega T/2) \tag{4.31}$$

The approximation $\hat{s}(t)$ is said to contain "$\sin x/x$ distortion," which occurs because $H_a(j\omega)$ is not an ideal low-pass filter. $H_a(j\omega)$ distorts the signal by causing a droop near the passband edge, as well as by passing high-frequency distortion terms which "leak" through the sidelobes of $H_a(j\omega)$. Therefore, a practical digital to analog converter is normally followed by an analog postfilter

$$H_p(j\omega) = \left[\begin{array}{ll} H_a^{-1}(j\omega), & 0 \le |\omega| < \pi/T \\ 0, & \omega \text{ otherwise} \end{array} \right] \tag{4.32}$$

which compensates for the distortion and produces the correct $\hat{s}(t)$, i.e., the correctly constructed CT output. Unfortunately, the postfilter $H_p(j\omega)$ cannot be implemented perfectly, and, therefore, the actual reconstructed signal always contains some distortion in practice that arises from errors in approximating the ideal postfilter. Figure 4.17 shows a digital processor, complete with analog to digital and digital to analog converters, and the accompanying analog pre- and postfilters necessary for proper operation.

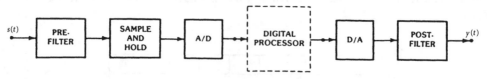

4.17   Analog pre- and postfilters required at the analog to digital and digital to analog interfaces.

4.8 Summary

This chapter presented many different Fourier transform concepts for both continuous time (CT) and discrete time (DT) signals and systems. Emphasis was placed on illustrating how these various forms of the Fourier transform relate to one another, and how they are all derived from more general complex transforms, the complex Fourier (or bilateral Laplace) transform for CT, and the bilateral z-transform for DT. It was shown that many of these transforms have similar properties which are inherited from their parent forms, and that a parallel hierarchy exists among Fourier transform concepts in the CT and the DT worlds. Both CT and DT sampling models were introduced as a means of representing sampled signals in these two different "worlds," and it was shown that the models are equivalent by virtue of having the same Fourier spectra when transformed into the Fourier domain with the appropriate Fourier transform. It was shown how Fourier analysis properly characterizes the relationship between the spectra of a CT signal and its DT counterpart obtained by sampling. The classical reconstruction formula was obtained as an outgrowth of this analysis. Finally, the discrete Fourier transform (DFT), the backbone for much of modern digital signal processing, was obtained from more classical forms of the Fourier transform by simultaneously discretizing the time and frequency domains. The DFT, together with the remarkable computational efficiency provided by the fast Fourier transform (FFT) algorithm, has contributed to the resounding success that engineers and scientists have experienced in applying digital signal processing to many practical scientific problems.

References

[1] VanValkenburg, M.E., *Network Analysis,* 3rd ed., Englewood Cliffs, NJ, Prentice-Hall, 1974.

[2] Oppenheim, A.V., Willsky, A.S., and Young, I.T., *Signals and Systems,* Englewood Cliffs, NJ, Prentice-Hall, 1983.

[3] Bracewell, R.N., *The Fourier Transform,* 2nd ed., New York, NY, McGraw-Hill, 1986.

[4] Oppenheim, A.V. and Schafer, R.W., *Discrete-Time Signal Processing,* Englewood Cliffs, NJ, Prentice-Hall, 1989.

[5] Jenkins, W.K. and Desai, M.D., "The discrete-frequency Fourier transform," *IEEE Trans. Circuits Syst.,* (CAS-33)7, 732–734, July 1986.

[6] Oppenheim, A.V. and Schafer, R.W., *Digital Signal Processing,* Englewood Cliffs, NJ, Prentice-Hall, 1975.

[7] Blahut, R.E., *Fast Algorithms for Digital Signal Processing,* Reading, MA, Addison-Wesley, 1985.

[8] Deller, J.R., Jr., "Tom, Dick, and Mary discover the DFT," *IEEE Signal Processing Mag.*, 11(2), pp. 36–50, April 1994.

[9] Burrus, C.S. and Parks, T.W., *DFT/FFT and Convolution Algorithms,* New York, John Wiley and Sons, 1985.

[10] Brigham, E.O., "The Fast Fourier Transform," Englewood Cliffs, NJ, Prentice-Hall, 1974.

5

z-Transform

Jelena Kovačević
AT&T Bell Laboratories

5.1 Introduction

When analyzing linear systems, one of the problems we often encounter is that of solving linear, constant-coefficient differential equations. A tool used for solving such equations is the Laplace transform. At the same time, to aid the analysis of linear systems, we extensively use Fourier-domain methods. With the advent of digital computers, it has become increasingly necessary to deal with discrete-time signals, or, sequences. These signals can be either obtained by sampling a continuous-time signal, or they could be inherently discrete. To analyze linear discrete-time systems, one needs a discrete-time counterpart of the Laplace transform (LT). Such a counterpart is found in the *z* transform, which similarly to the LT, can be used to solve linear constant-coefficient difference equations. In other words, instead of solving these equations directly, we transform them into a set of algebraic equations first, and then solve in this transformed domain. On the other hand, the *z*-transform can be seen as a generalization of the discrete-time Fourier transform (FT)

$$X(e^{j\omega}) = \sum_{n=-\infty}^{+\infty} x[n]e^{-j\omega n} \tag{5.1}$$

The above expression does not always converge, and thus, it is useful to have a representation which will exist for these nonconvergent instances. Furthermore, the use of the *z*-transform offers considerable notational simplifications. It also allows us to use the extensive body of work on complex variables to aid in analyzing discrete-time systems.

The z-transform, as pointed out by Jury in his classical text [3], is not new. It can be traced back to the early 18th century and the times of DeMoivre, who introduced the notion of the **generating function**, extensively used in probability theory

$$\Gamma(z) = \sum_{n=-\infty}^{+\infty} p[n]z^n \tag{5.2}$$

where $p[n]$ is the probability that the discrete random variable $\mathbf{n}$ will take a value n [8]. By comparing (5.2) and (5.3) below, we can see that the generating function $\Gamma(1/z)$ is the z-transform of the sequence $p[n] = p\{\mathbf{n} = n\}$. After these initial efforts, and due to the fast development of digital computers, a renewed interest in the z-transform occurred in the early 1950s, and the z-transform has been used for analyzing discrete-time systems ever since.

This section is intended as a brief introduction to the theory and application of the z-transform. For a rigorous mathematical treatment of the transform itself, the reader is referred to the book by one of the pioneers in the development of analysis of sampled data systems, Jury [3], and the references therein. For a more succinct account of the z-transform, its properties and use in discrete-time systems, consult, for example, [7]. A few other texts which contain parts on the z-transform include [1, 2, 5, 6, 10].

5.2   Definition of the z-Transform

Suppose we are given a discrete-time sequence $x[n]$, either inherently discrete time, or obtained by sampling a continuous-time signal $x_c(t)$, so that $x[n] = x_c(nT)$, $n \in \mathcal{Z}$, where T is the sampling period. Then the two-sided z-transform of $x[n]$ is defined as

$$X(z) = \sum_{n=-\infty}^{+\infty} x[n]z^{-n} \tag{5.3}$$

Here, z is a complex variable, and depending on its magnitude and the sequence $x[n]$, the above sum may or may not converge. The region in the z-plane where the sum does converge is called the **region of convergence** (ROC), and is discussed in more detail later.

Observe that in (5.3), n ranges from $-\infty$ to $+\infty$. That is why the z-transform defined in this way is called **two-sided**. One could define a **one-sided** z-transform, where n would range from 0 to $+\infty$. Obviously, the two definitions are equivalent only if the signal itself is one-sided, that is, if $x[n] = 0$, for $n < 0$. The advantage of using the one-sided z-transform is that it is useful in solving linear constant-coefficient difference equations with nonzero initial conditions and in the study of sampled-data feedback systems, discussed later. However, from now on, we deal mostly with the two-sided z-transform (see also Chapter 3 where the one-sided LT is used).

A power series given in (5.3) is a Laurent series, and thus for it to converge uniformly, it has to be absolutely summable, that is, the following must hold:

$$\sum_{n=-\infty}^{n=+\infty} |x[n]|\,|z|^{-n} < \infty \tag{5.4}$$

where $|z|$ is the magnitude of the complex variable z, i.e., $z = |z|e^{j\,\arg z}$. We can now see that if the z-transform converges for $z = z_1$, it will converge for all z such that $|z| = |z_1|$; that is, for all z on the circle $|z| = |z_1|$. In general, therefore, the ROC will be an annular region in the z-plane, and will be of the form

$$0 \le R_- < |z| < R_+ \le \infty \tag{5.5}$$

Here, R_- could be zero, and R_+ could be infinity, so that all of the cases given in Figure 5.1 are possible. Note that the points $z = 0$, or $z = \infty$ could be included in the region of convergence.

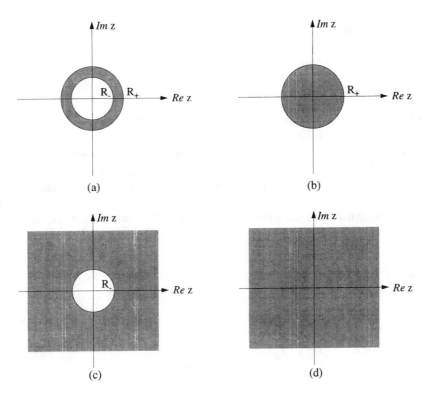

5.1 ROC of the z-transform. (a) General case: $0 \le R_- < |z| < R_+ \le \infty$. (b) ROC is the inside of the circle $|z| < R_+$ ($z = 0$ might be excluded). (c) ROC is the outside of the circle $R_- < |z|$ ($z = \infty$ might be excluded). (d) ROC is the whole z-plane (except possibly $z = 0$ or $z = \infty$).

A very important notion when talking about the ROC of the z-transform is that of the **unit circle**, because it makes the connection to the discrete-time FT. The unit circle is defined as the set of points in the z-plane for which $|z| = 1$. Let us evaluate the z-transform of a sequence on the unit circle. Because $|z| = 1$, it can be expressed as $z = e^{j\omega}$, where $\omega = \arg z$. Thus,

$$X(z)\big|_{|z|=1} = X(e^{j\omega}) = \sum_{n=-\infty}^{+\infty} x[n]e^{-j\omega n} \tag{5.6}$$

Note that here we use $X(e^{j\omega})$ rather than $X(\omega)$ to make it explicit that this is a function of $e^{j\omega}$. As can be seen from the equation, the z-transform evaluated on the unit circle is equivalent to the discrete-time FT given in (5.1). This immediately justifies the statement from the beginning of the section that the z-transform can be seen as a generalization of the FT. In other words, instances occur when the FT does not converge, while the z-transform does. One such instance is the unit-step function $u[n]$

$$u[n] = \begin{cases} 1, & n \geq 0, \\ 0, & \text{otherwise} \end{cases}$$

The sum

$$U(e^{j\omega}) = \sum_{n=-\infty}^{+\infty} u[n]e^{-j\omega n} = \sum_{n=0}^{+\infty} e^{-j\omega n}$$

is not absolutely summable and thus its FT does not converge, while

$$U(z) = \sum_{n=-\infty}^{+\infty} u[n]z^{-n} = \sum_{n=0}^{+\infty} z^{-n}$$

converges for $|z| > 1$. As a consequence, if the ROC includes the unit circle, then the discrete-time FT of a given sequence will exist, otherwise it will not.

The unit circle captures the periodicity of the discrete-time FT. If we start evaluating the z-transform on the unit circle at the point $(\text{Re } z, \text{Im } z) = (1, 0)$ corresponding to $\omega = 0$, going through $(\text{Re } z, \text{Im } z) = (0, j)$, $(\text{Re } z, \text{Im } z) = (-1, 0)$ which corresponds to $\omega = \pi$, and $(\text{Re } z, \text{Im } z) = (0, -j)$, back to $(\text{Re } z, \text{Im } z) = (1, 0)$ corresponding to $\omega = 2\pi$, we have evaluated one period of the FT and have returned to the same point. Thus, we are effectively warping the linear frequency axis into the unit circle (see Figure 5.2).

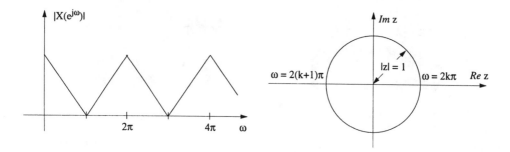

5.2   Warping of the linear frequency axis of the discrete-time FT into the unit circle of the z-transform. For example, the point $(\text{Re } z, \text{Im } z) = (1, 0)$ on the unit circle corresponds to points $\omega = 2k\pi$ on the ω axis.

We also mentioned that the z-transform is the discrete-time counterpart of the LT. Consider the function

$$x_{cs}(t) = \sum_{n=-\infty}^{+\infty} x_c(nT)\delta(t - nT) \tag{5.7}$$

or, the sampled version of the original continuous time function $x_c(t)$. Here, T is the sampling period, and $\delta(t)$ is the Dirac function. Taking the LT of $x_{cs}(t)$, we obtain

$$X_{cs}(s) = \sum_{n=-\infty}^{+\infty} x_c(nT)e^{-nTs} \tag{5.8}$$

If we now replace e^{sT} by z, we obtain the z-transform. Now, observe that $X_{cs}(s)$ in (5.8) is periodic, since

$$X_{cs}\left(s + j\frac{2\pi}{T}\right) = \sum_{n=-\infty}^{+\infty} x_c(nT)e^{-nT(s+j(2\pi)/T)},$$

$$= \sum_{n=-\infty}^{+\infty} x_c(nT)e^{-nTs}e^{-j2\pi n} = X_{cs}(s)$$

(5.9)

This means that $X_{cs}(s)$ is periodic along constant lines $\sigma = \sigma_{const}$ (parallel to the $j\omega$ axis). This further means that any line parallel to the $j\omega$ axis maps into a circle in the z-plane. It is easy to see that the $j\omega$ axis itself would map into the unit circle, while the left (or right) half-planes would map into the inside (or outside) of the unit circle, respectively.

Finally, let us say a few words about a very important class of signals, those whose z-transform is a rational function of z. They arise from systems that can be represented by linear constant-coefficient difference equations and are the signals with which we deal mostly in practice. If we represent such signals by

$$X(z) = \frac{N(z)}{D(z)}$$

(5.10)

then the zeroes of the numerator $N(z)$ are called **zeroes** of $X(z)$, while the zeroes of the denominator $D(z)$ are called **poles** of $X(z)$ (more precisely, a pole z_p will be a point at which $\lim_{z \to z_p} X(z)$ does not exist). How the poles can determine the region of convergence is the subject of a discussion later in the chapter.

5.3 Inverse z-Transform

We have seen that specifying the ROC when taking the z-transform of a sequence is an integral part of the process. For example, consider the following sequences: $x[n] = u[n]$ and $y[n] = -u[-n-1]$, where $u[n]$ is the unit-step function. Taking their z-transforms, we obtain

$$X(z) = \frac{1}{1-z^{-1}}, \qquad |z| > 1$$

(5.11)

$$Y(z) = \frac{1}{1-z^{-1}}, \qquad |z| < 1$$

(5.12)

They are identical except for their ROCs (see Figure 5.3). This tells us that without the ROC, our z-transform is not complete, and that if we are given $1/(1-z^{-1})$ as the z-transform of a sequence, taking the inverse is not uniquely specified unless we know the ROC.

We examine two ways of finding the inverse z-transform: first, and the formal one, by contour integration, and second, by partial fraction expansion. Note that the latter can be used only when $X(z)$ is a rational function of z; however, because most of the time we are dealing with rational functions of z, the partial fraction expansion method is used more often. We also mention two other informal techniques for determining the inverse z-transform.

Contour Integration

The formal inversion process for the z-transform is given by contour integration. It is obtained by taking the expression for the z-transform given by (5.3) and multiplying both sides by $(1/2\pi j)z^{k-1}$. Then the result is integrated counterclockwise along a closed contour C in the z-plane containing the origin, leading to

$$\oint_C \frac{1}{2\pi j} X(z)z^{k-1}dz = \oint_C \frac{1}{2\pi j}z^{k-1} \sum_{n=-\infty}^{+\infty} x[n]z^{-n} dz$$

(5.13)

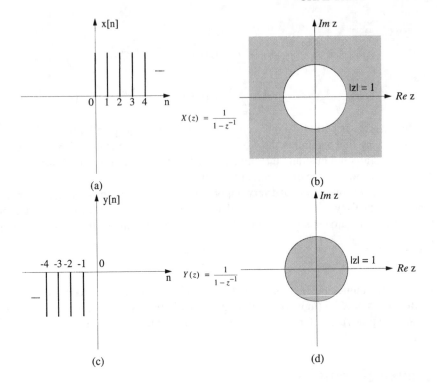

5.3 Two different sequences giving rise to the same z-transforms $1/(1 - z^{-1})$ except for their ROCs. (a) Sequence $x[n] = u[n]$ and (b) its ROC $|z| > 1$. (c) Sequence $y[n] = -u[-n-1]$ and (d) its ROC $|z| < 1$.

We choose the contour of integration to lie within the ROC, which will allow us to interchange the order of integration and summation on the right side of (5.13). This leads to

$$\frac{1}{2\pi j} \oint_C X(z) z^{k-1} \, dz = \sum_{n=-\infty}^{+\infty} x[n] \left(\frac{1}{2\pi j} \oint_C z^{-n+k-1} \, dz \right) \tag{5.14}$$

The integral in parentheses on the right side of (5.14) can be evaluated using Cauchy's integral formula, which states that if the contour of integration C contains the origin, and integration is performed counterclockwise, then

$$\frac{1}{2\pi j} \oint_C z^k \, dz = \begin{cases} 1, & k = -1, \\ 0, & \text{otherwise} \end{cases} \tag{5.15}$$

Substituting (5.15) in (5.14), we see that the integral is nonzero only for $n = k$, and thus (5.14) can be rewritten as

$$\frac{1}{2\pi j} \oint_C X(z) z^{k-1} \, dz = x[k] \tag{5.16}$$

The above equation is the inversion formula for the z-transform. To evaluate it for general functions can be quite difficult. However, in cases in which $X(z)$ is a rational function of z, we can make use of Cauchy's formula. It tells us that

$$\frac{1}{2\pi j} \oint_C \frac{F(z)}{z - z_p} \, dz = \begin{cases} F(z_p), & z_p \text{ inside } C, \\ 0, & \text{otherwise} \end{cases} \tag{5.17}$$

where C is a simple closed path and $F'(z)$, the derivative of $F(z)$, exists on and inside C. If we have a pole of multiplicity r enclosed in the contour C, and $F(z)$ and its $(r + 1)$ derivatives exist in and on C, then

$$\frac{1}{2\pi j} \oint_C \frac{F(z)}{(z - z_p)^r} \, dz = \begin{cases} \frac{1}{(r-1)!} \frac{d^{r-1}}{dz^{r-1}} F(z) \Big|_{z=z_p}, & z_p \text{ inside } C, \\ 0, & \text{otherwise} \end{cases} \tag{5.18}$$

Equations (5.17) and (5.18) are called *residues*. Use (5.17) and (5.18), and if we express

$$X(z)z^{k-1} = \frac{F(z)}{(z - z_p)^r} \tag{5.19}$$

where $F(z)$ has no poles at $z = z_p$, then Cauchy's residue theorem says that

$$x[k] = \frac{1}{2\pi j} \oint_C X(z)z^{k-1} \, dz = \sum_i R_i \tag{5.20}$$

where R_i are residues of $X(z)z^{k-1}$ at the poles inside the contour C. The poles outside the contour do not contribute to the sum. If no poles are inside the contour of integration for a certain k, then $x[k]$ is zero for that k. Do not ignore the fact that the contour of integration C must lie within the ROC.

In some instances it may be quite cumbersome to evaluate (5.20), for example, when we have a multiple-order pole at $z = 0$, whose order depends on k. In that case we can rewrite (5.20) as

$$x[k] = \sum_i R_i'$$

where R_i' is the residue of $X(1/z)z^{-k-1}$ at the poles inside the contour C', and C' is a circle of radius $1/s$ if C is a circle of radius s. For more details, see [7].

Partial Fraction Expansion

Another method of obtaining the inverse z-transform is by using partial fraction expansion. Note, however, that the partial fraction expansion method can be applied only to rational functions. Thus, suppose that $X(z)$ can be represented as in (5.10). We can then rewrite it as

$$X(z) = \frac{N_0}{D_0} \cdot \frac{\prod_{i=1}^{N}(1 - n_i z^{-1})}{\prod_{i=1}^{D}(1 - d_i z^{-1})} \tag{5.21}$$

where n_i are nontrivial zeroes of $N(z)$ and d_i are zeroes of $D(z)$. The partial fraction expansion of $X(z)$ can be written as

$$X(z) = \sum_{i=0}^{N-D} A_i z^{-i} + \sum_{i=1}^{D_s} \frac{B_i}{(1 - d_i z^{-1})} + \sum_{i=1}^{D_m} \sum_{m=1}^{P_i} \frac{C_{mi}}{(1 - d_i z^{-1})^m} \tag{5.22}$$

Here, if $N \geq D$, A_i can be obtained by long division of the numerator by the denominator; otherwise, that first sum in (5.22) disappears. In the second sum D_s denotes the number of single poles d_i of $X(z)$, and the coefficients B_i can be obtained as

$$B_i = (1 - d_i z^{-1})X(z) \Big|_{z=d_i} \tag{5.23}$$

The third sum (double sum) represents the part with multiple poles. D_m is the number of multiple poles d_i, and p_i are their respective multiplicities. The coefficients C_{mi} can be obtained as

$$C_{mi} = \frac{1}{(p_i - m)!(-d_i)^{p_i - m}} \frac{d^{p_i - m}}{dz^{p_i - m}} \left[(1 - d_i z)^{p_i} X(z^{-1})\right]\Bigg|_{z = d_i^{-1}} \qquad (5.24)$$

Once we have the expression (5.22), we can recognize each term as the z-transform of a known sequence. For example, $B_i/(1 - d_i z^{-1})$ will be the z-transform of either $B_i d_i^n u[n]$, or $-B_i d_i^n u[-n - 1]$, depending on whether $|z| > |d_i|$, or $|z| < |d_i|$.

Other Methods for Obtaining the Inverse z-Transform

While the two methods presented above will work in most cases, sometimes it can be more convenient to use simpler techniques.

One of these is the inspection method [7]. It consists of learning to recognize some often-used z-transform pairs. For example, if we are given $3/(1 - z^{-1})$ with the ROC $|z| < 1$, from (5.12) we can recognize it as the z-transform of $-3u[-n - 1]$. In this process the tables of z-transform pairs are an invaluable tool. An extensive list of z-transform pairs can be found in [3].

Another technique can be used if we are given a z-transform in the form of a power series expansion:

$$X(z) = \cdots + x[-1]z + x[0] + x[1]z^{-1} + \cdots$$

Then, we can identify each term with the appropriate power of z. For example, the coefficient with z^{-k} will be $x[k]$.

5.4   Properties of the z-Transform

While we can always obtain a z-transform of a sequence by directly applying its definition as given in (5.3), it is useful to have a list of properties at hand to help calculate a particular z-transform or inverse z-transform more easily. We divide these properties into two categories: properties of the ROC, and properties of the z-transform itself. In what follows R_x will denote the ROC of the signal $x[n]$, while R_{x-} and R_{x+} will denote its lower/upper bounds, respectively [as given in (5.5)].

Region of Convergence

The ROC is an integral part of the z-transform of a sequence. Thus, this section goes into more detail on some of the points touched upon earlier. These properties and the order in which they are presented follow those in [7]; therefore, for more details, see [7].

First, we said that the ROC is an annular region in the z-plane, i.e., $0 \leq R_- < |z| < R_+ \leq \infty$. This follows from the fact that if the z-transform converges for $z = z_1$, it will converge for all z such that $|z| = |z_1|$, that is, for all z on the circle $|z| = |z_1|$. Then, if we put $|z| = 1$ in (5.3), we obtain (5.6), or the discrete-time FT. Therefore, it is obvious that the FT of $x[n]$ converges iff the z-transform of $x[n]$ converges for $|z| = 1$, that is, iff the ROC of the z-transform contains the unit circle. Another useful property is that the ROC cannot contain any poles. This stems from the fact that if it did, the z-transform at the pole would be infinite and would not converge.

Consider now what happens if the sequence is of finite duration—it is zero except in a finite interval $-\infty < N_1 \leq n \leq N_2 < +\infty$. If all the values are finite, then the sequence is clearly absolutely summable and the z-transform will converge everywhere, except possibly at points $z = 0$ or $z = \infty$. Using the same type of arguments one can conclude that if the sequence is right-sided (it is zero for $n < N_1 < +\infty$), then the ROC will be the annular region outside of the finite pole of

$X(z)$ of the largest magnitude. Similarly, if the sequence is left-sided (it is zero for $n > N_2 > -\infty$), then the ROC will be the annular region inside the finite pole of $X(z)$ of the smallest magnitude. As a result, if a sequence is neither left- nor right-sided, the ROC will be an annular region bounded on the interior and the exterior by a pole.

Properties of the Transform

The sequences $x[n]$, $y[n]$, $\cdots$ will have associated z-transforms $X(z)$, $Y(z)$, ..., with ROCs R_x, R_y, ...,, in which each ROC will have its associated lower and upper bounds, as given in (5.5).

Linearity

$$ax[n] + by[n] \longleftrightarrow aX(z) + bY(z) \qquad \text{ROC} \supset R_x \cap R_y \tag{5.25}$$

To prove this, apply the definition given in (5.5). Note that the resulting ROC is at least as large as the intersection of the two starting ROCs. For example, if both $X(z)$ and $Y(z)$ are rational functions of z, and by adding $aX(z)$ to $bY(z)$ we introduce a zero that cancels one of the poles, the resulting ROC is larger than the intersection. If, on the other hand, no pole/zero cancellation exists, the resulting ROC is exactly the intersection.

Shift in Time

$$x[n - i] \longleftrightarrow z^{-i} X(z) \qquad \text{ROC} = R_x \tag{5.26}$$

The proof is straightforward and follows by the change of variables $k = n - i$ in (5.5). Note that the resulting ROC could gain or lose a few poles at $z = 0$ or $z = \infty$.

Time Reversal

$$x[-n] \longleftrightarrow X\left(\frac{1}{z}\right) \qquad \frac{1}{R_{x_+}} < |z| < \frac{1}{R_{x_-}} \tag{5.27}$$

Again, the proof is straightforward and follows by the change of variables $k = -n$ in (5.5).

Multiplication by an Exponential Sequence

One can easily show that the following holds:

$$a^n x[n] \longleftrightarrow X\left(\frac{z}{a}\right) \qquad |a|R_{x_-} < |z| < |a|R_{x_+} \tag{5.28}$$

Because z/a is the variable in the transform domain, we can see that all the poles of the original $X(z)$ have been scaled by a.

Multiplication by a Ramp

This property could also be called "differentiation of $X(z)$."

$$nx[n] \longleftrightarrow -z\frac{dX(z)}{dz}, \qquad \text{ROC} = R_x \tag{5.29}$$

To demonstrate this, differentiate both sides of (5.3) with respect to z and then multiply by $-z$ to obtain (5.29).

Convolution in Time

In transform domain convolution becomes simply the product of the two sequences. If we denote convolution by $*$,

$$x[n] * y[n] = \sum_{k=-\infty}^{+\infty} x[k]y[n-k] = \sum_{k=-\infty}^{+\infty} y[k]x[n-k] \tag{5.30}$$

then

$$x[n] * y[n] \longleftrightarrow X(z) \cdot Y(z), \qquad \text{ROC} \supset R_x \cap R_y \tag{5.31}$$

Although the proof is not difficult, we write it here, because this is one of the most useful properties. Thus, take the convolution in (5.30), multiply it by z^{-n} and sum over n

$$
\begin{aligned}
\sum_{n=-\infty}^{+\infty} \sum_{k=-\infty}^{+\infty} x[k]y[n-k]z^{-n} &= \sum_{k=-\infty}^{+\infty} x[k] \sum_{n=-\infty}^{+\infty} y[n-k]z^{-n}, \\
&= \sum_{k=-\infty}^{+\infty} x[k] \sum_{p=-\infty}^{+\infty} y[p]z^{-(p+k)}, \\
&= \sum_{k=-\infty}^{+\infty} x[k]z^{-k} \sum_{p=-\infty}^{+\infty} y[p]z^{-p}, \\
&= X(z)Y(z)
\end{aligned}
$$

where we have used change of variables $p = n - k$. As in the case of linearity, if it happens that a pole residing at the border of one of the ROCs is cancelled by a zero from the other transform, then the resulting ROC will be larger than the intersection of the individual ROCs; otherwise, it will be exactly their intersection.

Convolution in z-Domain

Convolution in z-domain is given by (for more details, refer to [3] or [7])

$$x[n] \cdot y[n] \longleftrightarrow \frac{1}{2\pi j} \oint_C X(\lambda)Y\left(\frac{z}{\lambda}\right)\lambda^{-1}d\lambda \qquad R_{x_-}R_{y_-} < |z| < R_{x_+}R_{y_+} \tag{5.32}$$

where C is a closed contour in the intersection of the ROCs of $X(\lambda)$ and $Y(z/\lambda)$, and integration is performed counterclockwise. This property is the generalization of the periodic convolution property of the FT. Suppose that the contour C is the unit circle and $\lambda = e^{j\omega}$, $z = e^{j\theta}$. Also, observe that $d\lambda = je^{j\omega}d\omega$, and that if λ goes around the unit circle, ω ranges from $-\pi$ to π. Then,

$$\frac{1}{2\pi j} \oint_C X(\lambda)Y\left(\frac{z}{\lambda}\right)\lambda^{-1}d\lambda = \frac{1}{2\pi} \int_{-\pi}^{\pi} X(e^{j\omega})Y(e^{j(\theta-\omega)})\,d\omega$$

This equation states that the product of two sequences has as its FT the periodic convolution of their FTs.

Conjugation

If we are given a complex sequence $x^*[n]$, then its z-transform pair is

$$x^*[n] \longleftrightarrow X^*(z^*) \qquad \text{ROC} = R_x \tag{5.33}$$

Real Part

This property and the next one use the conjugation property given in (5.33). Thus, if we express

$$\text{Re}\{x[n]\} = \frac{x[n] + x^*[n]}{2}$$

then by (5.33) and the linearity of the z-transform

$$\text{Re}\{x[n]\} \longleftrightarrow \frac{1}{2}\left[X(z) + X^*(z^*)\right] \qquad \text{ROC} \supset R_x \qquad (5.34)$$

Imaginary Part

Similarly to the previous property, if we express

$$\text{Im}\{x[n]\} = \frac{x[n] - x^*[n]}{2j}$$

then by (5.33) and the linearity of the z-transform

$$\text{Im}\{x[n]\} \longleftrightarrow \frac{1}{2j}\left[X(z) - X^*(z^*)\right] \qquad \text{ROC} \supset R_x \qquad (5.35)$$

Parseval's Relation

Parseval's relation is widely used in various transform domains, usually to find the energy of a signal. We start here with its more general formulation, and then reduce it to its usual form

$$\sum_{n=-\infty}^{+\infty} x[n]y^*[n] = \frac{1}{2\pi j} \oint_C X(\lambda)Y^*\left(\frac{1}{\lambda^*}\right)\lambda^{-1}d\lambda \qquad (5.36)$$

where C is a closed contour in the intersection of the ROCs of $X(\lambda)$ and $Y^*(1/\lambda^*)$, and integration is performed counterclockwise. Again, for the proof, we refer the reader to [3] or [7].

Suppose now that $y[n] = x[n]$. Then (5.36) reduces to

$$\sum_{n=-\infty}^{+\infty} |x[n]|^2 = \frac{1}{2\pi j} \oint_C X(\lambda)X^*\left(\frac{1}{\lambda^*}\right)\lambda^{-1}d\lambda \qquad (5.37)$$

If both $x[n]$ and $y[n]$ converge on the unit circle, i.e., their FTs exist, then, we can choose $\lambda = e^{j\omega}$ and (5.36) becomes

$$\begin{aligned}
\sum_{n=-\infty}^{+\infty} x[n]y^*[n] &= \frac{1}{2\pi j} \oint_C X(e^{j\omega})Y^*(e^{j\omega})e^{-j\omega}d(e^{j\omega}), \\
&= \frac{1}{2\pi} \int_{-\pi}^{\pi} X(e^{j\omega})Y^*(e^{j\omega})d\omega
\end{aligned}$$

which is the usual Parseval's relation in Fourier domain.

Initial Value Theorem

If $x[n]$ is causal (it is 0 for $n < 0$), then

$$x[0] = \lim_{z \to \infty} X(z) \qquad (5.38)$$

Final Value Theorem

If the poles of $X(z)$ are all inside the unit circle, then

$$\lim_{n \to \infty} x[n] = \lim_{z \to 1} \left[(1 - z^{-1}) X(z) \right] \tag{5.39}$$

5.5   Role of the z-Transform in Linear Time-Invariant Systems

The class of systems dealt with mostly are linear time-invariant systems. We discuss here the role of the z-transform in such systems. Recall that if we are given a system with input $x[n]$ and output $y[n]$, described by an operator $H[\cdot]$

$$y[n] = H[x[n]]$$

and $H[\cdot]$ is defined to be linear and time invariant, then

1. If inputs $x_1[n]$ and $x_2[n]$ produce outputs $y_1[n]$ and $y_2[n]$, then the input $ax_1[n] + bx_2[n]$ will produce the output $ay_1[n] + by_2[n]$.
2. If the input $x[n]$ produces output $y[n]$, then the input $x[n - k]$ will produce the output $y[n - k]$.

We recall a few properties of discrete linear time-invariant systems. For more details, refer to [4, 7]. Note first that $x[n]$ can be written as a superposition of unit samples as

$$x[n] = \sum_{k=-\infty}^{+\infty} x[k] \delta[n - k] \tag{5.40}$$

If this is the input and we want to find the corresponding output, then

$$\begin{aligned}
y[n] &= H[x[n]], \\
&= H \left[\sum_{k=-\infty}^{+\infty} x[k] \delta[n - k] \right], \\
&= \sum_{k=-\infty}^{+\infty} x[k] H[\delta[n - k]], \\
&= \sum_{k=-\infty}^{+\infty} x[k] h[n - k]
\end{aligned} \tag{5.41}$$

Here, $h[n]$ is called the **unit-sample response**, defined as the response of the system to the unit sample $\delta[n]$. Equation (5.41) is a convolution sum and thus it can be written as

$$y[n] = x[n] * h[n] \tag{5.42}$$

One of the most important properties of the z-transform is that the convolution in time has a product as its transform-domain pair. Therefore, using (5.31), we can write (5.42) in z-domain as

$$Y(z) = X(z) H(z) \tag{5.43}$$

Here, $H(z)$ is called the **system transfer function**

$$H(z) = \sum_{n=-\infty}^{+\infty} h[n] z^{-n} \tag{5.44}$$

We may also obtain this transfer function in another manner, if we assume that the systems we are dealing with are those that can be described by linear constant-coefficient difference equations (much in the same way as continuous time systems that can be represented by linear constant-coefficient differential equations). Along the way, the use of z-transform greatly simplifies analysis of such systems and the solutions to these equations reduce to solutions of algebraic equations. Hence, suppose that our system can be described by the following linear constant-coefficient difference equation:

$$\sum_{k=0}^{D} b_k y[n-k] = \sum_{k=0}^{N} a_k x[n-k] \tag{5.45}$$

This equation has many solutions. We assume that the system is causal and moreover, we assume that the initial conditions are satisfied so that the system is linear and time invariant. With these factors in mind, we can write the expression for the output of the system as

$$y[n] = -\sum_{k=1}^{D} \frac{b_k}{b_0} y[n-k] + \sum_{k=0}^{N} \frac{a_k}{b_0} x[n-k] \tag{5.46}$$

This means that the output at time n depends on the outputs from D previous instants as well as from the input at time n and at N previous instants. Taking the z-transform of both sides of (5.45) and using the linearity and shift in time properties of the z-transform, we obtain

$$Y(z) = \sum_{k=0}^{N} a_k z^{-k} \Big/ \sum_{k=0}^{D} b_k z^{-k} \, X(z) \tag{5.47}$$

Finally, using (5.43) we can identify $H(z)$ here as

$$H(z) = \sum_{k=0}^{N} a_k z^{-k} \Big/ \sum_{k=0}^{D} b_k z^{-k} \tag{5.48}$$

The function $H(z)$ is often referred to as a **filter**. If the denominator is a delay, i.e., $b_i z^{-i}$, such a filter will be a finite impulse response filter (FIR), as opposed to the infinite impulse response filter (IIR). Taking the inverse z-transform of (5.48) we may obtain the unit-sample response of the system $h[n]$.

Finally, let us see how one might obtain the **frequency response** of the system. Evaluate (5.44) on the unit circle (assuming that the ROC of $H(z)$ contains the unit circle)

$$H(z)\big|_{|z|=1} = H(e^{j\omega}) = \sum_{n=-\infty}^{+\infty} h[n] e^{-j\omega n} \tag{5.49}$$

Therefore, the frequency response of the system is the system transfer function evaluated on the unit circle:

$$H(e^{j\omega}) = H(z)\big|_{z=e^{j\omega}} \tag{5.50}$$

A consequence of this is that a linear time-invariant system is bounded-input–bounded-output stable (that is, bounded input produces bounded output), iff the ROC of the transfer function $H(z)$ contains the unit circle.

We mentioned earlier that we are using the double-sided z-transform, while the one-sided z-transform is most useful when solving linear constant-coefficient difference equations with nonzero initial conditions. To transform such an equation into an algebraic equation, we used the linearity and shift in time properties of the double-sided z-transform. We would like to be able to do the same with the one-sided z-transform; however, we must rederive the shift in time property in order to do that. Therefore, assume we are given a sequence $x[n-i]$ as in (5.26). Taking its one-sided z-transform, we obtain

$$
\begin{aligned}
\sum_{n=0}^{+\infty} x[n-i]z^{-n} &= \sum_{p=-i}^{+\infty} x[p]z^{-(p+i)}, \\
&= \sum_{p=-i}^{-1} x[p]z^{-(p+i)} + z^{-i} \sum_{p=0}^{+\infty} x[p]z^{-p}, \\
&= x[-i]+\cdots+x[-1]z^{-(i-1)} + z^{-i}X(z)
\end{aligned}
\tag{5.51}
$$

To solve (5.45), we can take the one-sided z-transform of both sides and use the linearity and shift in time properties to obtain

$$
\begin{aligned}
\sum_{k=0}^{D} b_k &\left(\sum_{p=-k}^{-1} y[p]z^{-(p+k)} + z^{-k}Y(z) \right) \\
&= \sum_{k=0}^{N} a_k \left(\sum_{p=-k}^{-1} x[p]z^{-(p+k)} + z^{-k}X(z) \right)
\end{aligned}
\tag{5.52}
$$

The output $y[n]$ for $n \geq 0$ can be obtained as the inverse z-transform of

$$
\begin{aligned}
Y(z) &= \left(\sum_{k=0}^{N} a_k z^{-k} \Big/ \sum_{k=0}^{D} b_k z^{-k} \right) X(z) \\
&+ \left(\sum_{k=0}^{N} a_k \sum_{p=-k}^{-1} x[p]z^{-(p+k)} - \sum_{k=0}^{D} b_k \sum_{p=-k}^{-1} y[p]z^{-(p+k)} \Big/ \sum_{k=0}^{D} b_k z^{-k} \right)
\end{aligned}
\tag{5.53}
$$

To solve for $Y(z)$ we need to know initial conditions $y[n]$, for $n = -D, \ldots, -1$, and for $x[n]$ for $n = -N, \ldots, -1$. Note that if $x[n]$ is causal, and the initial conditions are all zero, the above solution is the same as the one when the double-sided z-transform is used; i.e., it reduces to the first term on the right side of (5.53).

5.6   Variations on the z-Transform

Multidimensional z-Transform

Although the two-dimensional and multidimensional z-transforms are related to the one-dimensional z-transform, they are not straightforward generalizations of it. We give here a brief overview of the two-dimensional z-transform. For more details on it and the multidimensional z-transform, see [1, 2].

The two-dimensional z-transform of a sequence $x[n_1, n_2]$ is given by

$$
X(z_1, z_2) = \sum_{n_1=-\infty}^{+\infty} \sum_{n_2=-\infty}^{+\infty} x[n_1, n_2] z_1^{-n_1} z_2^{-n_2}
\tag{5.54}
$$

Here, z_1 and z_2 are complex variables and the region in the four-dimensional (z_1, z_2) space in which the above double sum converges is called the ROC. In one dimension, the ROC is an annular region in the z-plane, while here it is called the **Reinhardt domain**. The analog of the unit circle in one dimension is the **unit bicircle** for $|z_1| = 1$ and $|z_2| = 1$. On the unit bicircle, the two-dimensional z-transform becomes the two-dimensional discrete-time FT.

Properties of the two-dimensional z-transform are similar to those of the one-dimensional z-transform, and they can be found in [1, 2]. Here we list just two. First, for separable signals, the following holds:

$$x[n_1]y[n_2] \longleftrightarrow X(z_1)Y(z_2) \tag{5.55}$$

where (z_1, z_2) is in the ROC, if z_1 is in R_x and z_2 is in R_y. The differentiation property is as follows:

$$n_1 n_2 x[n_1, n_2] \longleftrightarrow z_1 z_2 \frac{\partial^2}{\partial z_1 \partial z_2} X(z_1, z_2), \qquad \text{ROC} = R_x \tag{5.56}$$

In analyzing two-dimensional systems it is useful to identify singularities. Given a two-dimensional linear shift-invariant system with an associated constant-coefficient difference equation

$$\sum_{k_1=0}^{D_1} \sum_{k_2=0}^{D_2} b_{k_1 k_2} y[n_1 - k_1, n_2 - k_2] = \sum_{k_1=0}^{N_1} \sum_{k_2=0}^{N_2} a_{k_1 k_2} x[n_1 - k_1, n_2 - k_2] \tag{5.57}$$

we can find the equivalent transfer function as

$$H(z_1, z_2) = \left(\sum_{k_1=0}^{N_1} \sum_{k_2=0}^{N_2} a_{k_1 k_2} z_1^{-k_1} z_2^{-k_2} \middle/ \sum_{k_1=0}^{D_1} \sum_{k_2=0}^{D_2} b_{k_1 k_2} z_1^{-k_1} z_2^{-k_2} \right) = \frac{N(z_1, z_2)}{D(z_1, z_2)} \tag{5.58}$$

Then, the zero of $H(z_1, z_2)$ is a point at which $A(z_1, z_2) = 0$ and $B(z_1, z_2) \neq 0$, while a pole is a point at which $B(z_1, z_2) = 0$. Note, however, that both zeroes and poles are continuous surfaces rather than a discrete set of points, as in one dimension. Note, also, that unlike in one dimension, no fundamental theorem of algebra tells us how to factorize a multidimensional polynomial into its factors, and thus, it is not easy to isolate poles and zeroes.

Modified z-Transform

The original, one-sided z-transform was developed to deal only with the signal at its sampling instants, and discard the rest. In many systems, particularly in mixed analog-digital systems, it is important to conserve the information about the signal between the sampling instants. Jury, among others, used the **modified z-transform** [3] to take care of this problem. The idea is to delay the continuous-time function $x_c(t)$ by a fictitious delay $(1 - \Delta)T$, where Δ varies from 0 to 1 (see Figure 5.4) in order to get all the values of $x_c(t)$ for $t = (n - 1 + \Delta)T, 0 \leq \Delta \leq 1, n \in \mathcal{Z}$; T is the sampling period. Then, the modified z-transform is defined as follows:

$$X_c(z, \Delta) = \sum_{n=-\infty}^{+\infty} x_c((n - 1 + \Delta)T)z^{-n}, \qquad 0 \leq \Delta \leq 1 \tag{5.59}$$

or, using the change of variable $k = n - 1$,

$$X_c(z, \Delta) = z^{-1} \sum_{k=-\infty}^{+\infty} x_c((k + \Delta)T)z^{-k}, \qquad 0 \leq \Delta \leq 1 \tag{5.60}$$

It is easy to see that the z-transform as defined in (5.3) can be obtained as a special case of the modified z-transform as

$$X_c(z) = X_c(z, \Delta)\big|_{\Delta=1} \tag{5.61}$$

Similarly to the z-transform, the modified z-transform possesses a number of useful properties. For these, the reader is referred to [3].

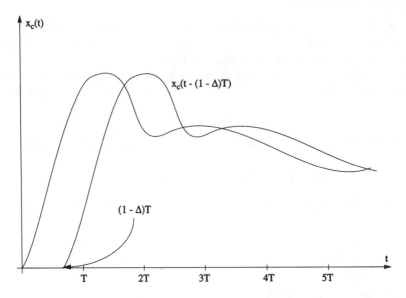

5.4   Modified z-transform takes into account values of the function between sampling instants by creating a fictitious delay $(1 - \Delta)T$.

Chirp z-Transform Algorithm

This is a brief discussion of an algorithm for computing the z-transform of a finite-length sequence, much as was done in the case of the DFT. Suppose we wanted to compute the z-transform on a circle concentric to the unit circle. We could use the DFT algorithm, with some minor modifications. However, the chirp z-transform algorithm used in radar systems [9] can be more efficient and allows one to compute the z-transform of a finite-length sequence on a spiral contour in the z-plane. The samples we compute are equally spaced in angle over some portion of the spiral. The algorithm employs the DFT as well, and has the complexity of

$$(N + M - 1) \log_2(N + M - 1)$$

where N is the number of nonzero values of the sequence $x[n]$, i.e., $x[0], \ldots, x[N - 1]$, and M is the number of points at which we evaluate the z-transform, $z_k, k = 1, \ldots, M$.

5.7   Concluding Remarks

This chapter developed the basics of the z-transform. We showed that it is a generalization of the discrete-time FT, and have explored the relationships between the two. We also discussed the connection to the LT. The inverse z-transform was presented, using both contour integration and partial fraction expansion. Although the partial fraction expansion method is easier, it can be used only for rational functions of z. We explored a number of properties of the ROC of the z-transform, as well as important properties of the transform itself. Finally, we showed how the z-transform is used in solving linear constant-coefficient difference equations. To conclude the section, we discussed the modified z-transform and the multidimensional z-transform. We also briefly mentioned the chirp z-transform algorithm, used for computing a few points of the z-transform of a finite-length sequence. For more details on the above topics, refer to [1, 2, 3, 5, 6, 7, 9, 10].

Acknowledgment

I would like to thank Professor Eli Jury for his encouragement and technical insight while writing this manuscript, as well as an anonymous reviewer for his useful remarks.

References

[1] Bose, N.K., *Applied Multidimensional System Theory,* New York, Van Nostrand Reinhold, 1982.

[2] Dudgeon, D.E. and Mersereau, R.M., *Multidimensional Digital Signal Processing,* Englewood Cliffs, NJ, Prentice-Hall, 1984.

[3] Jury, E.I., *Theory and Application of the z-Transform Method,* Malabar, FL, Robert E. Krieger, 1986.

[4] Kailath, T., *Linear Systems,* Englewood Cliffs, NJ, Prentice-Hall, 1980.

[5] Ludeman, L.C., *Fundamentals of DSP,* New York, Harper & Row, 1986.

[6] McGillem, C.D. and Cooper, G.R., *Continuous and Discrete Signal and System Analysis,* New York, Holt, Rinehart and Winston, 1974.

[7] Oppenheim, A.V. and Shafer, R.W., *Discrete-Time Signal Processing,* Englewood Cliffs, NJ, Prentice-Hall, 1989.

[8] Papoulis, A., *Probability, Random Variables and Stochastic Processes,* 2nd ed., New York, McGraw-Hill, 1984.

[9] Rabiner, L.R., Schafer, R.W., and Rader, C.M., "The chirp z-transform algorithm," *IEEE Trans. Audio Electroacoust.,* 17, 86–92, 1969.

[10] Vich, R., *z-Transform Theory and Applications,* Boston, D. Reidel, 1987.

6

Wavelet Transforms

P.P. Vaidyanathan
*California Institute of
Technology*

Igor Djokovic
*California Institute of
Technology*

6.1 Introduction

Transform techniques such as the Fourier and Laplace transforms and the z-transform have long been used in a wide variety of scientific and engineering disciplines [1, 2]. In a number of applications in which we require a joint time-frequency picture, it is necessary to consider other types of transforms or time-frequency representations. Many such methods have evolved. The wavelet transform technique [3]–[5] in particular has some unique advantages over other kinds of time-frequency representations such as the short-time Fourier transform. For historical developments as well as many technical details and original material see [5]. This chapter describes some of these representations, and explains the advantages of the wavelet transform, and the reason for its recent popularity.

A subclass of wavelet transforms [6] has an intimate connection with the theory of digital filter banks [7]–[10]. Filter banks have been known to the signal processing community for over 2 decades (see [7] and references therein). It is this relation that makes it possible to construct in a systematic way a wide family of wavelets with several desirable properties such as compact support (i.e., finite duration), smoothness, good time-frequency localization, and basis orthonormality (all these terms will be explained later).

The connection between wavelets and filter banks finds beautiful mathematical expression in the theory of multiresolution [11]. This enables us to compute the wavelet transform coefficients using

0-8493-0052-5/00/$0.00+$.50
© 2000 by CRC Press LLC

the so-called **fast wavelet transform (FWT)**, which is essentially a tree-structured filter bank. In addition to the practical value, many deep results from several disciplines find a unified home in the theory and development of the wavelet transform. This includes signal processing, circuit theory, communications, and mathematics. Our emphasis here is on this unification, and the beautiful big picture that it provides. Other tutorials on wavelets with different choices of emphasis can be found in [7], [12]–[14].

Scope and Outline

The literature on wavelets is enormous, and an attempt to do justice to everything would prove futile. Even a list of references that is fair to all contributors would be too long. We, therefore, restrict discussions to basic, core material. Sections 6.2 to 6.5 provide an overview, with the presentation given at a level that can be comprehended by most engineers. The more advanced results on wavelets, which brought them great attention in recent years, are presented in Sections 6.9 to 6.13. At the heart of these results lie several powerful mathematical tools, which are usually not familiar to engineers, and so we present a fairly extensive math review in three sections (Section 6.6 to 6.8). We suggest that the reader go through this review material once and then use it primarily as a reference.

The advanced Sections 6.9 to 6.13 are organized such that the main points, summarized as theorems for convenience of reference, can be appreciated even without the mathematical background material in Sections 6.6 to 6.8. The mathematical sections do, however, facilitate a deeper understanding. It is our hope that these sections will bring most readers to a point where they can pursue wavelet literature without difficulty.

Why Wavelets?

A commonly asked question is "why wavelets?," that is, "what are the advantages offered by wavelets over other types of transform techniques such as, the Fourier transform?" The answer to this question is fairly sophisticated, and also depends on the level at which we address the question. Several discussions addressing this question are scattered throughout this chapter. A convenient listing of the locations of these discussions is given in Section 6.14 under "Why Wavelets?"

General Notations and Acronyms

1. Boldfaced quantities represent matrices and vectors.
2. The notations $\mathbf{A}^T$, $\mathbf{A}^*$ and $\mathbf{A}^\dagger$ represent, respectively, the transpose, conjugate, and transpose-conjugate of the matrix $\mathbf{A}$.
3. The accent tilde is defined as follows: $\tilde{\mathbf{H}}(z) = \mathbf{H}^\dagger(1/z^*)$; thus, if $\mathbf{H}(z) = \sum_n \mathbf{h}(n)z^{-n}$, then $\tilde{\mathbf{H}}(z) = \sum_n \mathbf{h}^\dagger(-n)z^{-n}$. On the unit circle $\tilde{\mathbf{H}}(z) = \mathbf{H}^\dagger(z)$.
4. *Acronyms.* BIBO (bounded-input–bounded-output); FIR (finite impulse response); IIR (infinite impulse response); LTI (linear time invariant); PR (perfect reconstruction); STFT (short-time Fourier transform); WT (wavelet transform).
5. For LTI systems, "stability" stands for BIBO stability.
6. $\delta(n)$ denotes the unit pulse or discrete time impulse, defined such that $\delta(0) = 1$ and $\delta(n) = 0$ otherwise. This should be distinguished from the Dirac delta function [2], which is denoted as $\delta_a(t)$.
7. *Figures.* Sampled versions of continuous time signals are indicated with an arrow on the top [e.g., Fig. 6.10(a)]. The sampled versions are impulse trains of the form $\sum_n c(n)\delta_a(t - n)$, and are functions of continuous t.

6.2 Signal Representation Using Basis Functions

The electrical engineer is very familiar with the Fourier transform (FT) and its role in the study of linear time invariant (LTI) systems or **filters**. For example, the frequency response of an LTI system is the FT of its impulse response. The FT is also used routinely in the design and analysis of circuits. As a reminder, the FT of a signal $x(t)$ is given by the familiar integral $X(\omega) = \int_{-\infty}^{\infty} x(t)e^{-j\omega t}\, dt$ and the inverse transform by[1]

$$x(t) = \frac{1}{2\pi} \int_{-\infty}^{\infty} X(\omega)e^{j\omega t}\, d\omega \tag{6.1}$$

From this equation we can say that $x(t)$ has been expressed as a linear superposition (or linear combination) of an infinite number of functions $g_\omega(t) \overset{\triangle}{=} e^{j\omega t}$. Because the frequency ω is a continuous variable, *uncountably* many functions $g_\omega(t)$ are superimposed. Electrical engineers, in particular signal processors and communications engineers, are also familiar with two special classes of signals which can be regarded as a superposition of *countably* many functions:

$$x(t) = \sum_{n=-\infty}^{\infty} \alpha_n g_n(t) \tag{6.2}$$

where α_n are scalars (possibly complex) uniquely determined by $x(t)$. These two examples are time-limited signals for which we can find a Fourier series (FS), and band-limited signals which can be reconstructed from uniformly spaced samples by weighting them with shifted sinc functions (see below).

First consider a time-limited signal $x(t)$ with duration $0 \leq t \leq 1$ (Fig. 6.1). Under some mild conditions such a signal can be represented in the form (6.2) with $g_n(t) = e^{j2\pi nt}$. The expression (6.2) is then the FS of $x(t)$, and α_n are the Fourier coefficients. [In contrast we say that (6.1) is the **Fourier integral** of $x(t)$.] The **transform domain** signal $\{\alpha_n\}$ is a sequence, and the transform domain variable is discrete, namely, the frequencies $\omega_n \overset{\triangle}{=} 2\pi n$. Because $e^{j2\pi nt}$ is periodic in t with period one, the right-hand side of (6.2) is periodic, and it represents $x(t)$ only in $0 \leq t \leq 1$. It is sometimes convenient to replace the complex functions $e^{j2\pi nt}$ with the set of real functions $1, \sqrt{2}\cos(2\pi nt), \sqrt{2}\sin(2\pi nt), n > 0$, especially in circuit analysis.

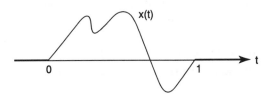

6.1 A finite duration signal, with support $0 \leq t \leq 1$.

Next, consider a band-limited signal $x(t)$ with FT $X(\omega)$ as demonstrated in Fig. 6.2. If we sample the signal at the Nyquist rate 2β rad/s (i.e., sampling period $T = \pi/\beta$), then multiple copies of the FT are generated [2], and we can recover $x(t)$ from the samples by use of an ideal low-pass filter

---

[1]At the moment it is not necessary to worry about the existence, invertibility, and the type (e.g., L^1 or L^2) of the FT. We return to the mathematical subtleties in Section 6.6.

$F(\omega)$ (Fig. 6.3). The impulse response of the filter is the **sinc function** $f(t) = \sin \beta t / \beta t$ so that the reconstruction formula is

$$x(t) = \sum_{n=-\infty}^{\infty} x(nT) f(t - nT) = \sum_{n=-\infty}^{\infty} x(nT) \frac{\sin \beta(t - nT)}{\beta(t - nT)}, \qquad T = \pi/\beta \qquad (6.3)$$

Comparing to (6.2) we see that the **transform domain coefficients** α_n can be regarded as the samples $x(nT)$, whereas the functions $g_n(t)$ are the shifted sinc functions.

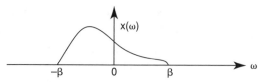

6.2   Fourier transform of a signal band-limited to $|\omega| < \beta$.

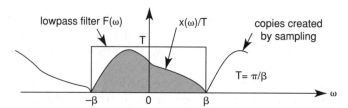

6.3   Use of low-pass filter $F(\omega)$ to recover $x(t)$ from its samples.

If a signal is time-limited or band-limited, we can express it as a countable linear combination of a set of fundamental functions (called **basis** functions, in fact an orthonormal basis; see below). If the signal is more arbitrary, i.e., not limited in time or bandwidth, can we still obtain such a countable linear combination? Suppose we restrict $x(t)$ to be a finite energy signal (i.e., $\int |x(t)|^2 dt < \infty$; also called L^2 signals, see below). Then this is possible. In fact, we can find an unusual kind of basis called the **wavelet basis,** fundamentally different from the Fourier basis. Representation of $x(t)$ using this basis has, in some applications, some advantages over the Fourier representation or the short-time (windowed) Fourier representation. Wavelet bases also exist for many other classes of signals, but this discussion is limited to the L^2 class of signals.

The most common kind of **wavelet representation** takes the form

$$x(t) = \sum_{k=-\infty}^{\infty} \sum_{n=-\infty}^{\infty} c_{kn} \underbrace{2^{k/2} \psi(2^k t - n)}_{\psi_{kn}(t)} \qquad (6.4)$$

The functions $\psi_{kn}(t)$ are typically (but not necessarily) linearly independent and form a basis for finite energy signals. The basis is very special in the sense that all the functions $\psi_{kn}(t)$ are derived from a single function $\psi(t)$ called the *wavelet,* by two operations: dilation ($t \rightarrow 2^k t$) and time shift ($t \rightarrow t - 2^{-k} n$). The advantage of such a basis is that it allows us to capture the details of a signal at various scales, while providing time-localization information for these "scales." Examples in future sections clarify this idea.

Why Worry About Signal Representations?

A common feature of all the above discussions is that we have taken a signal $x(t)$ and found an equivalent representation in terms of the transform domain quantity $\{\alpha_n\}$ in (6.2), or $\{c_{kn}\}$ in (6.4). If our only aim is to compute α_n from $x(t)$ and then recompute $x(t)$ from α_n, that would be a futile exercise. The motivation in practice is that the transform domain quantities are better suited in some sense. For example, in audio coding, decomposition of a signal into frequency components is motivated by the fact that the human ear perceives higher frequencies with less frequency resolution. We can use this information. We can also code the high-frequency components with relatively less precision, thereby enabling data compression. In this way we can take into account perceptual information during compression. Also, we can account for the fact that the error allowed by the human ear (due to quantization of frequency components) depends on the frequency masking property of the ear, and perform optimum-bit allocation for a given bit rate. Other applications of signal representations using wavelets include numerical analysis, solution of differential equations, and many others [5, 15, 16].

The main point, in any case, is that we typically perform certain manipulations with the transform domain coefficients α_n [or c_{kn} in (6.4)] before we recombine them to form an approximation of $x(t)$. Therefore, we really only have

$$\hat{x}(t) = \sum_n \hat{\alpha}_n g_n(t) \tag{6.5}$$

where $\{\hat{\alpha}_n\}$ approximates $\{\alpha_n\}$. This discussion gives rise to many questions: how best to choose the basis functions $g_n(t)$ for a given application? How to choose the compressed signal $\{\hat{\alpha}_n\}$ so that for a given data rate the reconstruction error is minimized? What, indeed, is the best way to define the reconstruction error?

These questions are deep and complicated and will take us too far afield. Our goal is to point out the basic advantages (sometimes) offered by the wavelet transform over other kinds of transforms (e.g., the FT).

Ideal Bandpass Wavelet

Consider a bandpass signal $x(t)$ with FT as shown in Fig. 6.4. Such signals arise in communication applications. The bandedges of the signal are ω_1 and ω_2 (and $-\omega_1$ and $-\omega_2$ on the negative side, which is natural if $x(t)$ is real). Viewed as a low-pass signal, the total bandwidth (counting negative frequencies also) is $2\omega_2$, but viewed as a bandpass signal, the total bandwidth is only 2β where $\beta = \omega_2 - \omega_1$. Does it mean that we can sample it at the rate 2β rad/s (which is the Nyquist rate for the low-pass case)?

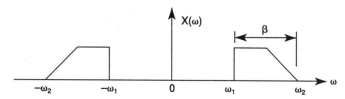

6.4   Fourier transform of a bandpass signal.

In the low-pass case sampling at Nyquist rate was enough to ensure that the copies of the spectrum created by the sampling did not overlap (Fig. 6.3). In the bandpass case we have two sets of such copies; one created by the positive half of the frequency $\omega_1 \le \omega \le \omega_2$ and the other by the negative half $-\omega_2 \le \omega \le -\omega_1$. This makes the problem somewhat more complicated. It can be shown that for sampling at the rate 2β no overlap of images exists iff one of the edges, ω_1 or ω_2, is a multiple of 2β. This is called the **bandpass sampling theorem**. The reconstruction of $x(t)$ from the samples proceeds exactly as in the low-pass case, except that the reconstruction filter $F(\omega)$ is now a bandpass filter (Fig. 6.5), occupying precisely the signal bandwidth. The first part of the expression (6.3), therefore, is still valid, i.e., $x(t) = \sum_n x(nT)f(t - nT)$, where $T = \pi/\beta$ again, but the sinc function is replaced with the bandpass impulse response $f(t)$.

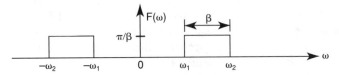

6.5   Bandpass filter to be used in the reconstruction of the bandpass signal from its samples.

Given a signal $x(t)$, imagine now that we have split its frequency axis into subbands in some manner (Fig. 6.6). Letting $y_k(t)$ denote the kth subband signal, we can write $x(t) = \sum_k y_k(t)$. This can be visualized as passing $x(t)$ through a bank of filters $\{H_k(\omega)\}$ [Fig. 6.7(a)], with responses as in Fig. 6.7(b). Note that each subband region is symmetric with respect to zero frequency, and therefore supports positive as well as negative frequencies. If the subband region $\omega_k \leq |\omega| < \omega_{k+1}$ satisfies the bandpass sampling condition, then the bandpass signal $y_k(t)$ can be expressed as a linear combination of its samples as before. Thus, $x(t) = \sum_k y_k(t) = \sum_k \sum_n y_k(nT_k)f_k(t - nT_k)$, where $T_k = \pi/\beta_k$. Here, $f_k(t)$ is the impulse response of the reconstruction filter (or synthesis filter) $F_k(\omega)$ shown in Fig. 6.7(c). Fig. 6.7(a) also shows this reconstruction schematic.

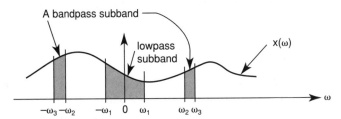

6.6   Splitting a signal into frequency subbands.

Figure 6.8 shows the set of synthesis filters $\{F_k(\omega)\}$ for two examples of frequency splitting arrangement, namely uniform splitting and nonuniform (octave) splitting. We will see later that the uniform splitting arrangement gives an example of the STFT representation (Sections 6.3 and 6.9). In this section we are interested in octave splitting. The bandedges of the filters here are $\omega_k = 2^k \pi$ ($k = \ldots -1, 0, 1, 2, \ldots$). The bandedges are such that $y_k(t)$ is a signal satisfying the bandpass sampling theorem. It has $\beta_k = 2^k \pi$, according to the notation of Fig. 6.7. It can be sampled at period $T_k = \pi/\beta_k = 2^{-k}$ without aliasing, and we can reconstruct it from samples as

$$y_k(t) = \sum_{n=-\infty}^{\infty} y_k(2^{-k}n) f_k(t - 2^{-k}n) \tag{6.6}$$

As k increases, the bandwidths of the filters increase so the sample spacing $T_k = 2^{-k}$ becomes finer.

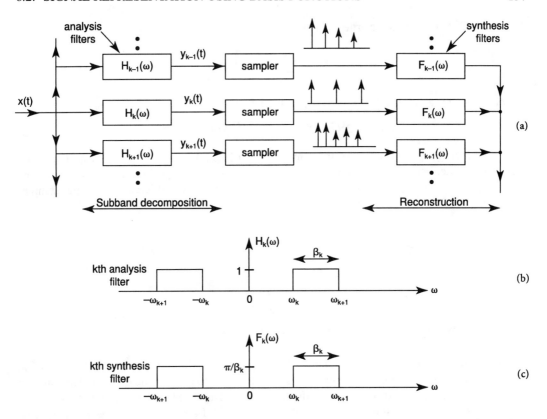

6.7 (a) Splitting a signal into subband signals, sampling, and then recombining; (b) response of the kth analysis filter; and (c) response of kth synthesis filter.

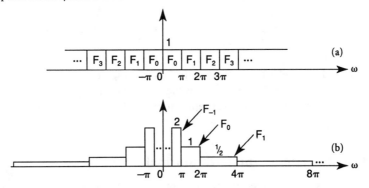

6.8 Two possible schemes to decompose a signal into frequency bands: (a) uniform splitting, and (b) octave-band splitting. The responses shown are those of synthesis filters.

Because $x(t) = \sum_k y_k(t)$ we see that $x(t)$ can be expressed as

$$x(t) = \sum_{k=-\infty}^{\infty} \sum_{n=-\infty}^{\infty} y_k(2^{-k}n) f_k(t - 2^{-k}n) \tag{6.7}$$

Our definition of the filters shows that the frequency responses are scaled versions of each other, i.e., $F_k(\omega) = 2^{-k} \Psi(2^{-k}\omega)$, with $\Psi(\omega)$ as in Fig. 6.9. The impulse responses are therefore related as

$f_k(t) = \psi(2^k t)$, and we can rewrite (6.7) as

$$x(t) = \sum_{k=-\infty}^{\infty} \sum_{n=-\infty}^{\infty} y_k(2^{-k}n)\psi(2^k t - n) \qquad (6.8)$$

We will write this as $x(t) = \sum_k \sum_n c_{kn}\psi_{kn}(t)$ by defining $c_{kn} = 2^{-k/2}y_k(2^{-k}n)$ and

$$\psi_{kn}(t) = 2^{k/2}\psi(2^k t - n) = 2^{k/2}\psi\left(2^k(t - 2^{-k}n)\right) \qquad (6.9)$$

Then the functions $\psi_{kn}(t)$ will have the same energy $\int |\psi_{kn}(t)|^2 dt$ for all k, n. From the analysis/synthesis filter bank point of view (Fig. 6.7) this is equivalent to making $H_k(\omega) = F_k(\omega)$ and rescaling, as shown in Fig. 6.10. With filters so rescaled, the wavelet coefficients c_{kn} are just samples of the outputs of the analysis filters $H_k(\omega)$.

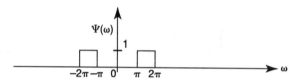

6.9   The fundamental bandpass function that generates a bandpass wavelet.

The function $\psi(2^k t)$ is a dilated version of $\psi(t)$ (squeezed version if $k > 0$ and stretched version if $k < 0$). The dilation factor 2^k is a power of 2, so this is said to be a dyadic dilation. The function $\psi(2^k(t - 2^{-k}n))$ is a shifted version of the dilated version. Thus, we have expressed $x(t)$ as a linear combination of shifted versions of (dyadic) dilated versions of a single function $\psi(t)$. The shifts $2^{-k}n$, are in integer multiples of 2^{-k}, where k governs the dilation. For completeness, note that the impulse response $\psi(t)$ corresponding to the function in Fig. 6.9 is given by

$$\psi(t) = \frac{\sin(\pi t/2)}{\pi t/2} \cos(3\pi t/2) \qquad \text{(ideal bandpass wavelet)} \qquad (6.10)$$

This is plotted in Fig. 6.11.

In (6.8) we obtained a wavelet representation for $x(t)$ [compare to (6.4)]. The function $\psi(t)$ is called the **ideal bandpass wavelet**, also known as the Littlewood-Paley wavelet. We now introduce some terminology for convenience and then return to more detailed definitions and discussions of the WT.

L^2 Spaces, Basis Functions, and Orthonormal Bases

Most of our discussions are restricted to the class of L^2 functions or square integrable functions, i.e., functions $x(t)$ for which $\int |x(t)|^2 dt$ exists and has a finite value. The norm, or L^2 norm, of such functions, denoted $\|x(t)\|_2$, is defined as $\|x(t)\|_2 = \left(\int |x(t)|^2 dt\right)^{1/2}$. The notation $L^2[a, b]$ stands for L^2 functions that are zero outside the interval $a \le t \le b$. The set $L^2(R)$ is the class of L^2 functions supported on the real line $-\infty < t < \infty$. We often abbreviate $L^2(R)$ as L^2.

The class of L^2 functions forms a (normed) linear vector space, i.e., any linear combination of functions in L^2 is still in L^2. In fact, it forms a special linear space such that a countable basis exists. That is, a sequence of linearly independent functions $\{g_n(t)\}$ exists in L^2 such that any L^2 function $x(t)$ can be expressed as $x(t) = \sum_n \alpha_n g_n(t)$, for a unique set of $\{\alpha_n\}$. We say that $g_n(t)$ are the basis functions. L^2 spaces have orthonormal bases. For such a basis, the basis functions satisfy

$$\langle g_k(t), g_m(t) \rangle = \delta(k - m) \qquad (6.11)$$

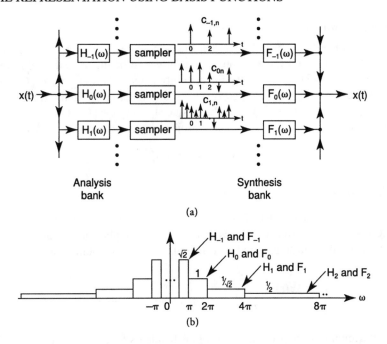

6.10 The octave-band splitting scheme. (a) The analysis bank, samplers, and synthesis bank; and (b) the filter responses.

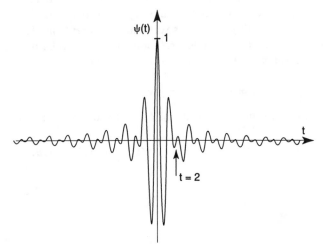

6.11 The ideal bandpass wavelet.

where the notation $\langle f(t), g(t) \rangle \overset{\Delta}{=} \int f(t) g^*(t) dt$ denotes the inner product between $f(t)$ and $g(t)$. For an orthonormal basis, the coefficients α_n in the expansion $x(t) = \sum_{n=-\infty}^{\infty} \alpha_n g_n(t)$ can thus be computed using the exceptionally simple relation

$$\alpha_n = \langle x(t), g_n(t) \rangle \qquad (6.12)$$

Two examples of orthonormal basis were shown above. The first is the FS expansion of a time-limited signal ($0 \le t \le 1$). Here, the basis functions $\{e^{j2\pi nt}\}$ are clearly orthonormal, with integrals going from 0 to 1. The second example is the expansion (6.3) of a band-limited signal; it can be shown that the shifted versions $f(t - nT)$ of the sinc functions form an orthonormal basis for band-limited signals (integrals going from $-\infty$ to ∞).

Orthogonal Projections

Suppose we consider a subset $\{g_{n_k}(t)\}$ of the orthonormal basis $\{g_n(t)\}$. Let S denote the subspace generated by $\{g_{n_k}(t)\}$ (an accurate statement would be that S is the "closure of the span of $\{g_{n_k}(t)\}$;" see Section 6.7). Consider the linear combination $y(t) = \sum_k \alpha_{n_k} g_{n_k}(t)$, where the α_{n_k} are evaluated as above, i.e., $\alpha_{n_k} = \langle x(t), g_{n_k}(t)\rangle$ for some signal $x(t)$. Then, $y(t) \in S$, and it can be shown that among all functions in S, $y(t)$ is the unique signal closest to $x(t)$ (i.e., $\|x(t) - y(t)\|_2$ is the smallest). We say that $y(t)$ is the orthogonal projection of $x(t)$ onto the subspace S, and write

$$y(t) = P_S[x(t)] \tag{6.13}$$

Wavelet Transforms

If a signal $x(t)$ is in L^2, then its FT $X(\omega)$ exists in the L^2 sense (see Section 6.6). We will see in Section 6.6 that the discussion which resulted in (6.8) is applicable for any signal $x(t)$ in L^2. Equation (6.8) means that the signal can be expressed as a linear combination of the form

$$x(t) = \sum_{k=-\infty}^{\infty} \sum_{n=-\infty}^{\infty} c_{kn} \underbrace{2^{k/2}\psi(2^k t - n)}_{\psi_{kn}(t)} \tag{6.14}$$

where $\psi(t)$ is the impulse response (Fig. 6.11) of the bandpass function $\Psi(\omega)$ in Fig. 6.9.[2] Because the frequency responses for two different values of k do not overlap, the functions $\psi_{kn}(t)$ and $\psi_{mi}(t)$ are orthogonal for $k \neq m$ (use Parseval's relation). For a given k, the functions $\psi_{kn}(t)$ are shifted versions of the impulse responses of the bandpass filter $F_k(\omega)$. From the ideal nature of this bandpass filter we can show that $\psi_{kn}(t)$ and $\psi_{km}(t)$ are also orthonormal for $n \neq m$. Thus, the set of functions $\{\psi_{kn}(t)\}$, with k and n ranging over all integers, forms an orthonormal basis for the class of L^2 functions, i.e., any L^2 function can be expressed as in (6.14) and furthermore,

$$\langle \psi_{kn}(t), \psi_{mi}(t)\rangle = \delta(k - m)\delta(n - i) \tag{6.15}$$

Because of this orthonormality, the coefficients c_{kn} are computed very easily as

$$c_{kn} = \langle x(t), \psi_{kn}(t)\rangle = \int_{-\infty}^{\infty} x(t)2^{k/2}\psi^*(2^k t - n)\,dt \tag{6.16}$$

Defining

$$\eta(t) = \psi^*(-t) \tag{6.17}$$

this takes the form

$$c_{kn} = \langle x(t), \eta_{kn}^*(-t)\rangle = \int_{-\infty}^{\infty} x(t)2^{k/2}\eta(n - 2^k t)\,dt \tag{6.18}$$

resembling a convolution.

Wavelet Transform Definitions

A set of basis functions $\psi_{kn}(t)$ derived from a single function $\psi(t)$ by dilations and shifts of the form

$$\psi_{kn}(t) = 2^{k/2}\psi(2^k t - n) \tag{6.19}$$

---

[2]The above equality and the convergence of the summation should be interpreted in the L^2 sense; see Section 6.6.

is said to be a **wavelet basis**, and $\psi(t)$ is called the **wavelet function**. The coefficients c_{kn} are the **wavelet transform coefficients**. The formula (6.16) that performs the transformation from $x(t)$ to c_{kn} is the **wavelet transform** of the signal $x(t)$. Equation (6.14) is the **wavelet representation** or the **inverse wavelet transform**. While this is only a special case of more general wavelet decompositions outlined at the end of this section, it is perhaps the most popular and useful.

Note that the kth dilated version $\psi(2^k t)$ has the shifted versions $\psi(2^k t - n) = \psi(2^k(t - 2^{-k}n))$, so the amount of shift is in integer multiples of 2^{-k}. Thus, the *stretched* versions are shifted by larger amounts and *squeezed* versions by smaller amounts. Even though we developed these ideas based on an example, the above definitions still hold generally for any orthonormal wavelet basis. For the ideal bandpass wavelet, the function $\psi(t)$ is real and symmetric [see (6.10)] so that $\eta(t) = \psi(t)$. For more general orthonormal wavelets we have the relation $\eta(t) = \psi^*(-t)$. We say that $\eta(t)$ is the **analyzing wavelet** [because of (6.18)] and $\psi(t)$ the **synthesis wavelet** [because of (6.14)]. For the nonorthonormal case we still have the transform and inverse transform equations as above, but the relation between $\psi(t)$ and $\eta(t)$ is not as simple as $\eta(t) = \psi^*(-t)$.

Before exploring the properties and usefulness of wavelets let us turn to a distinctly different example. This shows that unlike the Fourier basis functions $\{e^{j2\pi nt}\}$, the wavelet basis functions can be designed by the user. This makes them more flexible, interesting, and useful.

Haar Wavelet Basis

An orthonormal basis for L^2 functions was found by Haar [5] as early as 1910, which satisfies the definition of a wavelet basis given above. That is, the basis functions $\psi_{kn}(t)$ are derived from a single function $\psi(t)$ using dilations and shifts as in (6.19). To explain this system, first, consider a signal $x(t) \in L^2[0, 1]$. The Haar basis is built from two functions called $\phi(t)$ and $\psi(t)$, as described in Fig. 6.12. The basis function $\phi(t)$ is a constant in $[0, 1]$. The basis function $\psi(t)$ is constant on each half interval, and its integral is zero. After this, the remaining basis functions are obtained from $\psi(t)$ by dilations and shifts as indicated. It is clear from the figure that any two of these functions are mutually orthogonal. We have an orthonormal set, and it can be shown that this set of functions is an orthonormal basis for $L^2[0, 1]$. However, this is not exactly a wavelet basis yet because of the presence of $\phi(t)$.[3]

If we eliminate the requirement that $x(t)$ be supported or defined only on $[0, 1]$ and consider $L^2(R)$ functions then we can still obtain an orthonormal basis of the above form by including the shifted versions $\{\psi(2^k t - n)\}$ for *all* integer values of n, as well as the shifted versions $\{\phi(t - n)\}$. An alternative to the use of $\{\phi(t - n)\}$ would be to use stretched (i.e., $\psi(2^k t), k < 0$) as well as squeezed (i.e., $\psi(2^k t), k > 0$) versions of $\psi(t)$. The set of functions can thus be written as in (6.19), which has the form of a wavelet basis. It can be shown that this forms an orthonormal basis for $L^2(R)$. The FT of the Haar wavelet $\psi(t)$ is given by

$$\Psi(\omega) = je^{-j\omega/2}\frac{\sin^2(\omega/4)}{\omega/4} \qquad \text{(Haar wavelet)} \qquad (6.20)$$

The Haar wavelet has limited duration in time, whereas the ideal bandpass wavelet (6.10), being band-limited, has infinite duration in time.

---

[3]We see in Section 6.10 that the function $\phi(t)$ arises naturally in the context of the fundamental idea of multiresolution.

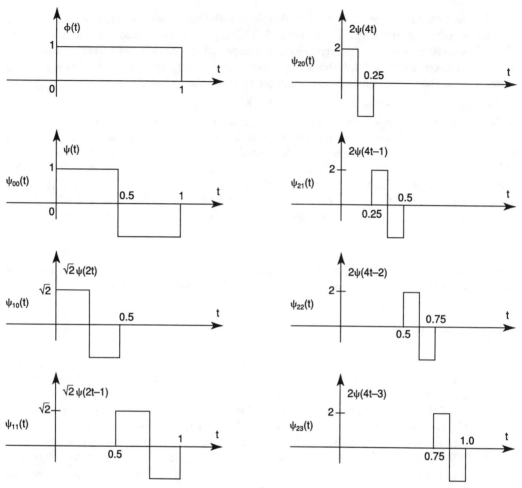

6.12   Examples of basis functions in the Haar basis for $L^2[0, 1]$.

Basic Properties of Wavelet Transforms

Based on the definitions and examples provided so far we can already draw some very interesting conclusions about wavelet transforms, and obtain a preliminary comparison to the FT.

1. **Concept of scale.** The functions $\psi_{kn}(t)$ are useful to represent increasingly finer "variations" in the signal $x(t)$ at various levels. For large k, the function $\psi_{kn}(t)$ looks like a "high frequency signal." This is especially clear from the plots of the Haar basis functions. (For the bandpass wavelets, see below.) Because these basis functions are not sinusoids, we do not use the term "frequency" but rather the term "scale." We say that the component $\psi_{kn}(t)$ represents a finer scale for larger k. Accordingly k (sometimes $1/k$) is called the scale variable. Thus, the function $x(t)$ has been represented as a linear combination of component functions that represent variations at different "scales." For instance, consider the Haar basis. If the signal expansion (6.14) has a relatively large value of $c_{4,2}$ this means that the component at scale $k = 4$ has large energy in the interval $[2/2^4, 3/2^4]$ (Fig. 6.14).

2. **Localized basis.** The above comment shows that if a signal has energy at a particular scale concentrated in a slot in the time domain, then the corresponding c_{kn} has large

value, i.e., $\psi_{kn}(t)$ contributes more to $x(t)$. The wavelet basis, therefore, provides *localization information* in time domain as well as in the *scale domain.* For example, if the signal is zero everywhere except in the interval $[2/2^4, 3/2^4]$ then the subset of the Haar basis functions which do not have their support in this interval are simply absent in this expansion.

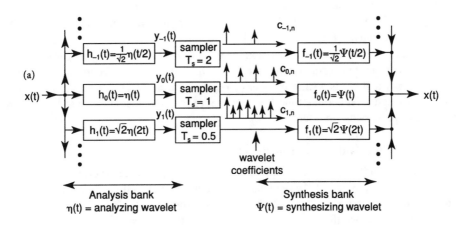

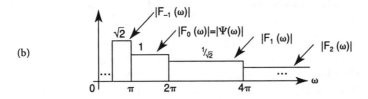

6.13 (a) Representing the dyadic WT as an analysis bank followed by samplers, and the inverse transform as a synthesis bank. For the orthonormal case, $\psi(t) = \eta^*(-t)$, and $f_k(t) = h_k^*(-t)$. (b) Filter responses for the example in which $\psi(t)$ is the ideal bandpass wavelet.

Note that the Haar wavelet has **compact support,** that is, the function $\psi(t)$ is zero everywhere outside a closed bounded interval ([0, 1] here). While the above discussions are motivated by the Haar basis, many of them are typically true, with some obvious modifications, for more general wavelets. Consider, for example, the ideal bandpass wavelet (Fig. 6.11) obtained from the bandpass filter $\Psi(\omega)$ in Fig. 6.9. In this case the basis functions do not have compact support, but are still locally concentrated around $t = 0$. Moreover, the basis functions for large k represent "fine" information, or the frequency component around the center frequency of the filter $F_k(\omega)$ (Fig. 6.10). The Haar wavelet and the ideal bandpass wavelet are two extreme examples (one is time limited and the other band-limited). Many intermediate examples can be constructed.

Filter Bank Interpretation and Time-Frequency Representation

We know that the wavelet coefficients c_{kn} for the ideal bandpass wavelet can be viewed as the sampled version of the output of a bandpass filter [Fig. 6.10(a)]. The same is true for any kind of WT. For this recall the expression (6.18) for the wavelet coefficients. This can be interpreted as the set of sampled output sequences of a bank of filters $H_k(\omega)$, with impulse response $h_k(t) = 2^{k/2}h_0(2^k t)$, where $h_0(t) = \eta(t)$. Thus the wavelet transform can be interpreted as a **nonuniform continuous time analysis filter bank,** followed by samplers. The Haar basis and ideal bandpass wavelet basis are two examples of the choice of these bandpass filters.

The wavelet coefficients c_{kn} for a given scale k are therefore obtained by sampling the output $y_k(t)$ of the bandpass filter $H_k(\omega)$, as indicated in Fig. 6.13(a). The first subscript k (the scale variable) represents the filter number. As k increases by 1, the center frequency ω_k increases by a factor of 2. The wavelet coefficients c_{kn} at scale k are merely the samples $y_k(2^{-k}n)$. As k increases, the filter bandwidth increases, and thus the samples are spaced by a proportionately finer amount 2^{-k}. The quantity $c_{kn} = y_k(2^{-k}n)$ measures the "amount" of the "frequency component" around the center frequency ω_k of the analysis filter $H_k(\omega)$, localized in *time* around $2^{-k}n$.

In wavelet transformation the transform domain is represented by the two integer variables k and n. This means that the transform domain is *two dimensional* (the time-frequency domain), and is *discretized.* We say that c_{kn} is a **time-frequency representation** of $x(t)$. Section 6.3 explains that this is an improvement over another time-frequency representation, the STFT, introducedmany years ago in the signal processing literature.

Synthesis Filter Bank and Reconstruction

The inner sum in (6.14) can be interpreted as follows: For each k, convert the sequence c_{kn} into an impulse train[4] $\sum_n c_{kn}\delta_a(t - 2^{-k}n)$ and pass it through a bandpass filter $F_k(\omega) = 2^{-k/2}\Psi(2^{-k}\omega)$ with impulse response $f_k(t) = 2^{k/2}\psi(2^k t)$. The outer sum merely adds the outputs of all these filters. Figure 6.13(a) shows this interpretation. Therefore, the reconstruction of the signal $x(t)$ from the wavelet coefficients c_{kn} is equivalent to the implementation of a **nonuniform continuous time synthesis filter bank,** with synthesis filters $f_k(t) = 2^{k/2}f_0(2^k t)$ generated by dilations of a single filter $f_0(t) \overset{\Delta}{=} \psi(t)$.

As mentioned earlier, the analyzing wavelet $\eta(t)$ and the synthesis wavelet $\psi(t)$ are related by $\eta(t) = \psi^*(-t)$ in the orthonormal case. Thus, the analysis and synthesis filters are related as $h_k(t) = f_k^*(-t)$; i.e., $H_k(\omega) = F_k^*(\omega)$. For the special case of the ideal bandpass wavelet (6.10), $\psi(t)$ is real and symmetric so that $f_k(t) = f_k^*(-t)$; i.e., $h_k(t) = f_k(t)$. Figure 6.13 summarizes the relations described in the preceding paragraphs.

Design of Wavelet Functions

Because all the filters in the analysis and synthesis banks are derived from the wavelet function $\psi(t)$, the quality of the frequency responses depends directly on $\Psi(\omega)$. In the time domain the Haar basis has poor smoothness (it is not even continuous), but it is well localized (compactly supported). Its FT $\Psi(\omega)$, given in (6.20), decays only as $1/\omega$ for large ω. The ideal bandpass wavelet, on the other hand, is poorly localized in time, but has very smooth behavior. In fact, because it is band-limited, $\psi(t)$ is infinitely differentiable, but it decays only as $1/t$ for large t. Thus, the Haar wavelet and the ideal bandpass wavelet represent two opposite extremes of the possible choices.

We could carefully design the wavelet $\psi(t)$ such that it is reasonably well localized in time domain, while at the same time sufficiently smooth or "regular." The term **regularity** is often used to quantify the degree of smoothness. For example, the number of times we can differentiate the wavelet $\psi(t)$ and the degree of continuity (so-called Hölder index) of the last derivative are taken as measures of regularity. We return to this in Sections 6.11 to 6.13, where we also present systematic procedures for design of the function $\psi(t)$. This can be designed in such a way that $\{2^{k/2}\psi(2^k t - n)\}$ forms an

---

[4]$\delta_a(t)$ is the Dirac delta function [2]. It is used here only as a schematic. The true meaning is that the output of $f_k(t)$ is $\sum_n c_{kn}f_k(t - 2^{-k}n)$.

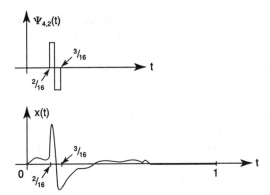

6.14 Example of an $L^2[0, 1]$ signal $x(t)$ for which the Haar component $\psi_{4,2}(t)$ dominates.

orthonormal basis with prescribed decay and regularity properties. It is also possible to design $\psi(t)$ such that we obtain other kinds of structures rather than an orthonormal basis, e.g., a Riesz basis or a frame (see Sections 6.7 and 6.8).

Wavelet Basis and Fourier Basis

Returning to the Fourier basis $g_k(t) = \{e^{j2\pi kt}\}$ for functions supported on $[0, 1]$, we see that $g_k(t) = g_1(kt)$, so that all the functions are dilated versions (dilations being integers rather than powers of integers) of $g_1(t)$. However, *these do not have the localization property of wavelets.* To understand this, note that $e^{j2\pi kt}$ has unit magnitude everywhere, and sines and cosines are nonzero almost everywhere. Thus, if we have a function $x(t)$ that is identically zero in a certain time slot (e.g., Fig. 6.14), then in order for the infinite series $\sum_n \alpha_n e^{j2\pi nt}$ to represent $x(t)$, extreme cancellation of terms must occur in that time slot. In contrast, if a compactly supported wavelet basis is used, it provides localization information as well as information about "frequency contents" in the form of "scales." The "transform domain" in traditional FT is represented by a single continuous variable ω. In the wavelet transform, where the transform coefficients are c_{kn}, the transform domain is represented by two integers k and n.

It is also clear that WT provide a great deal of flexibility because we can choose $\psi(t)$. With FTs, on the other hand, the basis functions (sines and cosines) are pretty much fixed (see, however, Section 6.3 on STFT).

More General Form of Wavelet Transformation

The most general form of the wavelet transform is given by

$$X(a, b) = \frac{1}{\sqrt{|a|}} \int_{-\infty}^{\infty} x(t)\psi\left(\frac{t-b}{a}\right) dt \qquad (6.21)$$

where a and b are real. This is called the continuous wavelet transform (CWT) because a and b are continuous variables. The transform domain is a two-dimensional domain (a, b). The restricted version of this, in which a and b take a discrete set of values $a = c^{-k}$ and $b = c^{-k}n$ (where k and n vary over the set of all integers), is called the discrete wavelet transform (DWT). The further special case, in which $c = 2$, i.e., $a = 2^{-k}$ and $b = 2^{-k}n$, is the WT discussed so far [see (6.16)] and is called the **dyadic DWT.** Expansions of the form (6.14) are also called **wavelet series expansions** by analogy with the FS expansion (a summation rather than an integral).

For fixed a, (6.21) is a convolution. Thus, if we apply the input signal $x(t)$ to a filter with impulse response $\psi(-t/a)/\sqrt{|a|}$, its output, evaluated at time b, will be $X(a, b)$. The filter has frequency response $\sqrt{|a|}\Psi(-a\omega)$. If we imagine that $\Psi(\omega)$ has a good bandpass response with center frequency ω_0, then the above filter is bandpass with center frequency $-a^{-1}\omega_0$; i.e., the wavelet transform $X(a, b)$, which is the output of the filter at time b, represents the "frequency content" of $x(t)$ around the frequency $-a^{-1}\omega_0$, "around" time b. Ignoring the minus sign [because $\psi(t)$ and $x(t)$ are typically real anyway], we see that the variable a^{-1} is analogous to frequency. In wavelet literature, the quantity $|a|$ is usually referred to as the "scale" rather than "inverse frequency."

For reasons which cannot be explained with our limited exposure thus far, the wavelet function $\psi(t)$ is restricted to be such that $\int \psi(t)dt = 0$. For the moment notice that this is equivalent to $\Psi(0) = 0$, which is consistent with the bandpass property of $\psi(t)$. In Section 6.10, where we generate wavelets systematically using multiresolution analysis, we see that this condition follows naturally from theoretical considerations.

6.3   The Short-Time Fourier Transform

In many applications we must accommodate the notion of frequency that evolves or changes with time. For example, audio signals are often regarded as signals with a time-varying spectrum, e.g., a sequence of short-lived pitch frequencies. This idea cannot be expressed with the traditional FT because $X(\omega)$ for each ω depends on $x(t)$ for all t.

The STFT was introduced as early as 1946 by Gabor [5] to provide such a time-frequency picture of the signal. Here, the signal $x(t)$ is multiplied with a window $v(t - \tau)$ centered or localized around time τ (Fig. 6.15) and the FT of $x(t)v(t - \tau)$ computed:

$$X(\omega, \tau) = \int_{-\infty}^{\infty} x(t)v(t - \tau)e^{-j\omega t}\, dt \qquad (6.22)$$

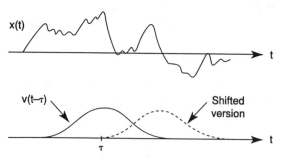

6.15   A signal $x(t)$ and the sliding window $v(t - \tau)$.

This is then repeated for shifted locations of the window, i.e., for various values of τ. That is, we compute not just one FT, but infinitely many. The result is a function of both time τ and frequency ω. If this must be practical we must make two changes: compute the STFT only for discrete values of ω, and use only a discrete number of window positions τ. In the traditional STFT both ω and τ are discretized on uniform grids:

$$\omega = k\omega_s \qquad \tau = nT_s. \qquad (6.23)$$

The STFT is thus defined as

$$X_{\text{STFT}}(k\omega_s, nT_s) = \int_{-\infty}^{\infty} x(t)v(t - nT_s)e^{-jk\omega_s t}\, dt, \qquad (6.24)$$

which we abbreviate as $X_{\text{STFT}}(k, n)$ when there is no confusion. Thus, the time domain is mapped into the **time-frequency domain**. The quantity $X_{\text{STFT}}(k\omega_s, nT_s)$ represents the FT of $x(t)$ "around time nT_s" and "around frequency $k\omega_s$." This, in essence, is similar to the WT: in both cases the transform domain is a two-dimensional discrete domain.

We compare wavelets and STFT on several grounds, giving a filter bank view and comparing time-frequency resolution and localization properties. Section 6.9 provides a comparison on deeper grounds: for example, when can we reconstruct a signal $x(t)$ from the STFT coefficients $X_{\text{STFT}}(k, n)$? Can we construct an orthonormal basis for L^2 signals based on the STFT? The advantage of WTs over the STFT will be clear after these discussions.

Filter Bank Interpretation

The STFT evaluated for some frequency ω_k can be rewritten as

$$X_{\text{STFT}}(\omega_k, \tau) = e^{-j\omega_k \tau} \int_{-\infty}^{\infty} x(t)v(t - \tau)e^{-j\omega_k(t-\tau)}\, dt \qquad (6.25)$$

The integral looks like a convolution of $x(t)$ with the filter impulse response

$$h_k(t) \overset{\triangle}{=} v(-t)e^{j\omega_k t} \qquad (6.26)$$

If $v(-t)$ has a FT looking like a low-pass filter then $h_k(t)$ looks like a bandpass filter with center frequency ω_k (Fig. 6.16). Thus, $X_{\text{STFT}}(\omega_k, \tau)$ is the output of this bandpass filter at time τ, down-shifted in frequency by ω_k. The result is a low-pass signal $y_k(t)$ whose output is sampled uniformly at time $\tau = nT_s$. For every frequency ω_k so analyzed, one such filter channel exists. With the frequencies uniformly located at $\omega_k = k\omega_s$, we get the analysis filter bank followed by downshifters and samplers as shown in Fig. 6.17.

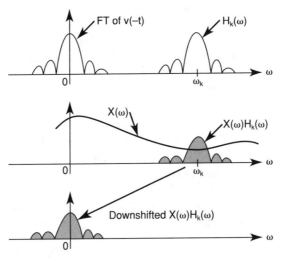

6.16 STFT viewed as a bandpass filter followed by a downshifter.

The STFT coefficients $X_{\text{STFT}}(k\omega_s, nT_s)$ therefore can be regarded as the uniformly spaced samples of the outputs of a bank of bandpass filters $H_k(\omega)$, all derived from one filter $h_0(t)$ by modulation: $h_k(t) = e^{jk\omega_s t}h_0(t)$; i.e., $H_k(\omega) = H_0(\omega - k\omega_s)$. (The filters are one sided in frequency so they have complex coefficients in the time domain, but ignore these details for now.) The output of $H_k(\omega)$ represents a portion of the FT $X(\omega)$ around the frequency $k\omega_s$. The downshifted version $y_k(t)$ is therefore a low-pass signal. In other words, it is a slowly varying signal, whose evolution as a function of t represents the evolution of the FT $X(\omega)$ around frequency $k\omega_s$. By sampling this slowly varying signal we can therefore compress the transform domain information.

If the window is narrow in the time domain, then $H_k(\omega)$ has large bandwidth. That is, good time resolution and poor frequency resolution are obtained. If the window is wide the opposite is true. Thus, if we try to capture the local information in time by making a narrow window, we get a fuzzy picture in frequency. Conversely, in the limit, as the filter becomes extremely localized in frequency, the window is very broad and STFT approaches the ordinary FT. That is, the time-frequency information collapses to the all-frequency information of ordinary FT. We see that time-frequency representation is inherently a compromise between time and frequency resolutions (or localizations). This is related to the uncertainty principle: as windows get narrow in time they have to get broad in frequency, and vice versa.

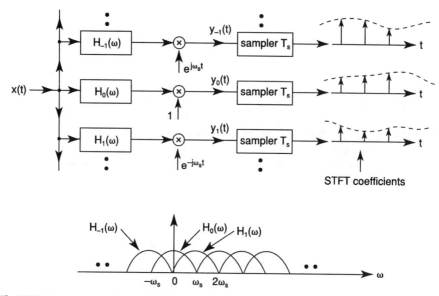

6.17   The STFT viewed as an analysis bank of uniformly shifted filters.

Optimal Time-Frequency Resolution: The Gabor Window

What is the best frequency resolution one can obtain for a given time resolution? For a given duration of the window $v(t)$ how small can the duration of $V(\omega)$ be? If we define duration according to common sense we are already in trouble because if $v(t)$ has finite duration then $V(\omega)$ has infinite duration. A more useful definition of duration is called the **root mean square** (RMS) duration. The RMS time duration D_t and the RMS frequency duration D_f for the window $v(t)$ are defined such that

$$D_t^2 = \frac{\int t^2 |v(t)|^2 \, dt}{\int |v(t)|^2 \, dt} \qquad D_f^2 = \frac{\int \omega^2 |V(\omega)|^2 \, d\omega}{\int |V(\omega)|^2 \, d\omega} \qquad (6.27)$$

Intuitively, D_t cannot be arbitrarily small for a specified D_f. The uncertainty principle says that $D_t D_f \geq 0.5$. Equality holds iff $v(t)$ has the shape of a Gaussian, i.e., $v(t) = Ae^{-\alpha t^2}$, $\alpha > 0$. Thus, the best joint time-frequency resolution is obtained by using the Gaussian window. This is also intuitively acceptable for the reason that the Gaussian is its own FT (except for scaling of variables and so forth). Gabor used the Gaussian window as early as 1946. Because it is of infinite duration, a truncated approximation is used in practice. The STFT based on the Gaussian is called the Gabor transform. A limitation of the Gabor transform is that it does not give rise to an orthonormal signal representation; in fact it cannot even provide a "stable basis." (Sections 6.7 and 6.9 explain the meaning of this.)

Wavelet Transform vs. Short-Time Fourier Transform

The STFT works with a fixed window $v(t)$. If a high frequency signal is being analyzed, many cycles are captured by the window, and a good estimate of the FT is obtained. If a signal varies very slowly with respect to the window, however, then the window is not long enough to capture it fully. From a filter bank viewpoint, notice that all the filters have identical bandwidths (Fig. 6.17). This means that the frequency resolution is uniform at all frequencies, i.e., the "percentage resolution" or accuracy is poor for low frequencies and becomes increasingly better at high frequencies. The STFT, therefore, does not provide uniform percentage accuracy for all frequencies; the computational resources are somehow poorly distributed.

Compare this to the WT which is represented by a nonuniform filter bank [Fig. 6.8(b)]. Here the frequency resolution gets poorer as the frequency increases, but the fractional resolution (i.e., the filter bandwidth $\Delta \omega_k$ divided by the center frequency ω_k) is constant for all k (the percentage accuracy is uniformly distributed in frequency). In the time domain this is roughly analogous to having a large library of windows; narrow windows are used to analyze high-frequency components and very broad windows are used to analyze low-frequency components. In electrical engineering language the filter bank representing WT is a **constant Q filter bank**, or an **octave band filter bank**. Consider, for example, the Haar wavelet basis. Here, the narrow basis functions $\psi_{2,n}(t)$ of Fig. 6.12 are useful to represent the highly varying components of the input, and are correspondingly narrower (have shorter support than the functions $\psi_{1,n}(t)$).

A second difference between the STFT and the WTs is the sampling rates at the outputs of the bandpass filters. These are identical for the STFT filters (since all filters have the same bandwidth). For the wavelet filters, these are proportional to the filter bandwidths, hence nonuniform [Fig. 6.10(a)]. This is roughly analogous to the situation that the narrower windows move in smaller steps compared to the wider windows. Compare again to Fig. 6.12 where $\psi_{2,n}(t)$ are moved in smaller steps as compared to $\psi_{1,n}(t)$ in the process of constructing the complete set of basis functions. The nonuniform (constant Q) filter stacking [Fig. 6.8(b)] provided by wavelet filters is also naturally suited for analyzing audio signals and sometimes even as components in the modeling of the human hearing system.

Time-Frequency Tiling

The fact that the STFT performs uniform sampling of time and frequency whereas the WT performs nonuniform sampling is represented by the diagram shown in Fig. 6.18. Here, the vertical lines represent time locations at which the analysis filter bank output is sampled, and the horizontal lines represent the center frequencies of the bandpass filters. The time frequency tiling for the STFT is a simple rectangular grid, whereas for the WT it has a more complicated appearance.

EXAMPLE 6.1:

Consider the signal $x(t) = \cos(10\pi t) + 0.5 \cos(5\pi t) + 1.2\delta_a(t - 0.07) + 1.2\delta_a(t + 0.07)$. It has impulses at $t = \pm 0.07$ in the time domain. Two impulses (or "lines") are found in the frequency domain, at $\omega_1 = 5\pi$ and $\omega_2 = 10\pi$. The function is shown in Fig. 6.19 with impulses replaced by narrow pulses. The aim is to try to compute the STFT or WT such that the impulses in time as well as those in frequency are resolved. Figure 6.20(a) to (c) shows the STFT plot for three widths of the window $v(t)$ and Fig. 6.20(d) shows the wavelet plot. The details of the window $v(t)$ and the

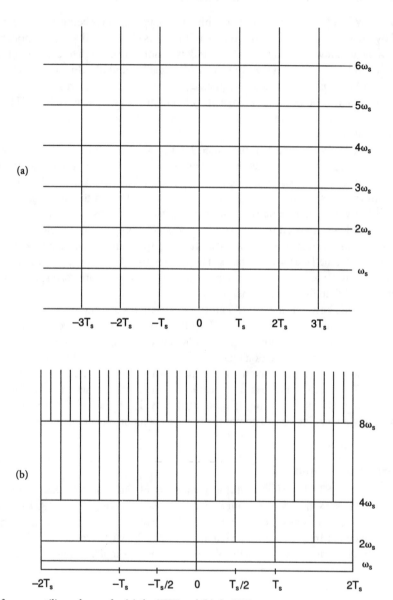

6.18 Time-frequency tiling schemes for (a) the STFT and (b) the WT.

wavelet $\psi(t)$ used for this example are described below, but first let us concentrate on the features of these plots.

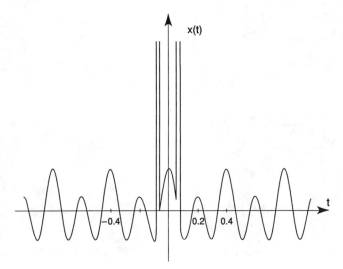

6.19   The signal to be analyzed by STFT and WT.

The STFT plots are time-frequency plots, whereas the wavelet plots are (a^{-1}, b) plots, where a and b are defined by (6.21). As explained in Section 6.2, the quantity a^{-1} is analogous to "frequency" in the STFT, and b is analogous to "time" in the STFT. The brightness of the plots in Fig. 6.20 is proportional to the magnitude of the STFT or WT, so the transform is close to zero in the dark regions. We see that for a narrow window with width equal to 0.1, the STFT resolves the two impulses in time reasonably well, but the impulses in frequency are not resolved. For a wide window with width equal to 1.0, the STFT resolves the "lines" in frequency very well, but not the time domain impulses. For an intermediate window width equal to 0.3, the resolution is poor in both time and frequency. The wavelet transform plot [Fig. 6.20(d)], on the other hand, simultaneously resolves both time and frequency very well. We can clearly see the locations of the two impulses in time, as well as the two lines in frequency.

The STFT for this example was computed using the Hamming window [2] defined as $v(t) = c[0.54 + 0.46\cos(\pi t/D)]$ for $-D \leq t \leq D$ and zero outside. The "widths" indicated in the figure correspond to $D = 0.1, 1.0$, and 0.3 (although the two-sided width is twice this). The wavelet transform was computed by using an example of the Morlet wavelet [5]. Specifically,

$$\psi(t) = e^{-t^2/16}(e^{j\pi t} - \alpha)$$

First, let us understand what this wavelet function is doing. The quantity $e^{-t^2/16}$ is the Gaussian (except for a constant scalar factor) with Fourier transform $4\sqrt{\pi}e^{-4\omega^2}$, which is again Gaussian, concentrated near $\omega = 0$. Thus, $e^{-t^2/16}e^{j\pi t}$ has a FT concentrated around $\omega = \pi$. Ignoring the second term α in the expression for $\psi(t)$, we see that the wavelet is a narrowband bandpass filter concentrated around π (Fig. 6.21).[5] If we set $a = 1$ in (6.21), then $X(1, b)$ represents the frequency

---

[5]The quantity α in the expression of $\psi(t)$ ensures that $\int \psi(t)dt = 0$ (Section 6.2). Because α is very small, it does not significantly affect the plots in Fig. 6.20.

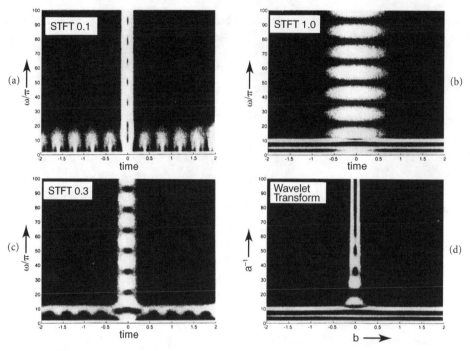

6.20   (a) to (c) STFT plots with window widths of 0.1, 1.0, and 0.3, respectively, and (d) WT plot.

contents around π. Thus, the frequencies $\omega_1 = 5\pi$ and $\omega_2 = 10\pi$ in the given signal $x(t)$ show up around points $a^{-1} = 5$ and $a^{-1} = 10$ in the WT plot, as seen from Fig. 6.20(d). In the STFT plots we have shown the frequency axis as ω/π so that the frequencies ω_1 and ω_2 show up at 5 and 10, making it easy to compare the STFT plots with the wavelet plot.

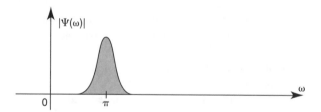

6.21   FT magnitude for the Morlet wavelet.

Mathematical Issues to be Addressed

While the filter bank viewpoint places wavelets and STFT on unified ground, several mathematical issues remain unaddressed. It is this deeper study that brings forth further subtle differences, giving wavelets a definite advantage over the STFT.

Suppose we begin from a signal $x(t) \in L^2$ and compute the STFT coefficients $X(k\omega_s, nT_s)$. How should we choose the sampling periods T_s and ω_s of the time and frequency grids so that we can reconstruct $x(t)$ from the STFT coefficients? (Remember that we are not talking about band-limited signals, and no sampling theorem is at work.) If the filters $H_k(\omega)$ are ideal one-sided bandpass filters with bandwidth ω_s, the downshifted low-pass outputs $y_k(t)$ (Fig. 6.16) can be sampled separately at the Nyquist rate ω_s or higher. This then tells us that $T_s \leq 2\pi/\omega_s$, that is,

$$\omega_s T_s \leq 2\pi \tag{6.28}$$

However, the use of ideal filters implies an impractical window $v(n)$.

If we use a practical window (e.g., one of finite duration), how should we choose T_s in relation to ω_s so that we can reconstruct $x(t)$ from the STFT coefficients $X(k\omega_s, nT_s)$? Is this a stable reconstruction? That is, if we make a small error in some STFT coefficient does it affect the reconstructed signal in an unbounded manner? Finally, does the STFT provide an orthonormal basis for L^2? These questions are deep and interesting, and require more careful treatment. We return to these in Section 6.9.

6.4 Digital Filter Banks and Subband Coders

Figure 6.22(a) shows a two-channel filter bank with input sequence $x(n)$ (a discrete-time signal). $G_a(z)$ and $H_a(z)$ are two digital filters, typically low-pass and high-pass. $x(n)$ is split into two subband signals, $x_0(n)$ and $x_1(n)$, which are then downsampled or decimated (see below for definitions). The total subband data rate, counting both subbands, is equal to the number of samples per unit time in the original signal $x(n)$. Digital filter banks provide a time-frequency representation for discrete time signals, similar to the STFT and WT for continuous time signals. The most common engineering application of the digital filter bank is in subband coding, which is used in audio, image, and video compression.

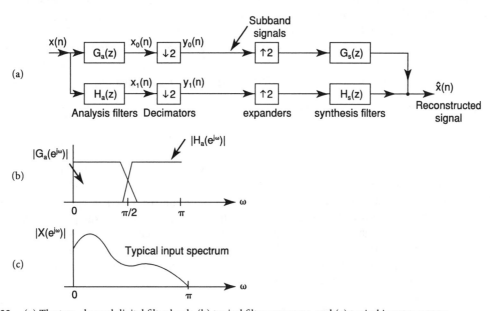

6.22   (a) The two-channel digital filter bank, (b) typical filter responses, and (c) typical input spectrum.

Neither subband coding nor such a time frequency representation is the main point of our discussion here. We are motivated by the fact that a deep mathematical connection exists between this digital filter bank and the continuous time WT. This fundamental relation, discovered by Daubechies [6], is fully elaborated in Section 6.10 to 6.13. This relation is what makes the WT so easy to design and attractive to implement in practice. Several detailed references on the topic of multirate systems and digital filter banks are available [7], so we will be brief.

The Multirate Signal Processing Building Blocks: The building blocks in the digital filter bank of Fig. 6.22(a) are digital filters, decimators, and expanders. The M-fold **decimator** or **downsampler** (denoted $\downarrow M$) is defined by the input-output relation $y(n) = x(Mn)$. The corresponding z-domain relation is $Y(z) = (1/M) \sum_{k=0}^{M-1} X(z^{1/M} e^{-j2\pi k/M})$. This relation is sometimes abbreviated by the notation $Y(z) = X(z)|_{\downarrow M}$ or $Y(e^{j\omega}) = X(e^{j\omega})|_{\downarrow M}$. The M-fold **expander** or **upsampler** (denoted $\uparrow M$) is defined by

$$y(n) = \begin{cases} x(n/M), & n = \text{multiple of } M, \\ 0, & \text{otherwise} \end{cases} \tag{6.29}$$

The transform domain relation for the expander is $Y(z) = X(z^M)$, i.e., $Y(e^{j\omega}) = X(e^{jM\omega})$.

Reconstruction from Subbands

In many applications it is desirable to reconstruct $x(n)$ from the decimated subband signals $y_k(n)$ (possibly after quantization). For this, we pass $y_k(n)$ through expanders and combine them with the synthesis filters $G_s(z)$ and $H_s(z)$. The system is said to have the **perfect reconstruction** (PR) property if $\hat{x}(n) = cx(n - n_0)$ for some $c \neq 0$ and some integer n_0. The PR property is not satisfied for several reasons. First, subband quantization and bit allocation are present, which are the keys to data compression using subband techniques. However, because since our interest here lies in the connection between filter banks and wavelets, we will not be concerned with subband quantization. Second, because the filters $G_a(z)$ and $H_a(z)$ are not ideal, aliasing occurs due to decimation. Using the above equations for the decimator and expander building blocks, we can obtain the following expression for the reconstructed signal: $\hat{X}(z) = T(z)X(z) + A(z)X(-z)$, where $T(z) = 0.5[G_a(z)G_s(z) + H_a(z)H_s(z)]$ and $A(z) = 0.5[G_a(-z)G_s(z) + H_a(-z)H_s(z)]$. The second term having $X(-z)$ is the aliasing term due to decimation. It can be eliminated if the filters satisfy

$$G_a(-z)G_s(z) + H_a(-z)H_s(z) = 0 \qquad \text{(alias cancellation)} \tag{6.30}$$

We can then obtain perfect reconstruction [$\hat{X}(z) = 0.5X(z)$] by setting

$$G_a(z)G_s(z) + H_a(z)H_s(z) = 1 \tag{6.31}$$

A number of authors have developed techniques to satisfy the PR conditions. In this chapter we are interested in a particular technique to satisfy (6.30) and (6.31), the conjugate quadrature filter (CQF) method, which was independently reported by Smith and Barnwell in 1984 [18] and by Mintzer in 1985 [19]. Vaidyanathan [20] showed that these constructions are examples of a general class of M channel filter banks satisfying a property called orthonormality or paraunitariness. More references can be found in [7]. The two-channel CQF solution was later rediscovered in the totally different contexts of multiresolution analysis [11] and compactly supported orthonormal wavelet construction [6]. These are discussed in future sections.

Conjugate Quadrature Filter (CQF) Solution

Suppose the low-pass filter $G_a(z)$ is chosen such that it satisfies the condition

$$\tilde{G}_a(z)G_a(z) + \tilde{G}_a(-z)G_a(-z) = 1 \qquad \text{for all } z \tag{6.32}$$

If we now choose the high-pass filter $H_a(z)$ and the two synthesis filters as

$$H_a(z) = z^{-1}\tilde{G}_a(-z) \qquad G_s(z) = \tilde{G}_a(z) \qquad H_s(z) = \tilde{H}_a(z) \tag{6.33}$$

then (6.30) and (6.31) are satisfied, and $\hat{x}(n) = 0.5x(n)$. In the time domain the above equations become

$$h_a(n) = -(-1)^n g_a^*(-n+1) \qquad g_s(n) = g_a^*(-n) \qquad h_s(n) = h_a^*(-n) \qquad (6.34)$$

The synthesis filters are **time-reversed conjugates** of the analysis filters. If we design a filter $G_a(z)$ satisfying the single condition (6.32) and determine the remaining three filters as above, then the system has the PR property. A filter $G_a(z)$ satisfying (6.32) is said to be **power symmetric**. Readers familiar with **half-band filters** will notice that the condition (6.32) says simply that $\tilde{G}_a(z)G_a(z)$ is half-band. To design a perfect reconstruction CQF system, we first design a low-pass half-band filter $G(z)$ with $G(e^{j\omega}) \geq 0$, and then extract a spectral factor $G_a(z)$. That is, find $G_a(z)$ such that $G(z) = \tilde{G}_a(z)G_a(z)$. The other filters can be found from (6.33).

Polyphase Representation

The polyphase representation of a filter bank provides a convenient platform for studying theoretical questions and also helps in the design and implementation of PR filter banks. According to this representation the filter bank of Fig. 6.22(a) can always be redrawn as in Fig. 6.23(a), which in turn can be redrawn as in Fig. 6.23(b) using standard multirate identities. Here, $\mathbf{E}(z)$ and $\mathbf{R}(z)$ are the "polyphase matrices," determined uniquely by the analysis and synthesis filters, respectively.

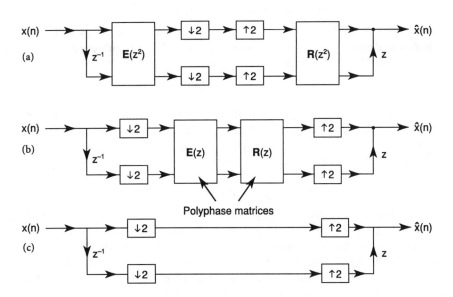

6.23 (a) The polyphase form of the filter bank, (b) further simplification, and (c) equivalent structure when $\mathbf{R}(z) = \mathbf{E}^{-1}(z)$.

If we impose the condition $\mathbf{R}(z)\mathbf{E}(z) = \mathbf{I}$, that is

$$\mathbf{R}(z) = \mathbf{E}^{-1}(z) \qquad (6.35)$$

the system reduces to Fig. 6.23(c), which is a perfect reconstruction system with $\hat{x}(n) = x(n)$. Equation (6.35) will be called the PR condition. Notice that insertion of arbitrary scale factors and delays to obtain $\mathbf{R}(z) = cz^{-K}\mathbf{E}^{-1}(z)$ does not affect the PR property.

Paraunitary Perfect Reconstruction System

A transfer matrix[6] $\mathbf{H}(z)$ is said to be paraunitary if $\mathbf{H}(e^{j\omega})$ is unitary; that is, $\mathbf{H}^\dagger(e^{j\omega})\mathbf{H}(e^{j\omega}) = \mathbf{I}$ (more generally $\mathbf{H}^\dagger(e^{j\omega})\mathbf{H}(e^{j\omega}) = c\mathbf{I}, c > 0$), for all ω. In all practical designs the filters are rational transfer functions so that the paraunitary condition implies $\tilde{\mathbf{H}}(z)\mathbf{H}(z) = \mathbf{I}$ for all z, where the notation $\tilde{\mathbf{H}}(z)$ was explained in Section 6.1. Note that $\tilde{\mathbf{H}}(z)$ reduces to transpose conjugation $\mathbf{H}^\dagger(e^{j\omega})$ on the unit circle. A filter bank in which $\mathbf{E}(z)$ is paraunitary and $\mathbf{R}(z) = \tilde{\mathbf{E}}(z)$ enjoys the PR property $\hat{x}(n) = cx(n), c \neq 0$. We often say that the analysis filter pair $\{G_a(z), H_a(z)\}$ is paraunitary instead of saying that the corresponding polyphase matrix is para-unitary.

The paraunitary property has played a fundamental role in electrical network theory [1, 21], and has a rich history (see references in Chapters 6 and 14 of [7]). Essentially, the scattering matrices of lossless (LC) multiports are paraunitary, i.e., unitary on the imaginary axis of the s-plane.

Properties of Paraunitary Filter Banks

Define the matrices $\mathbf{G}_a(z)$ and $\mathbf{H}_a(z)$ as follows:

$$\mathbf{G}_a(z) = \begin{bmatrix} G_a(z) & G_a(-z) \\ H_a(z) & H_a(-z) \end{bmatrix} \qquad \mathbf{G}_s(z) = \begin{bmatrix} G_s(z) & H_s(z) \\ G_s(-z) & H_s(-z) \end{bmatrix} \qquad (6.36)$$

Notice that these two matrices are fully determined by the analysis filters and synthesis filters, respectively. If $\mathbf{E}(z)$ and $\mathbf{R}(z)$ are paraunitary with $\tilde{\mathbf{E}}(z)\mathbf{E}(z) = 0.5\mathbf{I}$ and $\tilde{\mathbf{R}}(z)\mathbf{R}(z) = 0.5\mathbf{I}$, it can be shown that

$$\tilde{\mathbf{G}}_a(z)\mathbf{G}_a(z) = \mathbf{I} \qquad \tilde{\mathbf{G}}_s(z)\mathbf{G}_s(z) = \mathbf{I} \qquad (6.37)$$

In other words, the matrices $\mathbf{G}_a(z)$ and $\mathbf{G}_s(z)$ defined above are paraunitary as well.

Half-Band Property and Power Symmetry Property. The paraunitary property $\tilde{\mathbf{G}}_a(z)\mathbf{G}_a(z) = \mathbf{I}$ is also equivalent to $\mathbf{G}_a(z)\tilde{\mathbf{G}}_a(z) = \mathbf{I}$, which implies

$$\tilde{G}_a(z)G_a(z) + \tilde{G}_a(-z)G_a(-z) = 1 \qquad (6.38)$$

In other words, $G_a(z)$ is a power symmetric filter. A transfer function $G(z)$ satisfying $G(z) + G(-z) = 1$ is called a half-band filter. The impulse response of such $G(z)$ satisfies $g(2n) = 0$ for all $n \neq 0$ and $g(0) = 0.5$. We see that the power symmetry property of $G_a(z)$ says that $\tilde{G}_a(z)G_a(z)$ is a half-band filter. In terms of frequency response the power symmetry property of $G_a(z)$ is equivalent to

$$|G_a(e^{j\omega})|^2 + |G_a(-e^{j\omega})|^2 = 1 \qquad (6.39)$$

Imagine that $G_a(z)$ is a real-coefficient low-pass filter so that $|G_a(e^{j\omega})|^2$ has symmetry with respect to zero frequency. Then $|G_a(-e^{j\omega})|^2$ is as demonstrated in Fig. 6.24, and the power symmetry property means that the two plots in the Figure add up to unity. In this figure ω_p and ω_s are the bandedges, and δ_1 and δ_2 are the peak passband ripples of $G_a(e^{j\omega})$ (for definitions of filter specifications see [2] or [7]). Notice in particular that power symmetry of $G_a(z)$ implies that a symmetry relation exists between the passband and stopband specifications of $G_a(e^{j\omega})$. This is given by $\omega_s = \pi - \omega_p, \delta_2^2 = 1 - (1 - 2\delta_1)^2$.

Relation Between Analysis Filters. The property $\tilde{\mathbf{G}}_a(z)\mathbf{G}_a(z) = \mathbf{I}$ implies a relation between $G_a(z)$ and $H_a(z)$, namely $H_a(z) = e^{j\theta}z^N\tilde{G}_a(-z)$, where θ is arbitrary and N is an arbitrary odd integer. Let $N = -1$ and $\theta = 0$ for future simplicity. Then

$$H_a(z) = z^{-1}\tilde{G}_a(-z) \qquad (6.40)$$

---

[6]Transfer matrices are essentially transfer functions of multi-input multi-output systems. A review is found in Chapter 13 of [7].

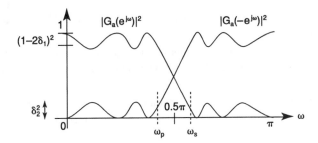

6.24 The magnitude responses $|G_a(e^{j\omega})|^2$ and $|G_a(-e^{j\omega})|^2$ for a real-coefficient power symmetric filter $G_a(z)$.

In particular, we have $|H_a(e^{j\omega})| = |G_a(-e^{j\omega})|$. Combining with the power symmetry property (6.39), we see that the two analysis filters are **power complementary**:

$$|G_a(e^{j\omega})|^2 + |H_a(e^{j\omega})|^2 = 1 \qquad (6.41)$$

for all ω. With $G_a(z) = \sum_n g_a(n)z^{-n}$ and $H_a(z) = \sum_n h_a(n)z^{-n}$ we can rewrite (6.40) as

$$h_a(n) = -(-1)^n g_a^*(-n + 1) \qquad (6.42)$$

Relation between Analysis and Synthesis Filters. If we use the condition $\mathbf{R}(z) = \tilde{\mathbf{E}}(z)$ in the definitions of $\mathbf{G}_s(z)$ and $\mathbf{G}_a(z)$ we obtain $\mathbf{G}_s(z) = \tilde{\mathbf{G}}_a(z)$, from which we conclude that the synthesis filters are given by $G_s(z) = \tilde{G}_a(z)$ and $H_s(z) = \tilde{H}_a(z)$. We can also rewrite these in the time domain; summarizing all this, we have

$$G_s(z) = \tilde{G}_a(z), \quad H_s(z) = \tilde{H}_a(z) \qquad g_s(n) = g_a^*(-n), \quad h_s(n) = h_a^*(-n) \qquad (6.43)$$

The synthesis filter coefficients are time-reversed conjugates of the analysis filters. Their frequency responses are conjugates of the analysis filter responses. In particular, $|G_s(e^{j\omega})| = |G_a(e^{j\omega})|$ and $|H_s(e^{j\omega})| = |H_a(e^{j\omega})|$. In view of the preceding relations the synthesis filters have all the properties of the analysis filters. For example, $G_s(e^{j\omega})$ is power symmetric, and the pair $\{G_s(e^{j\omega}), H_s(e^{j\omega})\}$ is power complementary. Finally, $H_s(z) = z\tilde{G}_s(-z)$, instead of (6.40).

Relation to Conjugate Quadrature Filter Design. The preceding discussions indicate that in a paraunitary filter bank the filter $G_a(z)$ is power symmetric, and the remaining filters are derived from $G_a(z)$ as in (6.40) and (6.43). This is precisely the CQF solution for PR, stated at the beginning of this section.

Summary of Filter Relations in a Paraunitary Filter Bank

If the filter bank of Fig. 6.22(a) is paraunitary, then the polyphase matrices $\mathbf{E}(z)$ and $\mathbf{R}(z)$ (Fig. 6.23) satisfy $\tilde{\mathbf{E}}(z)\mathbf{E}(z) = 0.5\mathbf{I}$ and $\tilde{\mathbf{R}}(z)\mathbf{R}(z) = 0.5\mathbf{I}$. Equivalently the filter matrices $\mathbf{G}_a(z)$ and $\mathbf{G}_s(z)$ satisfy $\tilde{\mathbf{G}}_a(z)\mathbf{G}_a(z) = \mathbf{I}$ and $\tilde{\mathbf{G}}_s(z)\mathbf{G}_s(z) = \mathbf{I}$. A number of properties follow from these:

1. All four filters, $G_a(z)$, $H_a(z)$, $G_s(z)$, and $H_s(z)$, are power symmetric. This property is defined, for example, by the relation (6.38). This means that the filters are spectral factors of half-band filters; for example, $\tilde{G}_s(z)G_s(z)$ is half-band.

2. The two analysis filters are related as in (6.40), so the magnitude responses are related as $|H_a(e^{j\omega})| = |G_a(-e^{j\omega})|$. The synthesis filters are time-reversed conjugates of the analysis filters as shown by (6.43). In particular, $G_s(e^{j\omega}) = G_a^*(e^{j\omega})$ and $H_s(e^{j\omega}) = H_a^*(e^{j\omega})$.

3. The analysis filters form a power complementary pair, i.e., (6.41) holds. The same is true for the synthesis filters.

4. Any two-channel paraunitary system satisfies the CQF equations (6.32, 6.33) (except for delays, constant scale factors, etc.). Conversely, any CQF design is a paraunitary filter bank.

5. The design procedure for two-channel paraunitary (i.e., CQF) filter banks is as follows: design a zero-phase low-pass half-band filter $G(z)$ with $G(e^{j\omega}) \geq 0$ and then extract a spectral factor $G_a(z)$. That is, find $G_a(z)$ such that $G(z) = \tilde{G}_a(z)G_a(z)$. Then choose the remaining three filters as in (6.33), or equivalently, as in (6.34).

Parametrization of Paraunitary Filter Banks

Factorization theorems exist for matrices which allow the expression of paraunitary matrices as a cascade of elementary paraunitary blocks. For example, let $\mathbf{H}(z) = \sum_{n=0}^{L} \mathbf{h}(n)z^{-n}$ be a 2×2 real causal FIR transfer matrix (thus, $\mathbf{h}(n)$ are 2×2 matrices with real elements). This is paraunitary iff it can be expressed as $\mathbf{H}(z) = \mathbf{R}_N \Lambda(z)\mathbf{R}_{N-1} \ldots \mathbf{R}_1 \Lambda(z)\mathbf{R}_0\mathbf{H}_0$ where

$$\mathbf{R}_m = \begin{bmatrix} \cos\theta_m & \sin\theta_m \\ -\sin\theta_m & \cos\theta_m \end{bmatrix} \quad \Lambda(z) = \begin{bmatrix} 1 & 0 \\ 0 & z^{-1} \end{bmatrix} \quad \mathbf{H}_0 = \begin{bmatrix} \alpha & 0 \\ 0 & \pm\alpha \end{bmatrix} \quad (6.44)$$

where α and θ_m are real. For a proof see [7]. The unitary matrix $\mathbf{R}_m$ is called a **rotation operator** or the **Givens rotation.** The factorization gives rise to a cascaded lattice structure that guarantees the paraunitary property structurally. This is useful in the design as well as the implementation of filter banks, as explained in [7]. Thus, if the polyphase matrix is computed using the cascaded structure, $G_a(z)$ is guaranteed to be power symmetric, and the relation $H_a(z) = z^{-1}\tilde{G}_a(-z)$ between the analysis filters automatically holds.

Maximally Flat Solutions

The half-band filter $G(z) \triangleq \tilde{G}_a(z)G_a(z)$ can be designed in many ways. One can choose to have equiripple designs or maximally flat designs [2]. An early technique for designing FIR maximally flat filters was proposed by Herrmann in 1971 [7]. This method gives closed form expressions for the filter coefficients and can be adapted easily for the special case of half-band filters. Moreover, the design automatically guarantees the condition $G(e^{j\omega}) \geq 0$, which in particular implies zero phase.

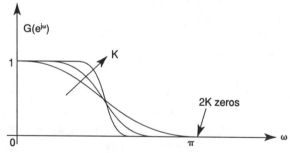

6.25   Maximally flat half-band filter responses with $2K$ zeroes at π.

The family of maximally flat half-band filters designed by Herrmann is demonstrated in Fig. 6.25. The transfer function has the form

$$G(z) = z^K \left(\frac{1 + z^{-1}}{2}\right)^{2K} \sum_{n=0}^{K-1} (-z)^n \binom{K + n - 1}{n} \left(\frac{1 - z^{-1}}{2}\right)^{2n} \tag{6.45}$$

The filter has order $4K - 2$. On the unit circle we find $2K$ zeroes, and all of these zeroes are concentrated at the point $z = -1$ (i.e., at $\omega = \pi$). The remaining $2K - 2$ zeroes are located in the z-plane such that $G(z)$ has the half-band property described earlier (i.e., $G(z) + G(-z) = 1$).

Section 6.13 explains that if the CQF bank is designed by starting from Herrmann's maximally flat half-band filter, then it can be used to design continuous time wavelets with arbitrary regularity (i.e., smoothness) properties.

Tree-Structured Filter Banks

The idea of splitting a signal $x(n)$ into two subbands can be extended by splitting a subband signal further, as demonstrated in Fig. 6.26(a). In this example the low-pass subband is split repeatedly. This is called a tree-structured filter bank. Each node of the tree is a two-channel analysis filter bank.

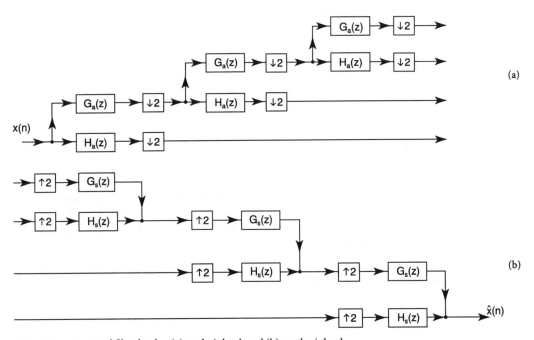

6.26 Tree-structured filter banks: (a) analysis bank and (b) synthesis bank.

The synthesis bank corresponding to Fig. 6.26(a) is shown in Fig. 6.26(b). We combine the signals in pairs in the same manner that we split them. It can be shown that if $\{G_a(z), H_a(z), G_s(z), H_s(z)\}$ is a PR system [i.e., satisfies $\hat{x}(n) = x(n)$ when connected in the form Fig. 6.22(a)] then the tree-structured analysis/synthesis system of Fig. 6.26 has PR $\hat{x}(n) = x(n)$.

The tree-structured system can be redrawn in the form shown in Fig. 6.27. For example, if we have a tree structure similar to Fig. 6.26 with three levels, we have $M = 4$, $n_0 = 2$, $n_1 = 4$, $n_2 = 8$, and $n_3 = 8$. If we assume that the responses of the analysis filters $G_a(e^{j\omega})$ and $H_a(e^{j\omega})$ are as in Fig. 6.28(a), the responses of the analysis filters $H_k(e^{j\omega})$ are as shown in Fig. 6.28(b). Note that

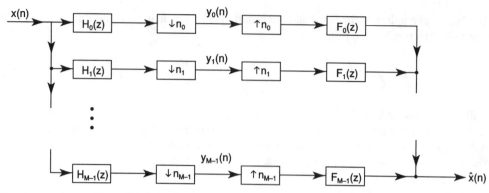

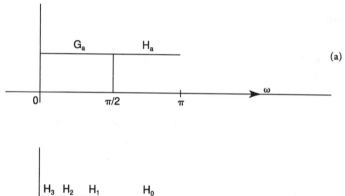

6.27   A general nonuniform digital filter bank.

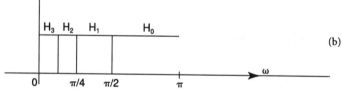

6.28   An example of responses: (a) $G_a(z)$ and $H_a(z)$, and (b) tree-structured analysis bank.

this resembles the wavelet transform [Fig. 6.8(b)]. The outputs of different filters are subsampled at different rates exactly as for wavelets. Thus, the tree-structured filter bank bears a close relationship to the WT. Sections 6.10 to 6.13 present the precise mathematical connection between the two, and the relation to multiresolution analysis.

Filter Banks and Basis Functions

Assuming PR [$\hat{x}(n) = x(n)$], we can express $x(n)$ as

$$x(n) = \sum_{k=0}^{M-1} \sum_{m=-\infty}^{\infty} y_k(m) \underbrace{f_k(n - n_k m)}_{\eta_{km}(n)} \qquad (6.46)$$

where $y_k(n)$ are the decimated subband signals, and $f_k(n)$ are the impulse responses of $F_k(z)$. Thus, the system is analogous to the filter bank systems which represented the continuous time STFT and WT in Sections 6.2 and 6.3. The collection of subband signals $y_k(m)$ can be regarded as a time-frequency representation for the sequence $x(n)$. As before, k denotes the **frequency index** and m the **time index** in the transform domain. If we have a PR filter bank we can recover $x(n)$ from this time-frequency representation using (6.46). The doubly indexed family of discrete time sequences $\{\eta_{km}(n)\}$ can be regarded as "basis functions" for the representation of $x(n)$.

To make things mathematically accurate, let $x(n) \in l^2$ (i.e., $\sum_n |x(n)|^2$ is finite). If the two-channel filter bank $\{G_a(z), H_a(z), G_s(z), H_s(z)\}$ which makes up the tree structure of Fig. 6.26 is paraunitary, it can be shown that $\eta_{km}(n)$ is an orthonormal basis for l^2. Orthonormality means

$$\sum_{n=-\infty}^{\infty} \eta_{k_1 m_1}(n) \eta_{k_2 m_2}^*(n) = \delta(k_1 - k_2)\delta(m_1 - m_2) \tag{6.47}$$

Notice that the basis functions (sequences) are not derived from a single function. Rather, they are derived from a *finite* number of filters $\{f_k(n)\}$ by time shifts of a specific form. The wavelet basis $\{2^{k/2}\psi(2^k t - n)\}$, on the other hand, is derived from a *single* wavelet function $\psi(t)$. We say that $\{\eta_{km}(n)\}$ is a *filter bank* type of basis for the space of l^2 sequences.

6.5 Deeper Study of Wavelets, Filter Banks, and Short-Time Fourier Transforms

We already know what the WT is and how it compares to the STFT, at least qualitatively. We are also familiar with time-frequency representations and digital filter banks. It is now time to fill in several important details, and generally be more quantitative. For example, we would like to mention some major technical limitations of the STFT which are not obvious from its definition, and explain that wavelets do not have this limitation.

For example, if the STFT is used to obtain an orthonormal basis for L^2 signals, the time-frequency RMS durations of the window $v(t)$ should satisfy $D_t D_f = \infty$. That is, either the time or the frequency resolution is very poor (Theorem 6.5). Also, if we have an STFT system in which the time-frequency sampling product $\omega_s T_s$ is small enough to admit redundancy (i.e., the vectors are not linearly independent as they would be in an orthonormal basis), the above difficulty can be eliminated (Section 6.9).

The Gabor transform, while admittedly a tempting candidate because of the optimal time-frequency resolution property ($D_t D_f$ minimized), has a disadvantage. For example, if we want to recover the signal $x(t)$ from the STFT coefficients, the reconstruction is *unstable* in the so-called critically sampled case (Section 6.9). That is, a small error in the STFT coefficients can lead to a large error in reconstruction.

The WT does not suffer from the above limitations of the STFT. Sections 6.11 to 6.13 show how to construct orthonormal wavelet bases with good time and frequency resolutions. We also show that we can start from a paraunitary digital filter bank and construct orthonormal wavelet bases for $L^2(R)$ very systematically (Theorem 6.13). Moreover, this can be done in such a way that many desired properties (e.g., compact support, orthonormality, good time frequency resolution, smoothness, and so forth) can be incorporated during the construction (Section 6.13). Such a construction is placed in evidence by the **theory of multiresolution**, which gives a unified platform for wavelet construction and filter banks (Theorems 6.6 and 6.7).

At this point, the reader may want to preview the above-mentioned theorems in order to get a flavor of things to come. However, to explain these results quantitatively, it is very convenient to review a number of mathematical tools. The need for advanced tools arises because of the intricacies associated with basis functions for infinite dimensional spaces, i.e., spaces in which the set of basis functions is an infinite set. (For finite dimensional spaces an understanding of elementary matrix theory would have been sufficient.) For example, a representation of the form $x(t) = \sum c_n f_n(t)$ in an infinite dimensional space could be *unstable* in the sense that a small error in the transform domain $\{c_n\}$ could be amplified in an unbounded manner during reconstruction. A special type of basis called the *Riesz basis* does not have this problem (orthonormal bases are special cases of these). Also, the so-called **frames** (Section 6.8) share many good properties of the Riesz bases, but may have redundant vectors (i.e., not a linearly independent set of vectors). For example, the concept of frames arises in the comparison of wavelets and the STFT. General STFT frames have an advantage over STFT bases. Frames also come into consideration when the connection between wavelets and paraunitary digital filter banks is explained in Section 6.11. When describing the connection between wavelets and nonunitary filter banks, one again encounters Riesz bases and the idea of **biorthogonality**.

Because it is difficult to find all the mathematical background material in one place, we review a carefully selected set of topics in the next few sections. These are very useful for a deeper understanding of wavelets and STFT. The material in Section 6.6 is fairly standard (Lebesgue integrals, L^p spaces, L^1 and L^2 FTs). The material in Sections 6.7 and 6.8 (Riesz bases and frames) are less commonly known among engineers, but play a significant role in wavelet theory. The reader may want to go through these review sections (admittedly dense), once during first reading and then use them as a reference. Following this review we return to our discussions of wavelets, STFT, and filter banks.

6.6   The Space of L^1 and L^2 Signals

We developed the wavelet representation in Section 6.2 based on the framework of a bank of bandpass filters. To make everything mathematically meaningful it becomes necessary to carefully specify the types of signals, types of FTs, etc. For example, the concept of ideal bandpass filtering is appealing to engineers, but a difficulty arises. An ideal bandpass filter $H(\omega)$ is not stable, that is, $\int |h(t)| \, dt$ does not exist [2]. In other words $h(t)$ does not belong to the space L^1 (see below).

Why should this matter if we are discussing theory? The frequency domain developments based on Fig. 6.2, which finally give rise to the time domain expression (6.8), implicitly rely on the convolution theorem (*convolution in time implies multiplication in frequency*). However, the convolution theorem is typically proved only for L^1 signals and bounded L^2 signals; it is not valid for arbitrary signals. Therefore, care must be exercised when using these familiar engineering notions in a mathematical discussion.

Lebesgue Integrals

In most engineering discussions we think of the integrals as Riemann integrals, but in order to handle several convergence questions in the development of Fourier series, convolution theorems, and wavelet transforms, it is necessary to use Lebesgue integration. Lebesgue integration theory has many beautiful results which are not true for the Riemann integral under comparable assumptions about signals. This includes theorems that allow us to interchange limits, integrals, and infinite sums freely.

All integrals in this chapter are Lebesgue integrals. A review of Lebesgue integration is beyond the scope of this chapter, although many excellent references, for example, [22], are available. A few elementary comparisons between Riemann and Lebesgue integrals are given next.

1. If $x(t)$ is Riemann integrable on a bounded interval $[a, b]$ then it is also Lebesgue integrable on $[a, b]$. The converse is not true, however. For example, if we define $x(t) = -1$ for all rationals and $x(t) = 1$ for all irrationals in $[0,1]$, then $x(t)$ is not Riemann integrable in $[0,1]$, but it is Lebesgue integrable, and $\int_0^1 x(t) \, dt = 1$.

2. A similar statement is not true for the unbounded interval $(-\infty, \infty)$. For the unbounded interval $(-\infty, \infty)$ the Riemann integral is defined only as a limit called the **improper**

integral.[7] Consider the sinc function defined as: $s(t) = \sin t / t$ for $t \neq 0$, and $s(0) = 1$. This has improper Riemann integral $= \pi$, but is not Lebesgue integrable.

3. If $x(t)$ is Lebesgue integrable, so is $|x(t)|$. The same is not true for Riemann integrals, as demonstrated by the sinc function $s(t)$ of the preceding paragraph.

4. If $|x(t)|$ is Lebesgue integrable, so is $x(t)$ as long as it is measurable.[8] This, however, is not true for Riemann integrals. If we define $x(t) = -1$ for all rationals and 1 for all irrationals in $[0,1]$, it is not Riemann integrable in $[0,1]$ although $|x(t)|$ is.

5. If $x(t)$ is (measurable and) bounded by a non-negative Lebesgue integrable function $g(t)$ [i.e., $|x(t)| \leq g(t)$] then $x(t)$ is Lebesgue integrable.

Sets of Measure Zero

A subset S of real numbers is said to have measure zero if, given $\epsilon > 0$, we can find a countable union $\cup_i I_i$ of open intervals I_i [intervals of the form (a_i, b_i), i.e., $a_i < x < b_i$] such that $S \subset \cup_i I_i$ and the total length of the intervals $< \epsilon$. For example, the set of all integers (in fact, any countable set of real numbers, e.g., rationals) has measure zero. Uncountable sets of real numbers exist which have measure zero, a famous example being the Cantor set [22].

When something is said to be true "almost everywhere" (abbreviated a.e.) or "for almost all t" it means that the statement holds everywhere, except possibly on a set of measure zero. For example, if $x(t) = y(t)$ everywhere except for integer values of t, then $x(t) = y(t)$ a.e. An important fact in Lebesgue integration theory is that if two Lebesgue integrable functions are equal a.e., then their integrals are equal. In particular, if $x(t) = 0$ a.e., the Lebesgue integral $\int x(t)dt$ exists and is equal to zero.

Convergence Theorems

What makes the Lebesgue integral so convenient is the existence of some powerful theorems which allow us to interchange limits with integrals and summations under very mild conditions. These theorems have been at the center of many beautiful results in Fourier and wavelet transform theory.

Let $\{g_k(t)\}$, $1 \leq k \leq \infty$ be a sequence of Lebesgue integrable functions. In general, this sequence may not have a limit, and even if it did, the limit may not be integrable. Under some further mild postulates, we can talk about limits and their integrals. In what follows we often say "$g(t)$ is a pointwise limit a.e. of the sequence $\{g_k(t)\}$," or "$g_k(t)$ converges to $g(t)$ a.e." This means that for any chosen value of t (except possibly in a set of measure zero), we have $g_k(t) \to g(t)$ as $k \to \infty$.

Monotone Convergence Theorem: *Suppose $\{g_k(t)\}$ is* **nondecreasing** *a.e. (i.e., for almost all values of t, $g_k(t)$ is nondecreasing in k) and $\int g_k(t)\, dt$ is a* **bounded sequence.** *Then $\{g_k(t)\}$ converges a.e. to a Lebesgue integrable function $g(t)$ and $\lim_k \int g_k(t)\, dt = \int \lim_k g_k(t)\, dt$, i.e., $\lim_k \int g_k(t)\, dt = \int g(t)\, dt$. That is, we can* **interchange the limit with the integral**.

Dominated Convergence Theorem: *Suppose $\{g_k(t)\}$ is dominated by a non-negative Lebesgue integrable function $f(t)$, i.e., $|g_k(t)| \leq f(t)$ a.e., and $\{g_k(t)\}$ converges to a limit $g(t)$ a.e. Then*

---

[7]Essentially we consider $\int_{-a}^{b} x(t)dt$ and let a and b go to ∞ *separately*. This limit, the improper Riemann integral, should not be confused with the **Cauchy principal value**, which is the limit of $\int_{-a}^{a} x(t)\, dt$ as $a \to \infty$. The function $x(t) = t$ has Cauchy principal value $= 0$, but the improper Riemann integral does not exist.

[8]The notion of a measurable function is very subtle. Any continuous function is measurable, and any Lebesgue integrable function is measurable. In fact, examples of nonmeasureable functions are so rare and so hard to construct that practically no danger exists that we will run into one. We take measurability for granted and never mention it.

the limit $g(t)$ is Lebesgue integrable and $\lim_k \int g_k(t)dt = \int \lim_k g_k(t)\, dt$, *i.e.,* $\lim_k \int g_k(t)dt = \int g(t)\, dt$. *That is, we can* **interchange the limit with the integral**.

Levi's Theorem: *Suppose* $\int \sum_{k=1}^m |g_k(t)|\, dt$ *is a bounded sequence in* m. *Then* $\int \sum_{k=1}^\infty g_k(t)\, dt = \sum_{k=1}^\infty \int g_k(t)\, dt$. *This means, in particular, that* $\sum_{k=1}^\infty g_k(t)$ *converges a.e. to a Lebesgue integrable function. This theorem permits us to* **interchange infinite sums with the integrals**.

Fatou's Lemma: *Let (a)* $g_k(t) \geq 0$ *a.e., (b)* $g_k(t) \to g(t)$ *a.e., and (c)* $\int g_k(t)\, dt \leq A$ *for some* $0 < A < \infty$. *Then the limit* $g(t)$ *is Lebesgue integrable and* $\int g(t) \leq A$. (Stronger versions of this result exist [23], but we shall not require them here.)

L^p Signals

Let p be an integer such that $1 \leq p < \infty$. A signal $x(t)$ is said to be an L^p signal if it is measurable, and if $\int |x(t)|^p dt$ exists. We define the L^p norm of $x(t)$ as $\|x(t)\|_p = [\int |x(t)|^p dt]^{1/p}$. For fixed p the set of L^p signals forms a vector space. It is a normed linear vector space, with norm defined as above. The term "linear" means that if $x(t)$ and $y(t)$ are in L^p, then $\alpha x(t) + \beta y(t)$ is also in L^p for any complex α and β. Because any two signals $x(t)$ and $y(t)$ that are equal a.e. cannot be distinguished (i.e., $\|x(t) - y(t)\|_p = 0$), each element in L^p is in reality "a set of functions that are equal a.e." Each such set becomes an "equivalence class" in mathematical language. For $p = 2$ the quantity $\|x(t)\|_p^2$ is equal to the **energy** of $x(t)$, as defined in signal processing texts. Thus, an L^2 signal is a finite-energy (or square-integrable) signal. For $p = \infty$ the above definitions do not make sense, and we simply define L^∞ to be the space of **essentially bounded signals.** A signal $x(t)$ is said to be essentially bounded if there exists a number $B < \infty$ such that $|x(t)| \leq B$ a.e. We often omit the term "essential" for simplicity; it arises because of the a.e. in the inequality. The norm $\|x(t)\|_\infty$ is taken as essential supremum of $|x(t)|$ over all t. That is, $\|x(t)\|_\infty$ is the smallest number such that $|x(t)| \leq \|x(t)\|_\infty$ a.e.

L^1, L^2, and L^∞ functions are particularly interesting for engineers. Note that neither L^1 nor L^2 contains the other. However, bounded L^1 functions are in L^2, and L^2 functions on bounded intervals are in L^1. That is,

$$L^1 \cap L^\infty \subset L^2 \quad \text{and} \quad L^2[a, b] \subset L^1[a, b] \tag{6.48}$$

Thus, L^2 is already bigger than bounded L^1 functions. Moreover,

$$x(t) \in L^1 \cap L^\infty \quad \Rightarrow \quad x(t) \in L^p \text{ for all } p > 1$$

This follows because $|x(t)|^p \leq |x(t)|\, \|x(t)\|_\infty^{p-1}$. Thus, $|x(t)|^p$ is (measurable and) bounded by a Lebesgue integrable function (because $|x(t)|$ is integrable), and is therefore integrable.

Orthonormal Signals in L^2

The inner product $\langle x(t), y(t) \rangle = \int x(t) y^*(t)\, dt$ always exists for any $x(t)$ and $y(t)$ in L^2. Thus, the product of two L^2 functions is an L^1 function. If $\langle x(t), y(t) \rangle = 0$ we say that $x(t)$ and $y(t)$ are orthogonal. Clearly, $\|x(t)\|_2^2 = \langle x(t), x(t) \rangle$. Consider a sequence $\{g_n(t)\}$ of signals such that any pair of these are orthogonal, and $\|g_n(t)\|_2 = 1$ for all n. This is said to be an orthonormal sequence. The following two results are fundamental.

THEOREM 6.1 *Let* $\{g_n(t)\}, 1 \leq n \leq \infty$ *be an orthonormal sequence in* L^2. *Define* $c_n = \langle x(t), g_n(t) \rangle$ *for some* $x(t) \in L^2$. *Then the sum* $\sum_n |c_n|^2$ *converges, and* $\sum_n |c_n|^2 \leq \|x(t)\|^2$.

THEOREM 6.2 *(Riesz-Fischer Theorem) Let* $\{g_n(t)\}, 1 \leq n \leq \infty$ *be an orthonormal sequence in* L^2 *and let* $\{c_n\}$ *be a sequence of complex numbers such that* $\sum_n |c_n|^2$ *converges. Then there exists*

$x(t) \in L^2$ such that $c_n = \langle x(t), g_n(t) \rangle$, and $x(t) = \sum_n c_n g_n(t)$ (with equality interpreted in the L^2 sense; see below).

The space L^2 is more convenient to work with than L^1. For example, the inner product and the concept of orthonormality are undefined in L^1. Moreover (see following section), the FT in L^2 has more time-frequency symmetry than in L^1. In Section 6.7 we will define **unconditional bases**, which have the property that any rearrangement continues to be a basis. It turns out that any orthonormal basis in L^2 is unconditional, whereas the L^1 space does not even have an unconditional basis.

Equality and Convergence in L^p Sense

Let $x(t)$ and $y(t)$ be L_p functions ($p < \infty$). Then $\|x(t) - y(t)\|_p = 0$ iff $x(t) = y(t)$ a.e. For example, if $x(t)$ and $y(t)$ differ only for every rational t we still have $\|x(t) - y(t)\|_p = 0$. Whenever $\|x(t) - y(t)\|_p = 0$, we say that $x(t) = y(t)$ in L^p sense. Now consider a statement of the form

$$x(t) = \sum_{n=1}^{\infty} c_n g_n(t) \tag{6.49}$$

for $p < \infty$, where $g_n(t)$ and $x(t)$ are in L^p. This means that the sum converges to $x(t)$ in the L^p sense; that is, $\|x(t) - \sum_{n=1}^{N} c_n g_n(t)\|_p$ goes to zero as $N \to \infty$. If we modify the limit $x(t)$ by adding some number to $x(t)$ for all rational t, the result is still a limit of $\sum_{n=1}^{N} c_n g_n(t)$ in the L^p sense. L^p limits are unique only in the a.e. sense. We omit the phrase "in the L^p sense" whenever it is clear from the context.

l^p Spaces

Let p be an integer with $1 \le p \le \infty$. The collection of all sequences $x(n)$ such that $\sum_n |x(n)|^p$ converges to a finite value is denoted l^p. This is a linear space with norm $\|x(n)\|_p$ defined so that $\|x(n)\|_p = \left(\sum_n |x(n)|^p \right)^{1/p}$. Unlike L^p spaces, the l^p spaces satisfy the following inclusion rule:

$$l^1 \subset l^2 \subset l^3 \subset \ldots l^\infty \tag{6.50}$$

The spaces l^1 and l^2 are especially interesting in circuits and signal processing. If $h(n) \in l^1$, $\sum_n |h(n)| < \infty$. This is precisely the condition for the BIBO (bounded-input–bounded-output) stability of a linear time invariant system with impulse response $h(n)$ [2].

Continuity of Inner Products

If $\{x_n(t)\}$ is a sequence in L^2 and has an L^2 limit $x(t)$, then for any $y(t) \in L^2$,

$$\lim_{n \to \infty} \langle x_n(t), y(t) \rangle = \langle \lim_{n \to \infty} x_n(t), y(t) \rangle = \langle x(t), y(t) \rangle \tag{6.51}$$

with the second limit interpreted in the L^2 sense. Thus, limits can be interchanged with inner product signs. Similarly, infinite summation signs can be interchanged with the inner product sign, that is, $\sum_{n=1}^{\infty} \langle \alpha_n x_n(t), y(t) \rangle = \langle \sum_{n=1}^{\infty} \alpha_n x_n(t), y(t) \rangle$, provided the second summation is regarded as an L^2 limit. These follow from the fundamental property that *inner products are continuous* [23].

Next suppose $\{x_n(t)\}$ is a sequence of functions in L^p for some integer $p \ge 1$, and suppose $x_n(t) \to x(t)$ in the L^p sense. Then $\|x_n(t)\|_p \to \|x(t)\|_p$ as well. We can rephrase this as

$$\lim_{n \to \infty} \|x_n(t)\|_p = \left\| \lim_{n \to \infty} x_n(t) \right\|_p = \|x(t)\|_p \tag{6.52}$$

Thus, the limit sign can be interchanged with the norm sign, where the limit in the second expression is in the L^p sense. This follows because

$$\Big| \|x_n(t)\|_p - \|x(t)\|_p \Big| \leq \|x_n(t) - x(t)\|_p \to 0 \quad \text{as} \quad n \to \infty$$

Fourier Transforms

The Fourier transform is defined for L^1 and L^2 signals in different ways. The properties of these two types of FT are significantly different. In the signal processing literature, in which we ultimately seek engineering solutions (such as filter approximation with rational transfer functions), this distinction often is not necessary. However, when we try to establish that a certain set of signals is a basis for a certain class, we must be careful, especially if we use tools such as the FT, convolution theorem, etc. (as we implicitly did in Section 6.2). Detailed references for this section include [15, 22], and [23].

L^1 Fourier Transform

Given a signal $x(t) \in L^1$, its FT $X(\omega)$ (the L^1 FT) is defined in a manner that is familiar to engineers:

$$X(\omega) = \int_{-\infty}^{\infty} x(t)e^{-j\omega t} \, dt \tag{6.53}$$

The existence of this integral is assured by the fact that $x(t)$ is in L^1.[9] In fact the above integral exists iff $x(t) \in L^1$. The L^1 FT has the following properties:

1. $X(\omega)$ is a continuous function of ω.
2. $X(\omega) \to 0$ as $|\omega| \to \infty$. This is called the **Riemann-Lebesgue lemma.**
3. $X(\omega)$ is bounded, and $|X(\omega)| \leq \|x(t)\|_1$.

In engineering applications we often draw the ideal low-pass filter response ($F(\omega)$ in Fig. 6.3) and consider it to be the FT of the impulse response $f(t)$, but this frequency response is discontinuous and already violates property 1. This is because $f(t)$ is not in L^1 and $F(\omega)$ is not the L^1-FT of $f(t)$. That $f(t)$ is not in L^1 is consistent with the fact that the ideal filter is not BIBO stable (i.e., a bounded input may not produce bounded output because $\int |f(t)| \, dt$ is not finite).

Inverse Fourier Transform. The FT $X(\omega)$ of an L^1 signal generally is not in L^1. For example, if $x(t)$ is the rectangular pulse, then $X(\omega)$ is the sinc function which is not absolutely integrable. Thus, the familiar inverse transform formula

$$x(t) = \frac{1}{2\pi} \int_{-\infty}^{\infty} X(\omega)e^{j\omega t} \, d\omega \tag{6.54}$$

does not make sense in general. However, because $X(\omega)$ is continuous and bounded, it is integrable on any bounded interval, so $\int_{-c}^{c} X(\omega)e^{j\omega t} \, d\omega / 2\pi$ exists for any finite c. This quantity may even have a limit as $c \to \infty$, even if the Lebesgue integral or improper Rieman integral, does not exist. Such a limit (the Cauchy principal value) does represent the original function $x(t)$ under some conditions. Two such cases are outlined next.

Case 1. Thus, suppose $x(t) \in L^1$ and suppose that it is of **bounded variation** in an interval $[a, b]$; that is, it can be expressed as the difference of two nondecreasing functions [22]. Then we can show

---

[9]Because $x(t)$ is Lebesgue integrable (hence, measurable) the product $x(t)e^{-j\omega t}$ is measurable, and it is bounded by the integrable function $|x(t)|$. Thus, $x(t)e^{-j\omega t}$ is integrable.

that the above Cauchy principal value exists, and

$$\frac{x(t^+) + x(t^-)}{2} = \lim_{c \to \infty} \frac{1}{2\pi} \int_{-c}^{c} X(\omega)e^{j\omega t} \, d\omega \qquad (6.55)$$

for every $t \in (a, b)$. The notations $x(t^-)$ and $x(t^+)$ are the left-hand limit and the right-hand limit, respectively, of $x(\cdot)$ at t; for functions of bounded variation, these limits can be shown to exist. If $x(\cdot)$ is continuous at t, then $x(t^-) = x(t^+) = x(t)$, and the above reduces to the familiar inversion formula.

Case 2. Suppose now that $x(t) \in L^1$ and $X(\omega) \in L^1$ as well. Then the integral $y(t) \stackrel{\Delta}{=} \int_{-\infty}^{\infty} X(\omega)e^{j\omega t} d\omega / 2\pi$ exists as a Lebesgue integral, and $y(t) = x(t)$ a.e. [23]. In particular, if $x(\cdot)$ is continuous at t, $x(t) = \int_{-\infty}^{\infty} X(\omega)e^{j\omega t} d\omega / 2\pi$.

If $x(t)$ and $X(\omega)$ are both in L^1 they are both in L^2 as well. This is shown as follows: because $x(t) \in L^1$ implies that $X(\omega)$ is bounded, we see that $X(\omega) \in L^1 \cap L^\infty$. So $X(\omega) \in L^p$ for all integer p (See previous section). In particular, $X(\omega) \in L^2$, so $x(t) \in L^2$ as well (by Parseval's relation; see below).

The L^2 Fourier Transform

The L^1 Fourier transform lacks the convenient property of time-frequency symmetry. For example, even though $x(t)$ is in L^1, $X(\omega)$ may not be in L^1. Also, even though $x(t)$ may not be continuous, $X(\omega)$ is necessarily continuous. The space L^2 is much easier to work with. Not only can we talk about inner products and orthonormal bases, perfect symmetry also exists between time and frequency domains. We must define the L^2-FT differently because the usual definition (6.53) is meaningful only for L^1 signals. Suppose $x(t) \in L^2$ and we truncate it to the interval $[-n, n]$. This truncated version is in L^1 because of (6.48), and its L^1 FT exists:

$$X_n(\omega) = \int_{-n}^{n} x(t)e^{-j\omega t} \, dt \qquad (6.56)$$

It can be shown that $X_n(\omega)$ is in L^2 and that the sequence $\{X_n(\omega)\}$ has a limit in L^2. That is, there exists an L^2 function $X(\omega)$ such that

$$\lim_{n \to \infty} \|X_n(\omega) - X(\omega)\|_2 = 0 \qquad (6.57)$$

This limit $X(\omega)$ is defined to be the L^2 FT of $x(t)$. Some of the properties are listed next:

1. $X(\omega)$ is in L^2, and we can compute $x(t)$ from $X(\omega)$ in an entirely analogous manner, namely the L^2 limit of $\int_{-n}^{n} X(\omega)e^{j\omega t} d\omega / 2\pi$.

2. If $x(t)$ is in L^1 and L^2, then the above computation gives the same answer as the L^1-FT (6.53) a.e. For example, consider the rectangular pulse $x(t) = 1$ in $[-1, 1]$ and zero otherwise. This is in L^1 and L^2, and the FT using either definition is $X(\omega) = 2\sin\omega/\omega$. This answer is in L^2, but not in L^1. The inverse L^2-FT of $X(\omega)$ is the original $x(t)$.

3. If $x(t) \in L^2$ and $X(\omega) \in L^1$ then the Lebesgue integral $\int_{-\infty}^{\infty} X(\omega)e^{j\omega t} d\omega / 2\pi$ exists, and equals $x(t)$ a.e.

4. Parseval's relation holds, i.e., $\sqrt{2\pi}\|x(t)\|_2 = \|X(\omega)\|_2$. Thus, the FT is a linear transformation from L^2 to L^2, which preserves norms except the scale factor $\sqrt{2\pi}$. (Note that this would not make sense if $x(t)$ were only in L^1.) In particular, it is a bounded transformation because the norm $\|X(\omega)\|_2$ in the transform domain is bounded by the norm $\|x(t)\|_2$ in the original domain.

5. Unlike the L^1-FT, the L^2-FT $X(\omega)$ need not be continuous. For example, the impulse response of an ideal low-pass filter (sinc function) is in L^2 and its FT is not continuous.

6. Let $\{f_n(t)\}$ be a sequence in L^2 and let $x(t) = \sum_n c_n f_n(t)$ be a convergent summation (in the L^2 sense). With upper case letters denoting the L^2-FTs, $X(\omega) = \sum_n c_n F_n(\omega)$. This result is obvious for finite summations because of the linearity of the FT. For infinite summations this follows from the property that the L^2-FT is a continuous mapping from L^2 to L^2. (This in turn follows from the result that it is a bounded linear transformation). The continuity allows us to move the FT operation inside the infinite summation.

Thus, complete symmetry exists between the time and frequency domains. The L^2-FT is a one-to-one mapping from L^2 onto L^2. Moreover, because $\sqrt{2\pi}\|x(t)\|_2 = \|X(\omega)\|_2$, it is a norm preserving mapping—one says that the L^2-FT is an isometry from L^2 to L^2.

l^1 Fourier Transform

If a sequence $x(n) \in l^1$ its discrete-time FT $X(e^{j\omega}) = \sum_n x(n)e^{-j\omega n}$ exists, and is the l^1-FT of $x(n)$. It can be shown that $X(e^{j\omega})$ is a continuous function of ω and that $|X(e^{j\omega})|$ is bounded.

Convolutions

Suppose $h(t) \in L^1$ and $x(t) \in L^p$ for some p in $1 \le p \le \infty$. The familiar convolution integral defined by $(x * h)(t) = \int x(\tau)h(t - \tau)d\tau$ exists for almost all t [23]. If we define a function $y(t)$ to be $x * h$ where it exists and to be zero elsewhere, the result is, in fact, an L^p function. We simply say that the convolution of an L^1 function with an L^p function gives an L^p function. By recalling that an LTI system is stable (i.e., BIBO stable), iff its impulse response is in L^1, we have the following examples:

1. If an L^1 signal is input to a stable LTI system, the output is in L^1. Because the convolution of two L^1 signals is in L^1, the cascade of two stable LTI systems is stable, a readily accepted fact in engineering.

2. If an L^2 signal (finite energy input) is input to a stable LTI system, the output is in L^2.

3. If an L^∞ signal is input to a stable LTI system, the output is in L^∞ (i.e., bounded inputs produce bounded outputs).

If $x(t)$ and $h(t)$ are both in L^1, their convolution $y(t)$ is in L^1, and all three signals have L^1-FT. The **convolution theorem** [23] says that these three are related as $Y(\omega) = H(\omega)X(\omega)$. When signals are not necessarily in L^1 we cannot in general write this, even if convolution might itself be well defined.

Convolution Theorems for L^2 Signals

For all our discussions in the preceding sections, the signals were restricted to be in L^2, but not necessarily in L^1. In fact, even the filters are often only in L^2. For example, ideal bandpass filters (Fig. 6.8) are unstable, and therefore only in L^2. For arbitrary L^2 signals $x(t)$ and $h(t)$, the convolution theorem does not hold. We therefore need to better understand L^2 convolution.

Assume that $x(t)$ and $h(t)$ are both in L^2. Their convolution $y(t) = \int x(\tau)h(t - \tau)d\tau$ exists for all t, as the integral is only an inner product in L^2. Using Schwartz inequality [23], we also have $|y(t)| \le \|x(t)\|_2\|h(t)\|_2$, that is, $y(t) \in L^\infty$. Suppose the filter $h(t)$ has the further property that the frequency response $H(\omega)$ is bounded, i.e., $|H(\omega)| \le B$ a.e., for some $B < \infty$. Then we can show that $y(t) \in L^2$, and that the convolution theorem holds ($Y(\omega) = H(\omega)X(\omega)$). To prove this, note that

$$y(t) = \int x(\tau)h(t-\tau)d\tau = \frac{1}{2\pi} \int X(\omega)H(\omega)e^{j\omega t} d\omega \qquad (6.58)$$

from Parseval's relation which holds for L^2 signals [23]. If $|H(\omega)| \leq B$, then $|X(\omega)H(\omega)|^2 \leq B^2|X(\omega)|^2$. Therefore, $|X(\omega)H(\omega)|^2$ is bounded by the integrable function $|X(\omega)|^2$, and is therefore integrable. Thus, $X(\omega)H(\omega) \in L^2$, and the preceding equation establishes that $y(t) \in L^2$. The equation also shows that $y(t)$ and $H(\omega)X(\omega)$ form an L^2-FT pair, so $Y(\omega) = H(\omega)X(\omega)$.

Bounded L^2 Filters

Filters for which $h(t) \in L^2$ and $H(\omega)$ bounded are called **bounded L^2 filters.** The preceding discussion shows that bounded L^2 filters admit the convolution theorem, although arbitrary L^2 filters do not. Another advantage of bounded L^2 filters is that a cascade of two bounded L^2 filters, $h_1(t)$ and $h_2(t)$, is a bounded L^2 filter, just as a cascade of two stable filters would be stable. To see this, note that the cascaded impulse response is the convolution $h(t) = (h_1 * h_2)(t)$. By the preceding discussion, $h(t) \in L^2$, and moreover, $H(\omega) = H_1(\omega)H_2(\omega)$. Clearly, $H(\omega)$ is still bounded. Bounded L^2 filters are therefore very convenient to work with. Fortunately, all filters in the discussion of wavelets and filter banks are bounded L^2 filters, even though they may not be BIBO stable (as are the ideal bandpass filters in Fig. 6.8). We summarize the preceding discussions as follows.

THEOREM 6.3 *(Convolution of L^2 functions) We say that $h(t)$ is a bounded L^2 filter if $h(t) \in L^2$ and $|H(\omega)| \leq B < \infty$ a.e.*

1. *Let $x(t) \in L^2$, and let $h(t)$ be a bounded L^2 filter. Then $y(t) = (x * h)(t)$ exists for all t and $y(t) \in L^2$. Moreover, $Y(\omega) = H(\omega)X(\omega)$.*
2. *If $h_1(t)$ and $h_2(t)$ are bounded L^2 filters, then their cascade $h(t) = (h_1 * h_2)(t)$ is a bounded L^2 filter, and $H(\omega) = H_1(\omega)H_2(\omega)$.*

6.7 Riesz Basis, Biorthogonality, and Other Fine Points

In a finite dimensional space, such as the space of all N-component Euclidean vectors, the ideas of basis and orthonormal basis are easy to appreciate. When we extend these ideas to infinite dimensional spaces (i.e., where the basis $\{g_n(t)\}$ has infinite number of functions), a number of complications and subtleties arise. Our aim is to point these out. References for this section include [5, 15], and [24].

Readers familiar with **Hilbert spaces** will note that the L^2 space is a Hilbert space; all our developments here are valid for any Hilbert space $\mathcal{H}$. Elements in $\mathcal{H}$ (vectors) are typically denoted x, y, etc. When we deal with the Hilbert space L^2, the vectors are functions and are denoted as $x(t)$, $y(t)$, etc. for clarity. Similarly, for the special case of Euclidean vectors we use boldface, e.g., **x**, **y**, etc. The reader not familiar with Hilbert spaces can assume that all discussions are in L^2 and that x is merely a simplification of the notation $x(t)$.

Finite Dimensional Vector Spaces

We first look at the finite dimensional case and then proceed to the infinite dimensional case. Consider an $N \times N$ matrix $\mathbf{F} = [\mathbf{f}_1 \ \mathbf{f}_2 \ \dots \ \mathbf{f}_N]$. We assume that this is nonsingular, that is, the columns $\mathbf{f}_n$ are linearly independent. These column vectors form a basis for the N-dimensional Euclidean space $\mathcal{C}^N$ of complex N-component vectors. This space is an example of a finite dimensional Hilbert space, with inner product defined as $\langle \mathbf{x}, \mathbf{y} \rangle = \mathbf{y}^\dagger \mathbf{x} = \sum_{n=1}^{N} x_n y_n^*$. The norm $\|\mathbf{x}\|$ induced by this inner product is defined as $\|\mathbf{x}\| = \sqrt{\langle \mathbf{x}, \mathbf{x} \rangle}$. Thus $\|\mathbf{x}\|^2 = \mathbf{x}^\dagger \mathbf{x} = \sum_{n=1}^{N} |x_n|^2$.

Any vector $\mathbf{x} \in \mathcal{C}^N$ can be expressed as $\mathbf{x} = \sum_{n=1}^{N} c_n \mathbf{f}_n$ for some uniquely determined set of scalars c_n. We can abbreviate this as $\mathbf{x} = \mathbf{Fc}$, where $\mathbf{c} = [c_1 \ c_2 \ \ldots \ c_N]^T$. The matrix $\mathbf{F}$ can be regarded as a linear transformation from $\mathcal{C}^N$ to $\mathcal{C}^N$. The nonsingularity of $\mathbf{F}$ means that for every $\mathbf{x} \in \mathcal{C}^N$ we can find a unique $\mathbf{c}$ such that $\mathbf{x} = \mathbf{Fc}$.

Boundedness of F and Its Inverse

In practice we have a further requirement that if the norm $\|\mathbf{c}\|$ is "small" then $\|\mathbf{x}\|$ should also be "small," and vice versa. This requirement implies, for example, that if a small error occurs in the transmission or estimate of the vector $\mathbf{c}$, the corresponding error in $\mathbf{x}$ is also small. From the relation $\mathbf{x} = \mathbf{Fc}$ we obtain

$$\|\mathbf{x}\|^2 = \mathbf{x}^\dagger \mathbf{x} = \mathbf{c}^\dagger \mathbf{F}^\dagger \mathbf{Fc} \tag{6.59}$$

Letting λ_M and λ_m denote the maximum and minimum eigenvalues of $\mathbf{F}^\dagger \mathbf{F}$ it then follows that $\|\mathbf{x}\|^2 \geq \lambda_m \|\mathbf{c}\|^2$ and that $\|\mathbf{x}\|^2 \leq \lambda_M \|\mathbf{c}\|^2$. That is,

$$\lambda_m \|\mathbf{c}\|^2 \leq \|\mathbf{x}\|^2 \leq \lambda_M \|\mathbf{c}\|^2 \tag{6.60}$$

with $0 < \lambda_m \leq \lambda_M < \infty$, where $0 < \lambda_m$ follows from the nonsingularity of $\mathbf{F}$. Thus, the transformation $\mathbf{F}$, which converts $\mathbf{c}$ into $\mathbf{x}$, has an amplification factor bounded by λ_M in the sense that $\|\mathbf{x}\|^2 \leq \lambda_M \|\mathbf{c}\|^2$. Similarly, the inverse transformation $\mathbf{G} = \mathbf{F}^{-1}$, which converts $\mathbf{x}$ into $\mathbf{c}$, has amplification bounded by $1/\lambda_m$. Because λ_M is finite, we say that $\mathbf{F}$ is a bounded linear transformation, and because $\lambda_m \neq 0$ we see that the inverse transformation is also bounded.

Using $\mathbf{x} = \sum_n c_n \mathbf{f}_n$ and $\|\mathbf{c}\|^2 = \sum_n |c_n|^2$ we can rewrite the preceding inequality as

$$A \sum_n |c_n|^2 \leq \left\| \sum_n c_n \mathbf{f}_n \right\|^2 \leq B \sum_n |c_n|^2 \tag{6.61}$$

where $A = \lambda_m > 0$ and $B = \lambda_M < \infty$, and all summations are for $1 \leq n \leq N$. Readers familiar with the idea of a Riesz basis in infinite dimensional Hilbert spaces will notice that the above is in the form that agrees with that definition. We will return to this issue later.

Biorthogonality

With $\mathbf{F}^{-1}$ denoted as $\mathbf{G}$, let $\mathbf{g}_n^\dagger$ denote the rows of $\mathbf{G}$:

$$\mathbf{G} = \begin{bmatrix} \mathbf{g}_1^\dagger \\ \mathbf{g}_2^\dagger \\ \vdots \\ \mathbf{g}_N^\dagger \end{bmatrix}, \quad \mathbf{F} = \begin{bmatrix} \mathbf{f}_1 & \mathbf{f}_2 & \ldots & \mathbf{f}_N \end{bmatrix} \tag{6.62}$$

The property $\mathbf{GF} = \mathbf{I}$ implies $\mathbf{g}_k^\dagger \mathbf{f}_n = \delta(k-n)$:

$$\langle \mathbf{f}_n, \mathbf{g}_k \rangle = \delta(k-n) \tag{6.63}$$

for $1 \leq k, n \leq N$. Equivalently, $\langle \mathbf{g}_k, \mathbf{f}_n \rangle = \delta(k-n)$.

Two sets of vectors, $\{\mathbf{f}_n\}$ and $\{\mathbf{g}_k\}$, satisfying (6.63) are said to be **biorthogonal**. Because $\mathbf{c} = \mathbf{F}^{-1}\mathbf{x} = \mathbf{Gx}$ we can write the elements of $\mathbf{c}$ as $c_n = \mathbf{g}_n^\dagger \mathbf{x} = \langle \mathbf{x}, \mathbf{g}_n \rangle$. Then $\mathbf{x} = \sum_n c_n \mathbf{f}_n = \sum_n \langle \mathbf{x}, \mathbf{g}_n \rangle \mathbf{f}_n$. Next, $\mathbf{G}^\dagger$ is a nonsingular matrix, therefore, we can use its columns $\mathbf{g}_n$, instead of the columns of $\mathbf{F}$, to obtain a similar development, and express the arbitrary vector $\mathbf{x} \in \mathcal{C}^N$ as $\mathbf{x} = \sum_n \langle \mathbf{x}, \mathbf{f}_n \rangle \mathbf{g}_n$. Summarizing, we have

$$\mathbf{x} = \sum_n \langle \mathbf{x}, \mathbf{g}_n \rangle \mathbf{f}_n = \sum_n \langle \mathbf{x}, \mathbf{f}_n \rangle \mathbf{g}_n \qquad (6.64)$$

where the summations are for $1 \le n \le N$. By using the expressions $c_n = \langle \mathbf{x}, \mathbf{g}_n \rangle$ and $\mathbf{x} = \sum_n c_n \mathbf{f}_n$ we can rearrange the inequality (6.61) into $B^{-1} \|\mathbf{x}\|^2 \le \sum_n |\langle \mathbf{x}, \mathbf{g}_n \rangle|^2 \le A^{-1} \|\mathbf{x}\|^2$. With the columns $\mathbf{g}_n$ of $\mathbf{G}^\dagger$, rather than the columns of $\mathbf{F}$, used as the basis for $\mathcal{C}^N$ we obtain similarly

$$A\|\mathbf{x}\|^2 \le \sum_n |\langle \mathbf{x}, \mathbf{f}_n \rangle|^2 \le B\|\mathbf{x}\|^2 \qquad (6.65)$$

where $1 \le n \le N$, and $A = \lambda_m$, $B = \lambda_M$ again. Readers familiar with the idea of a frame in an infinite dimensional Hilbert space will recognize that the above inequality defines a frame $\{\mathbf{f}_n\}$.

Orthonormality

The basis $\mathbf{f}_n$ is said to be orthonormal if $\langle \mathbf{f}_k, \mathbf{f}_n \rangle = \delta(k-n)$, i.e., $\mathbf{f}_n^\dagger \mathbf{f}_k = \delta(k-n)$. Equivalently, $\mathbf{F}$ is unitary, i.e., $\mathbf{F}^\dagger \mathbf{F} = \mathbf{I}$. In this case the rows of the inverse matrix $\mathbf{G}$ are the quantities $\mathbf{f}_n^\dagger$. But $\mathbf{F}^\dagger \mathbf{F} = \mathbf{I}$, and we have $\lambda_m = \lambda_M = 1$, or $A = B = 1$. With this, (6.60) becomes $\|\mathbf{c}\| = \|\mathbf{x}\|$, or $\sum_n |c_n|^2 = \left\| \sum_n c_n \mathbf{f}_n \right\|^2$. This shows that equation (6.61) is a generalization of the orthonormal situation. Similarly, biorthogonality (6.63) is a generalization of orthonormality.

Basis in Infinite Dimensional Spaces

When the simple idea of a basis in a finite dimensional space (e.g., the Euclidean space $\mathcal{C}^N$) is extended to infinite dimensions, several new issues arise which make the problem nontrivial. Consider the sequence of functions $\{f_n\}$, $1 \le n \le \infty$ in a Hilbert space $\mathcal{H}$. Because of the infinite range of n, linear combinations of the form $\sum_{n=1}^\infty c_n f_n$ must now be considered. The problem that immediately arises is one of convergence. For arbitrary sequences c_n this sum does not converge, so the statement "all linear combinations" must be replaced with something else.[10]

Closure of Span

First define the set of all *finite* linear combinations of the form $\sum_{n=1}^N c_n f_n$, where N varies over all integers ≥ 1. This is called the **span of** $\{f_n\}$. Now suppose $x \in \mathcal{H}$ is a vector not necessarily in the span of $\{f_n\}$, but can be approximated as closely as we wish by vectors in the span. In other words, given an $\epsilon > 0$ we can find N and the sequence of constants c_{nN} such that

$$\left\| x - \sum_{n=1}^N c_{nN} f_n \right\| < \epsilon \qquad (6.66)$$

where $\|x\|$ is the norm defined as $\|x\| = \sqrt{\langle x, x \rangle}$. If we append all such vectors x to the span of $\{f_n\}$ we get the **closure of the span** of $\{f_n\}$.[11] Note that c_{nN} in general depends on ϵ because N depends on ϵ.

---

[10] Our review uses $1 \le n \le \infty$ to be consistent with standard math texts, but all the crucial results hold for doubly infinite sequences and summations, i.e., for the case $-\infty \le n \le \infty$. This is what we need in the case of Fourier and wavelet bases; see for example, (6.3) and (6.4).

[11] The term "closure" has its origin from the theory of metric spaces, more generally, topological vector spaces. We do not require the deeper, more general meaning here.

Completeness

A sequence of vectors $\{f_n\}$ is said to be **complete** in $\mathcal{H}$ if the closure of the linear span of $\{f_n\}$ equals $\mathcal{H}$. Therefore, any $x \in \mathcal{H}$ can be approximated, as closely as we wish, by finite linear combinations of f_n in the sense of (6.66). This is also expressed by saying that the linear span of $\{f_n\}$ is **dense** in $\mathcal{H}$. Completeness of $\{f_n\}$ in a Hilbert space is equivalent to the statement that the only vector orthogonal to all f_n is the zero vector.

Infinite Summations

When we write $x = \sum_{n=1}^{\infty} c_n f_n$ we mean that the infinite summation converges to x in the norm of $\mathcal{H}$. In other words, given $\epsilon > 0$, there exists n_0 such that

$$\left\| x - \sum_{n=1}^{N} c_n f_n \right\| < \epsilon \qquad \text{for all } N \geq n_0 \tag{6.67}$$

This statement is stronger than saying that x is in the closure of the linear span of $\{f_n\}$. The latter statement only requires (6.66), where N and hence c_{nN} depends on ϵ. In (6.67) $\{c_n\}$ is a fixed sequence.

Linear Independence

Let $\{f_n\}, n = 1, 2, \ldots$ be a sequence of vectors in an infinite dimensional Hilbert space $\mathcal{H}$. Unlike in a finite dimensional space, one must distinguish between several types of linear independence.

Type 1: $\{f_n\}$ has *finite linear independence* if $\sum_{n=1}^{N} c_n f_n = 0$, for any finite N implies $c_n = 0, 1 \leq n \leq N$.

Type 2: $\{f_n\}$ is *ω-independent* if $\sum_{n=1}^{\infty} c_n f_n = 0$ implies $c_n = 0$ for all n (where the infinite sum is interpreted as explained above).

Type 3: $\{f_n\}$ is *minimal* if none of the f_m is in the closure of the span of the remaining set of f_n.

Type 3 independence implies type 2, which in turn implies type 1. Thus, type 3 is the strongest kind of linear independence. The reason that it is stronger than type 2 is that type 2 implies we cannot have $f_m = \sum_{n \neq m} c_n f_n$. However, for a type 2 independent sequence $\{f_n\}$, it is possible that we can make

$$\left\| f_m - \sum_{\substack{n=1 \\ n \neq m}}^{N} c_{nN} f_n \right\| < \epsilon \tag{6.68}$$

for any given $\epsilon > 0$ by choosing N and c_{nN} properly.[12] Type 3 linear independence prohibits even this. The comments following Example 6.3 make this distinction more clear. (see p. 174).

Basis or Schauder Basis

A sequence of vectors $\{f_n\}$ in $\mathcal{H}$ is a Schauder basis for $\mathcal{H}$ if any $x \in \mathcal{H}$ can be expressed as $x = \sum_{n=1}^{\infty} c_n f_n$, and the sequence of scalars $\{c_n\}$ is unique for a given x. The second condition can

---

[12] As we make ϵ increasingly smaller, we may need to change N and *all coefficients* c_{kN}. Therefore, this does not imply $f_m = \sum_{n \neq m} c_n f_n$ for fixed $\{c_n\}$.

be replaced with the statement that $\{f_n\}$ is ω-independent. A subtle result for Hilbert spaces [24] is that a Schauder basis automatically satisfies minimality (i.e., type 3 independence).

A Schauder basis is ω-independent and complete in the sense defined above. Conversely, ω-independence and completeness do not imply that $\{f_n\}$ is a Schauder basis; completeness only means that we may approximate any vector as closely as we wish in the sense of (6.66), where c_{kN} *depend on N*. In this chapter "independence" (or linear independence) stands for ω-independence. Similarly "basis" stands for Schauder basis unless qualified otherwise.

Riesz Basis

Any basis $\{\mathbf{f}_n\}$ in a finite dimensional space satisfies (6.61), which in turn ensures that the transformation from $\mathbf{x}$ to $\{c_n\}$ and that from $\{c_n\}$ to $\mathbf{x}$ are stable. For a basis in an infinite dimensional space, (6.61) is not automatically guaranteed, as shown by the following example.

EXAMPLE 6.2:

Let $\{e_n\}$, $1 \leq n \leq \infty$ be an orthonormal basis in a Hilbert space $\mathcal{H}$, and define the sequence $\{f_n\}$ by $f_n = e_n/n$. Then we can show that f_n is still a basis, i.e., it satisfies the definition of a Schauder basis. Suppose we pick $x = \epsilon e_k$ for some k. Then, $x = \sum_n c_n f_n$, with $c_k = \epsilon k$, and $c_n = 0$ for all other n. Thus, $\sum_n |c_n|^2 = \epsilon^2 k^2$ and grows as k increases, although $\|x\| = \epsilon$ for all k. That is, a "small" error in x can become amplified in an unbounded manner. Recall that this could never happen in the finite dimensional case because $A > 0$ in (6.61). For our basis $\{f_n\}$, we can indeed show that no $A > 0$ satisfies (6.61). To see this let $c_n = 0$ for all n, except that $c_k = 1$. Then, $\sum_n c_n f_n = f_k = e_k/k$ and has norm $1/k$. So (6.61) reads $A \leq 1/k^2 \leq B$ for all $k \geq 1$. This is not possible with $A > 0$.

If $\{e_n\}$, $1 \leq n \leq \infty$ is an orthonormal basis in an infinite dimensional Hilbert space $\mathcal{H}$, then any vector $x \in \mathcal{H}$ can be expressed uniquely as $x = \sum_{n=1}^{\infty} c_n e_n$ where

$$\|x\|^2 = \sum_{n=1}^{\infty} |c_n|^2.$$

This property automatically ensures the stability of the transformations from x to $\{c_n\}$ and vice versa. The Riesz basis is defined such that this property is made more general.[13]

DEFINITION 6.1 **Definition of a Riesz basis.** A sequence $\{f_n\}$, $1 \leq n \leq \infty$ in a Hilbert space $\mathcal{H}$ is a Riesz basis if it is *complete* and constants A and B exist such that $0 < A \leq B < \infty$ and

$$A \sum_{n=1}^{\infty} |c_n|^2 \leq \left\| \sum_{n=1}^{\infty} c_n f_n \right\|^2 \leq B \sum_{n=1}^{\infty} |c_n|^2 \qquad (6.69)$$

for all choice of c_n satisfying $\sum_n |c_n|^2 < \infty$.

In a finite dimensional Hilbert space, A and B come from the extreme eigenvalues of a nonsingular matrix $\mathbf{F}^\dagger \mathbf{F}$, so automatically $A > 0$ and $B < \infty$. In other words, any basis in a finite dimensional space is a Riesz basis. As Example 6.2 shows, this may not be the case in infinite dimensions.

---

[13]For readers familiar with bounded linear transformations in Hilbert spaces, we state that a basis is a Riesz basis iff it is related to an orthonormal basis via a bounded linear transformation with bounded inverse.

Unconditional Basis

It can be shown that the Riesz basis is an **unconditional basis**, that is, any reordering of $\{f_n\}$ is also a basis (and the new c_n are the correspondingly reordered versions). This is a nontrivial statement; an arbitrary (Schauder) basis is not necessarily unconditional. In fact, the space of L^1 functions (which is a Banach space, not a Hilbert space) does not have an unconditional basis.

Role of the Constants A and B

1. *Strongest linear independence.* The condition $A > 0$ means, in particular, that $\sum_n c_n f_n \neq 0$ unless c_n is zero for all n. This is just ω-independence. Actually the condition $A > 0$ means that the vectors $\{f_n\}$ are independent in the strongest sense (type 3), that is, $\{f_n\}$ is minimal. To see this, assume this is not the case by supposing some vector f_m is in the closure of the span of the others. Then, given arbitrary $\epsilon > 0$ we can find N and c_{nN} satisfying (6.66) with $x = f_m$. Defining $c_n = -c_{nN}$ for $n \neq m$ and $c_m = 1$, (6.69) implies $A\left(1 + \sum_{n \neq m} |c_{nN}|^2\right) \leq \epsilon^2$. Because ϵ is arbitrary, this is not possible for $A > 0$.

2. *Distance between vectors.* The condition $A > 0$ also implies that no two vectors in $\{f_n\}$ can become "arbitrarily close." To see this, choose $c_k = -c_m = 1$ for some k, m and $c_n = 0$ for all other n. Then (6.69) gives $2A \leq \|f_k - f_m\|^2 \leq 2B$. Thus, the distance between any two vectors is at least $\sqrt{2A}$, at most $\sqrt{2B}$.

3. *Bounded basis.* A Riesz basis is a bounded basis in the sense that $\|f_n\|$ cannot get arbitrarily large. In fact, by choosing $c_n = 0$ for all but one value of n, we can see that $0 < A \leq \|f_n\|^2 \leq B < \infty$. That is, the norms of the vectors in the basis cannot become arbitrarily small or large. Note that the basis in Example 6.2 violates this, because $\|f_n\| = 1/n$. Therefore, Example 6.2 is only a Schauder basis and not a Riesz basis.

4. *Stability of basis.* The condition $A > 0$ yields $\sum_n |c_n|^2 \leq A^{-1}\|x\|^2$, where $x = \sum_n c_n f_n$. This means that the transformation from the vector x to the sequence $\{c_n\}$ is bounded, so a small error in x is not amplified in an unbounded manner. Similarly, the inequality $\|x\|^2 \leq B \sum_n |c_n|^2$ shows that the role of B is to ensure that the inverse transformation from c_n to x is bounded. Summarizing, the transformation from x to $\{c_n\}$ is numerically stable (i.e., small errors not severely amplified) because $A > 0$, and the reconstruction of x from $\{c_n\}$ is numerically stable because $B < \infty$.

5. *Orthonormality.* For a Riesz basis with $A = B = 1$ the condition (6.69) reduces to $\sum_n |c_n|^2 = \|\sum_n c_n f_n\|^2$. It can be shown that such a Riesz basis is simply an orthonormal basis. The properties listed above show that the Riesz basis is as good as an orthonormal basis in most applications. Any Riesz basis can be obtained from an orthonormal basis by means of a bounded linear transformation with a bounded linear inverse.

EXAMPLE 6.3: Mishaps with a System That Is Not a Riesz Basis.

Let us modify Example 6.2 to $f_n = (e_n/n) + e_1$, $n \geq 1$, where $\{e_n\}$ is an orthonormal basis. As $n \to \infty$ the vectors f_n move arbitrarily closer together (although $\|f_n\|$ approaches unity from above). Formally, $f_n - f_m = (e_n/n) - (e_m/m)$, so $\|f_n - f_m\|^2 = (1/n^2) + (1/m^2)$, which goes to zero as $n, m \to \infty$. Thus there does not exist $A > 0$ satisfying (6.69) (because of comment 2 above). This, then, is not a Riesz basis; in fact, this is not even a Schauder basis (see below). This example also has $B = \infty$. To see this let $c_n = 1/n$, then $\sum_n |c_n|^2$ converges, but $\|\sum_{n=1}^{N} c_n f_n\|^2$ does not converge as $N \to \infty$ (as we can verify), so (6.69) is not satisfied for finite B. *Such mishaps cannot occur with a Riesz basis.*

In this example $\{f_n\}$ is not minimal (which is type 3 independence). Note that $\|f_1 - f_n\|$ gets arbitrarily small as n increases to infinity, therefore, f_1 is in the closure of the span of $\{f_n\}$, $n \neq 1$. However, $\{f_n\}$ is ω-independent; no sequence $\{c_n\}$ exists such that $\|\sum_{n=1}^{N} c_n f_n\| \to 0$ as $N \to \infty$. In any case, the fact that $\{f_n\}$ is not minimal (i.e., not independent in the strongest sense) shows that it is not even a Schauder basis.

Biorthogonal Systems, Riesz Bases, and Inner Products

When discussing finite dimensional Hilbert spaces we found that given a basis $\mathbf{f}_n$ (columns of a nonsingular matrix) we can express any vector $\mathbf{x}$ as a linear combination $\mathbf{x} = \sum_n \langle \mathbf{x}, \mathbf{g}_n \rangle \mathbf{f}_n$, where $\mathbf{g}_n$ is such that the biorthogonality property $\langle \mathbf{f}_m, \mathbf{g}_n \rangle = \delta(m - n)$ holds. A similar result is true for infinite dimensional Hilbert spaces.

THEOREM 6.4 *(Biorthogonality and Riesz Basis) Let $\{f_n\}$ be a basis in a Hilbert space $\mathcal{H}$. Then there exists a unique sequence $\{g_n\}$ biorthogonal to $\{f_n\}$, that is,*

$$\langle f_m, g_n \rangle = \delta(m - n) \quad \text{(biorthogonality)} \tag{6.70}$$

Moreover, the unique expansion of any $x \in \mathcal{H}$ in terms of the basis $\{f_n\}$ is given by

$$x = \sum_{n=1}^{\infty} \langle x, g_n \rangle f_n \tag{6.71}$$

It is also true that the biorthogonal sequence $\{g_n\}$ is a basis and that $x = \sum_{n=1}^{\infty} \langle x, f_n \rangle g_n$. Moreover, if $\{f_n\}$ is a Riesz basis, then $\sum_n |\langle x, g_n \rangle|^2$ and $\sum_n |\langle x, f_n \rangle|^2$ are finite, and we have

$$A\|x\|^2 \leq \sum_{n=1}^{\infty} |\langle x, f_n \rangle|^2 \leq B\|x\|^2 \tag{6.72}$$

where A and B are the same constants as in the definition (6.69) of a Riesz basis.

This beautiful result resembles the finite dimensional version, where f_n corresponds to the column of a matrix and g_n corresponds to the rows (conjugated) of the inverse matrix. In this sense we can regard the biorthogonal pair of sequences $\{f_n\}$, $\{g_n\}$ as inverses of each other. Both are bases for $\mathcal{H}$. A proof of the above result can be obtained by combining the ideas on pp. 28 to 32 of [24]. The theorem implies, in particular, that if $\{f_n\}$ is a Riesz basis, then any vector in the space can be written in the form $\sum_{n=1}^{\infty} c_n f_n$, where $c_n \in l^2$.

Summary of Riesz Basis

The Riesz basis $\{f_n\}$ in a Hilbert space is a complete set of vectors, linearly independent in the strongest sense (i.e., type 3 or minimal). It is a bounded basis with bounded inverse. Any two vectors are separated by at least $\sqrt{2A}$, that is, $\|f_n - f_m\|^2 \geq 2A$. The norm of each basis vector is bounded as $\|f_n\| \leq \sqrt{B}$. In the expression $x = \sum_n c_n f_n$ the computation of x from c_n as well as the computation of c_n from x are numerically stable because $B < \infty$ and $A > 0$, respectively. A Riesz basis with $A = B = 1$ is an orthonormal basis. In fact, any Riesz basis can be obtained from an orthonormal basis via a bounded linear transformation with a bounded inverse. Given any basis $\{f_n\}$ in a Hilbert space, a unique biorthogonal sequence $\{g_n\}$ exists such that we can express any $x \in \mathcal{H}$ as $x = \sum_{n=1}^{\infty} \langle x, g_n \rangle f_n$ as well as $x = \sum_{n=1}^{\infty} \langle x, f_n \rangle g_n$; if this basis is also a Riesz basis then $\sum_n |\langle x, f_n \rangle|^2$ and $\sum_n |\langle x, g_n \rangle|^2$ are finite. If $\{f_n\}$ is a Riesz basis, then any vector $x \in \mathcal{H}$ can be written in the form $x = \sum_{n=1}^{\infty} c_n f_n$, where $c_n \in l^2$.

6.8   Frames in Hilbert Spaces

A frame in a Hilbert space $\mathcal{H}$ is a sequence of vectors $\{f_n\}$ with certain special properties. While a frame is not necessarily a basis, it shares some properties of a basis. For example, we can express any vector $x \in \mathcal{H}$ as a linear combination of the frame elements, i.e., $x = \sum_n c_n f_n$. However, frames generally have redundancy—the frame vectors are not necessarily linearly independent, even in the weakest sense defined in Section 6.7. The Riesz basis (hence, any orthonormal basis) is a special case of frames. The concept of a frame is useful when discussing the relation between wavelets, STFTs, and filter banks. Frames were introduced by Duffin and Schaeffer [25], and used in the context of wavelets and STFT by Daubechies [5]. Excellent tutorials can be found in [12] and [24].

Definition of a Frame

A sequence of vectors $\{f_n\}$ in a (possibly infinite dimensional) Hilbert space $\mathcal{H}$ is a frame if there exist constants A and B with $0 < A \leq B < \infty$ such that for any $x \in \mathcal{H}$ we have

$$A\|x\|^2 \leq \sum_{n=1}^{\infty} |\langle x, f_n \rangle|^2 \leq B\|x\|^2 \tag{6.73}$$

The constants A and B are called frame bounds.

In Section 6.7 we saw that a Riesz basis, which by definition satisfies (6.69), also satisfies (6.72), which is precisely the frame definition. A Riesz basis is, therefore, also a frame, but it is a special case of a frame, where the set of vectors is minimal.

Any frame is complete. That is, if a vector $x \in \mathcal{H}$ is orthogonal to all elements in $\{f_n\}$, then $x = 0$, otherwise $A > 0$ is violated. Thus any $x \in \mathcal{H}$ is in the closure of the span of the frame. In fact, we will see that more is true; for example, we can express $x = \sum c_n f_n$, although $\{c_n\}$ may not be unique. The frame elements are not necessarily linearly independent, as demonstrated by examples below. A frame, then, is not necessarily a basis. Compare (6.73) to the Riesz basis definition (6.69), where the left inequality forced the vectors f_n to be linearly independent (in fact, minimal). The left inequality for a frame only ensures completeness, not linearindependence.

Representing Arbitrary Vectors in Terms of Frame Elements

We will see that, given a frame $\{f_n\}$ we can associate with it another sequence $\{g_n\}$ called the dual frame, such that any element $x \in \mathcal{H}$ can be represented as $x = \sum_{n=1}^{\infty} \langle x, g_n \rangle f_n$. We also can write $x = \sum_{n=1}^{\infty} \langle x, f_n \rangle g_n$. This representation in terms of $\{f_n\}$ and $\{g_n\}$ resembles the biorthogonal system discussed in Section 6.7, but some differences are pointed out later.

Stability of Computations

To obtain the representation $x = \sum_{n=1}^{\infty} \langle x, f_n \rangle g_n$ we compute (at least conceptually) the coefficients $\langle x, f_n \rangle$ for all n. This computation is a linear transformation from $\mathcal{H}$ to the space of sequences. The inverse transform computes x from this sequence by using the formula $x = \sum_{n=1}^{\infty} \langle x, f_n \rangle g_n$. The condition $B < \infty$ in the frame definition ensures that the transformation from x to $\langle x, f_n \rangle$ is bounded. Similarly, the condition $A > 0$ ensures that the inverse transformation from $\langle x, f_n \rangle$ to x is bounded. The conditions $A > 0$ and $B < \infty$, therefore, ensure stability; small errors in one domain are not arbitrarily amplified in the other domain. A similar advantage was pointed out earlier for the Riesz basis—for arbitrary bases in infinite dimensional spaces such an advantage cannot be claimed (Example 6.4).

If we wish to use the dual representation $x = \sum_{n=1}^{\infty} \langle x, g_n \rangle f_n$ instead of $x = \sum_{n=1}^{\infty} \langle x, f_n \rangle g_n$ we must compute $\langle x, g_n \rangle$, etc.; the roles of A and B are taken up by $1/B$ and $1/A$, respectively, and similar discussions hold. This is summarized in Fig. 6.29.

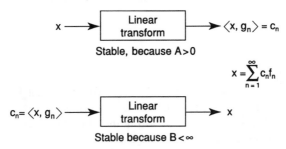

6.29 Representation of x using frame elements $\{f_n\}$. The transformation from x to $\{c_n\}$ and vice versa are stable.

Exact Frames, Tight Frames, Riesz Bases, and Orthonormal Bases

The resemblance between a Riesz basis and a frame is striking. Compare (6.69) to (6.73). One might wonder what the precise relation is. Thus far, we know that a Riesz basis is a frame. To go deeper, we need a definition: a frame $\{f_n\}$ which ceases to be a frame if any element f_k is deleted is said to be an **exact frame**. Such a frame has no redundancy. A frame with $A = B$ is said to be a **tight frame**. The defining property reduces to $\|x\|^2 = A^{-1} \sum_n |\langle x, f_n \rangle|^2$, resembling Parseval's theorem for an orthonormal basis. A frame is *normalized* if $\|f_n\| = 1$ for all n. The following facts concerning exact frames and tight frames are fundamental:

1. A tight frame with $A = B = 1$ and $\|f_n\| = 1$ for all n (i.e., a normalized tight frame with frame bound $= 1$) is an orthonormal basis [5].
2. $\{f_n\}$ is an exact frame iff it is a Riesz basis [24]. Moreover, if a frame is not exact then it cannot be a basis [12]. *Thus, if a frame is a basis it is certainly a Riesz basis.*
3. Because an orthonormal basis is a Riesz basis, a normalized tight frame with frame bound equal to 1 is automatically an exact frame.

Some examples follow which serve to clarify the preceding concepts and definitions. In these examples the sequence $\{e_n\}$, $n \geq 1$ is an orthonormal basis for $\mathcal{H}$. Thus, $\{e_n\}$ is a tight frame with $A = B = 1$, and $\|e_n\| = 1$.

EXAMPLE 6.4:

Let $f_n = e_n/n$ as in Example 6.2. Then $\{f_n\}$ is still a (Schauder) basis for $\mathcal{H}$, but it is not a frame. In fact, this satisfies (6.73) only with $A = 0$; i.e., the inverse transformation (reconstruction) from $\langle x, f_n \rangle$ to x is not bounded. To see why $A = 0$, note that if we let $x = e_k$ for some $k > 0$ then $\|x\| = 1$, whereas $\sum_n |\langle x, f_n \rangle|^2 = 1/k^2$. The first inequality in the frame definition becomes $A \leq 1/k^2$, which cannot be satisfied for all k unless $A = 0$. In this example a finite B works because $|\langle x, f_n \rangle| = |\langle x, e_n \rangle|/n$ for each n. Therefore, $\sum |\langle x, f_n \rangle|^2 \leq \sum |\langle x, e_n \rangle|^2 = \|x\|^2$.

EXAMPLE 6.5:

Suppose we modify the above example as follows: define $f_n = (e_n/n) + e_1$. We know that this is no longer a basis (Example 6.3). We now have $B = \infty$ in the frame definition, so this is not a frame. To verify this, let $x = e_1$ so $\|x\| = 1$. Then $\langle x, f_n \rangle = 1$ for all $n > 1$, so $\sum_n |\langle x, f_n \rangle|^2$ does not converge to a finite value.

EXAMPLE 6.6:

Consider the sequence of vectors $\{e_1, e_1, e_2, e_2, \ldots\}$ This is a tight frame with frame bounds $A = B = 2$. Note that even though the vectors are normalized and the frame is tight this is not an orthonormal basis. This has a redundancy of two in the sense that each vector is repeated twice. This frame is not even a basis, therefore, it is not a Riesz basis.

EXAMPLE 6.7:

Consider the sequence of vectors $\{e_1, (e_2/\sqrt{2}), (e_2/\sqrt{2}), (e_3/\sqrt{3}), (e_3/\sqrt{3}), (e_3/\sqrt{3}), \ldots\}$. Again, redundancy occurs, so it is not a basis. It is a tight frame with $A = B = 1$, but not an exact frame, and clearly not a basis. It has redundancy (repeated vectors).

Frame Bounds and Redundancy

For a tight frame with unit norm vectors f_n, the frame bound measures the redundancy. In Example 6.6 the redundancy is two (every vector repeated twice), and indeed $A = B = 2$. In Example 6.7, where we still have redundancy, the frame bound $A = B = 1$ does not indicate it. The frame bound of a tight frame measures redundancy only if the vectors f_n have unit norm as in Example 6.6.

Frame Operator, Dual Frame, and Biorthogonality

The frame operator $\mathcal{F}$ associated with a frame $\{f_n\}$ in a Hilbert space $\mathcal{H}$ is a linear operator defined as

$$\mathcal{F}x = \sum_{n=1}^{\infty} \langle x, f_n \rangle f_n \tag{6.74}$$

The summation can be shown to be convergent by using the definition of the frame. The frame operator $\mathcal{F}$ takes a vector $x \in \mathcal{H}$ and produces another vector in $\mathcal{H}$. The norm of $\mathcal{F}x$ is bounded as follows:

$$A\|x\| \le \|\mathcal{F}x\| \le B\|x\| \tag{6.75}$$

The frame operator is a **bounded linear operator** (because $B < \infty$; hence it is a continuous operator [12]. Its inverse is also a bounded linear operator because $A > 0$).

From (6.74) we obtain $\langle \mathcal{F}x, x \rangle = \sum_n |\langle x, f_n \rangle|^2$ by interchanging the inner product with the infinite summation. This is permitted by the continuity of the operator $\mathcal{F}$ and the continuity of inner products; (see Section 6.6). Because $\{f_n\}$ is complete, the right-hand side is positive for $x \ne 0$. Thus, $\langle \mathcal{F}x, x \rangle > 0$ unless $x = 0$, that is, $\mathcal{F}$ is a positive definite operator. The realness of $\langle \mathcal{F}x, x \rangle$ also means that $\mathcal{F}$ is self-adjoint, or $\langle \mathcal{F}x, y \rangle = \langle x, \mathcal{F}y \rangle$ for any $x, y \in \mathcal{H}$.

The importance of the frame operator arises from the fact that if we define $g_n = \mathcal{F}^{-1} f_n$, any $x \in \mathcal{H}$ can be expressed as

$$x = \sum_{n=1}^{\infty} \langle x, g_n \rangle f_n = \sum_{n=1}^{\infty} \langle x, f_n \rangle g_n \qquad (6.76)$$

The sequence $\{g_n\}$ is itself a frame in $\mathcal{H}$ called the **dual frame**. It has frame bounds B^{-1} and A^{-1}. Among all representations of the form $x = \sum_n c_n f_n$, the representation $x = \sum_n \langle x, g_n \rangle f_n$ possesses the special property that the energy of the coefficients is minimized, i.e., $\sum_n |\langle x, g_n \rangle|^2 \le \sum_n |c_n|^2$ with equality iff $c_n = \langle x, g_n \rangle$ for all n [12]. As argued earlier, the computation of $\langle x, f_n \rangle$ from x and the inverse computation of x from $\langle x, f_n \rangle$ are numerically stable operations because $B < \infty$ and $A > 0$, respectively.

For the special case of a tight frame ($A = B$), the frame operator is particularly simple. We have $\mathcal{F}x = Ax$, and so $g_n = \mathcal{F}^{-1} f_n = f_n/A$. Any vector $x \in \mathcal{H}$ can be expressed as

$$x = \frac{1}{A} \sum_{n=1}^{\infty} \langle x, f_n \rangle f_n \quad \text{(tight frames)} \qquad (6.77)$$

Notice also that (6.73) gives

$$\sum_{n=1}^{\infty} |\langle x, f_n \rangle|^2 = A \|x\|^2 \quad \text{(tight frames)} \qquad (6.78)$$

For a tight frame with $A = 1$, the above equations resemble the representation of x using an orthonormal basis, even though such a tight frame is not necessarily a basis because of possible redundancy (Example 6.7).

Exact Frames and Biorthogonality

For the special case of an exact frame (i.e., a Riesz basis) the sequence $\{f_n\}$ is minimal, and it is biorthogonal to the dual frame sequence $\{g_n\}$. This is consistent with our observation at the end of Section 6.7.

Summary of Frames

A sequence of vectors $\{f_n\}$ in a Hilbert space $\mathcal{H}$ is a frame if there exist constants $A > 0$ and $B < \infty$ such that (6.73) holds for every vector $x \in \mathcal{H}$. Frames are complete (because $A > 0$), but not necessarily linearly independent. The constants A and B are called the frame bounds. A frame is *tight* if $A = B$. A tight frame with $A = B = 1$ and with normalized vectors ($\|f_n\| = 1$) is an orthonormal basis. For a tight frame with $\|f_n\| = 1$, the frame bound A measures redundancy. Any vector $x \in \mathcal{H}$ can be expressed in either of the two ways shown in (6.76). Here, $g_n = \mathcal{F}^{-1} f_n$, where $\mathcal{F}$ is the *frame operator* defined in (6.74). The frame operator is a bounded linear operator and is self-adjoint (in fact, positive). The sequence $\{g_n\}$ is the *dual frame* and has frame bounds B^{-1} and A^{-1}. For a tight frame the frame representation reduces to (6.77). A frame is *exact* if deletion of any vector f_m destroys the frame property. A sequence $\{f_n\}$ is an exact frame iff it is a Riesz basis. An exact frame $\{f_n\}$ is biorthogonal to the dual frame $\{g_n\}$.

Figure 6.30 is a Venn diagram, which shows the classification of frames and bases and the relationship between these.

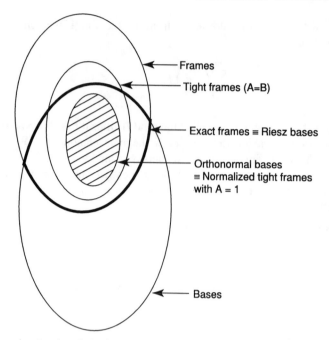

6.30   A Venn diagram showing the relation between frames and bases in a Hilbert space.

6.9   Short-Time Fourier Transform: Invertibility, Orthonormality, and Localization

In Section 6.8 we saw that a vector x in an infinite dimensional Hilbert space (e.g., a function $x(t)$ in L^2) can be expanded in terms of a sequence of vectors $\{f_n\}$ called a frame, that is $x = \sum_{n=1}^{\infty} \langle x, g_n \rangle f_n$. One of the most important features of frames is that the construction of the expansion coefficients $\langle x, g_n \rangle$ from x as well as the reconstruction of x from these coefficients are numerically stable operations because $A > 0$ and $B < \infty$ (see Section 6.8). Riesz and orthonormal bases, which are special cases of a frame (Fig. 6.30), also share this numerical stability.

In Section 6.3 we attempted to represent an L^2 function in terms of the short time Fourier transform (STFT). The STFT coefficients are constructed using the integral (6.24). Denote for simplicity

$$g_{kn}(t) = v^*(t - nT_s)e^{jk\omega_s t} \tag{6.79}$$

The computation of the STFT coefficients can be written as

$$X_{\text{STFT}}(k\omega_s, nT_s) = \Big\langle x(t), g_{kn}(t) \Big\rangle \tag{6.80}$$

This is a linear transformation which converts $x(t)$ into a two-dimensional sequence because k and n are integers. Our hope is to be able to reconstruct $x(t)$ using an inverse linear transformation (inverse STFT) of the form

$$x(t) = \sum_{k=-\infty}^{\infty} \sum_{n=-\infty}^{\infty} X_{\text{STFT}}(k\omega_s, nT_s) f_{kn}(t) \tag{6.81}$$

We know that this can be done in a numerically stable manner if $\{g_{kn}(t)\}$ is a frame in L^2 and $\{f_{kn}(t)\}$ is the dual frame. The fundamental questions, then, are under what conditions does $\{g_{kn}(t)\}$ constitute a frame? Under what further conditions does this become a Riesz basis, better still, an orthonormal basis? With such conditions, what are the time-frequency localization properties of the resulting STFT? The answers depend on the window $v(t)$ and the sample spacings ω_s and T_s.

We first construct a very simple example which shows the existence of orthonormal STFT bases, and indicate a fundamental disadvantage in the example. We then state the answers to the above general questions without proof. Details can be found in [5, 12], and [16].

EXAMPLE 6.8: Orthonormal Short-Time Fourier Transform Basis.

Suppose $v(t)$ is the rectangular window shown in Fig. 6.31, applied to an L^2 function $x(t)$. The product $x(t)v(t)$ therefore has finite duration. If we sample its FT at the rate $\omega_s = 2\pi$ we can recover $x(t)v(t)$ from these samples (this is like a Fourier series of the finite duration waveform $x(t)v(t)$). Shifting the window by successive integers (i.e., $T_s = 1$), we can in this way recover successive pieces of $x(t)$ from the STFT, with sample spacing $\omega_s = 2\pi$ in the frequency domain. Thus, the choice $T_s = 1$ and $\omega_s = 2\pi$ (so, $\omega_s T_s = 2\pi$) leads to an STFT $X_{\text{STFT}}(k\omega_s, nT_s)$, from which we can reconstruct $x(t)$ for all t. The quantity $g_{kn}(t)$ becomes

$$g_{kn}(t) = v(t-n)e^{jk\omega_s t} = v(t-n)e^{j2\pi kt} \qquad (6.82)$$

Because the successive shifts of the window do not overlap, the functions $g_{kn}(t)$ are orthonormal for different values of n. The functions are also orthonormal for different values of k. Summarizing, the rectangular window of Fig. 6.31, with the time-frequency sampling durations $T_s = 1$ and $\omega_s = 2\pi$, produces an orthonormal STFT basis for L^2 functions.

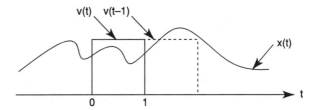

6.31 The rectangular window in STFT.

This example is reminiscent of the Nyquist sampling theorem in the sense that we can reconstruct $x(t)$ from (time-frequency) samples, but the difference is that $x(t)$ is an L^2 signal, not necessarily band-limited. Note that T_s and ω_s cannot be arbitrarily interchanged (even if $\omega_s T_s = 2\pi$ is preserved). Thus, if we had chosen $T_s = 2$ and $\omega_s = \pi$ (preserving the product $\omega_s T_s$) we would not have obtained a basis because two successive positions of the window would be spaced too far apart and we would miss 50% of the signal $x(t)$.

Time-Frequency Sampling Density for Frames and Orthonormal Bases

Let us assume that $v(t)$ is normalized to have unit energy, i.e., $\int |v(t)|^2 dt = 1$ so that $\|g_{kn}(t)\| = 1$ for all k, n. If we impose the condition that $g_{kn}(t)$ be a frame, then it can be shown that the frame bounds satisfy the condition

$$A \le \frac{2\pi}{\omega_s T_s} \le B \qquad (6.83)$$

regardless of how $v(t)$ is chosen. As an orthonormal basis is a tight frame with $A = B = 1$, *an STFT orthonormal basis must therefore have $\omega_s T_s = 2\pi$.*

It can further be shown that if $\omega_S T_S > 2\pi$, $\{g_{kn}(t)\}$ cannot be a frame. For $\omega_s T_s < 2\pi$ we can find frames (but not orthonormal basis) by appropriate choice of window $v(t)$. The critical time-frequency sampling density is $(\omega_s T_s)^{-1} = (2\pi)^{-1}$. If the density is smaller we cannot have frames, and if it is larger we cannot have orthonormal basis, only frames.

Orthonormal Short-Time Fourier Transform Bases have Poor Time-Frequency Localization

If we wish to have an orthonormal STFT basis, the time-frequency density is constrained so that $\omega_s T_s = 2\pi$. Under this condition suppose we choose $v(t)$ appropriately to design such a basis. The time-frequency localization properties of this system can be judged by computing the mean square durations D_t^2 and D_f^2 defined in (6.27). It has been shown by Balian and Low [5, 16] that one of these is necessarily infinite no matter how $v(t)$ is designed. Thus, an orthonormal STFT basis always satisfies $D_t D_f = \infty$. That is, either the time localization or the frequency resolution is very poor. This is summarized in the following theorem.

THEOREM 6.5 *Let the window $v(t)$ be such that $\{g_{kn}(t)\}$ in (6.79) is an orthonormal basis for L^2 (which means, in particular that $\omega_s T_s = 2\pi$). Define the RMS durations D_t and D_f for the window $v(t)$ as usual (6.27). Then, either $D_t = \infty$ or $D_f = \infty$.*

Return now to Example 6.8, where we constructed an orthonormal STFT basis using the rectangular window of Fig. 6.31. Here, $T_s = 1$ and $\omega_s = 2\pi$ so that $\omega_s T_s = 2\pi$. The window $v(t)$ has finite mean square duration D_t^2. Its FT $V(\omega)$ has magnitude $|V(\omega)| = |\sin(\omega/2)/(\omega/2)|$ so that $\int \omega^2 |V(\omega)|^2 d\omega$ is not finite. This demonstrates the result of Theorem 6.5. One can try to replace the window $v(t)$ with something for which $D_t D_f$ is finite, but this cannot be done without violating orthonormality.

Instability of the Gabor Transform

Gabor constructed the STFT using the Gaussian window $v(t) = ce^{-t^2/2}$. In this case the sequence of functions $\{g_{kn}(t)\}$ can be shown to be complete in L^2 (in the sense defined in Section 6.7) as long as $\omega_s T_s \leq 2\pi$. However, if $\omega_s T_s = 2\pi$ the system is not a frame because it can be shown that $A = 0$ in (6.73). Thus, the reconstruction of $x(t)$ from $X_{STFT}(k\omega_s, nT_s)$ is unstable if $\omega_s T_s = 2\pi$ (see Section 6.8), even though $\{g_{kn}(t)\}$ is complete. Although the Gabor transform has the ideal time frequency localization (minimum $D_t D_f$), it cannot provide a *stable basis*, hence, it is certainly not an orthonormal basis, whenever $\omega_s T_s = 2\pi$.

Because orthonormal STFT basis is not possible if $\omega_s T_s \neq 2\pi$, this shows that an orthonormal basis can never be achieved with the Gabor transform (Gaussian windowed STFT), no matter how we choose ω_s and T_s. The Gabor example also demonstrates the fact that even if we successfully construct a complete set of functions (not necessarily a basis) to represent $x(t)$, it may not be useful because of the instabilty of reconstruction. If we construct Riesz bases (e.g., orthonormal bases) or more generally frames, this disadvantage disappears. For example, with the Gabor transform if we let $\omega_s T_s < 2\pi$ then all is well. We obtain a frame [so $A > 0$ and $B < \infty$ in (6.73)]; we have stable reconstruction and good time frequency localization, but not orthonormality. Figure 6.32 summarizes these results pertaining to the time-frequency product $\omega_s T_s$ in the STFT.

A major advantage of the WT over the STFT is that it is free from the above difficulties. For example, we can obtain an orthonormal basis for L^2 with excellent time-frequency localization (finite, controllable $D_t D_f$). We will also see how to constrain such a wavelet $\psi(t)$ to have the additional property of **regularity** or smoothness. Regularity is a property which is measured by the continuity and differentiability of $\psi(t)$. More precisely, it is quantified by the Hölder index (defined in Section 6.13). In the next few sections where we construct wavelets based on paraunitary filter banks, we will see how to achieve all this systematically.

6.10 Wavelets and Multiresolution

Section 6.11 to 6.13 show how to construct compactly supported wavelets systematically to obtain orthonormal bases for L^2. The construction is such that excellent time-frequency localization is possible. Moreover, the smoothness or regularity of the wavelets can be controlled. The construction is based on the two-channel paraunitary filter bank described in Section 6.4. In that section, we denoted the synthesis filters as $G_s(z)$ and $H_s(z)$, with impulse responses $g_s(n)$ and $h_s(n)$, respectively.

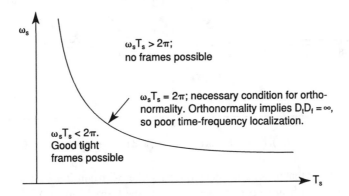

6.32 Behavior of STFT representations for various regions of time-frequency sampling product $\omega_s T_s$. The curve $\omega_s T_s = 2\pi$ is critical; see text.

All constructions are based on obtaining the wavelet $\psi(t)$ and an auxiliary function $\phi(t)$, called the scaling function, from the impulse response sequences $g_s(n)$ and $h_s(n)$. We do this by using time domain recursions of the form

$$\phi(t) = 2 \sum_{n=-\infty}^{\infty} g_s(n)\phi(2t - n) \qquad \psi(t) = 2 \sum_{n=-\infty}^{\infty} h_s(n)\phi(2t - n) \qquad (6.84)$$

called **dilation equations.** Equivalently, in the frequency domain

$$\Phi(\omega) = G_s(e^{j\omega/2})\Phi(\omega/2) \qquad \Psi(\omega) = H_s(e^{j\omega/2})\Phi(\omega/2) \qquad (6.85)$$

If $\{G_s(z), H_s(z)\}$ is a paraunitary pair with further mild conditions (e.g., that the low-pass filter $G_s(e^{j\omega})$ has a zero at π and no zeroes in $[0, \pi/3]$) the preceding recursions can be solved to obtain $\psi(t)$, which gives rise to an orthonormal wavelet basis $\{2^{k/2}\psi(2^k t - n)\}$ for L^2. By constraining $G_s(e^{j\omega})$ to have a sufficient number of zeroes at π we can further control the Hölder index (or regularity) of $\psi(t)$ (see Section 6.13).

Our immediate aim is to explain the occurrence of the function $\phi(t)$, and the curious recursions (6.84) called the *dilation equations* or **two-scale equations.** These have origin in the beautiful theory of multiresolution for L^2 spaces [4, 11]. Because multiresolution theory lays the foundation for the construction of the most practical wavelets to date, we give a brief description of it here.

The Idea of Multiresolution

Return to Fig. 6.13(a), where we interpreted the wavelet transformation as a bank of continuous time analysis filters followed by samplers, and the inverse transformation as a bank of synthesis filters. Assume for simplicity the filters are ideal bandpass. Figure 6.13(b) is a sketch of the frequency responses. The bandpass filters $F_k(\omega) = 2^{-k/2}\Psi(\omega/2^k)$ become increasingly narrow as k decreases (i.e., as k becomes more and more negative). Instead of letting k be negative, suppose we keep only $k \geq 0$ and include a low-pass filter $\Phi(\omega)$ to cover the low frequency region. Then we get the picture of Fig. 6.33. This is analogous to Fig. 6.12, where we used the pulse function $\phi(t)$ instead of using negative k in $\psi(2^k t - n)$.

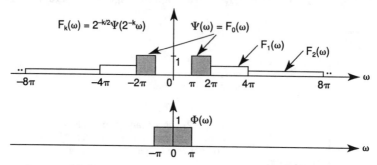

6.33   The low-pass function $\Phi(\omega)$, bandpass function $\Psi(\omega)$, and the stretched bandpass filters $F_k(\omega)$.

Imagine for a moment that $\Phi(\omega)$ is an ideal low-pass filter with cutoff $\pm\pi$. Then we can represent any L^2 function $F(\omega)$ with support restricted to $\pm\pi$ in the form $F(\omega) = \sum_{n=-\infty}^{\infty} a_n \Phi(\omega) e^{-j\omega n}$. This is simply the FS expansion of $F(\omega)$ in $[-\pi, \pi]$, and it follows that $\sum_n |a_n|^2 < \infty$ (Theorem 6.1). In the time domain this means

$$f(t) = \sum_{n=-\infty}^{\infty} a_n \phi(t - n) \tag{6.86}$$

Let us denote by V_0 the closure of the span of $\{\phi(t-n)\}$. Thus, V_0 is the class of L^2 signals that are band-limited to $[-\pi, \pi]$. Because $\phi(t)$ is the sinc function, the shifted functions $\{\phi(t-n)\}$ form an orthonormal basis for V_0.

Consider now the subspace $W_0 \subset L^2$ of bandpass functions band-limited to $\pi < |\omega| \leq 2\pi$. The bandpass sampling theorem (Section 6.2) allows us to reconstruct such a bandpass signal $g(t)$ from its samples $g(n)$ by using the ideal filter $\Psi(\omega)$. Denoting the impulse response of $\Psi(\omega)$ by $\psi(t)$ we see that $\{\psi(t-n)\}$ spans W_0. It can be verified that $\{\psi(t-n)\}$ is an orthonormal basis for W_0. Moreover, as $\Psi(\omega)$ and $\Phi(\omega)$ do not overlap, it follows from Parseval's theorem that W_0 is orthogonal to V_0.

Next, consider the space of all signals of the form $f(t) + g(t)$, where $f(t) \in V_0$ and $g(t) \in W_0$. This space is called the direct sum (in this case, orthogonal sum) of V_0 and W_0, and is denoted as $V_1 = V_0 \oplus W_0$. It is the space of all L^2 signals band-limited to $[-2\pi, 2\pi]$. We can continue in this manner and define the spaces V_k and W_k for all k. Then, V_k is the space of all L^2 signals band-limited to $[-2^k\pi, 2^k\pi]$, and W_k is the space of L^2 functions band-limited to $2^k\pi < |\omega| \leq 2^{k+1}\pi$. The general recursive relation is $V_{k+1} = V_k \oplus W_k$. Figure 6.34 demonstrates this for the case in which the filters are ideal bandpass. Only the positive half of the frequency axis is shown for simplicity.

It is clear that we could imagine V_0 itself to be composed of subspaces V_{-1} and W_{-1}. Thus, $V_0 = V_{-1} \oplus W_{-1}$, $V_{-1} = V_{-2} \oplus W_{-2}$, and so forth. In this way we have defined a sequence of spaces $\{V_k\}$ and $\{W_k\}$ for all integers k such that the following conditions are true:

$$V_{k+1} = V_k \oplus W_k \qquad W_k \perp W_m, \quad k \neq m \tag{6.87}$$

where $\perp$ means "orthogonal," i.e., the functions in W_k are orthogonal to those in W_m. It is clear that $V_k \subset V_{k+1}$.

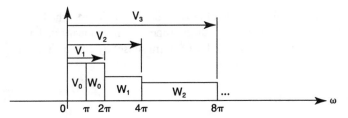

6.34   Toward multiresolution analysis. The spaces $\{V_k\}$ and $\{W_k\}$ spanned by various filter responses.

We will see later that *even if the ideal filters* $\Phi(\omega)$ *and* $\Psi(\omega)$ *are replaced with nonideal approximations,* we can sometimes define sequences of subspaces V_k and W_k satisfying the above conditions. The importance of this observation is that *whenever* $\Psi(\omega)$ *and* $\Phi(\omega)$ *are such that we can construct such a subspace structure, the impulse response* $\psi(t)$ *of the filter* $\Psi(\omega)$ *can be used to generate an orthonormal wavelet basis.* While this might seem too complicated and convoluted, we will see that the construction of the function $\phi(t)$ is quite simple and elegant, and simplifies the construction of orthonormal wavelet bases. A realization of these ideas based on paraunitary filter banks is presented in Section 6.11. It is now time to be more precise with definitions as well as statements of the results.

Definition of Multiresolution Analysis

Consider a sequence of closed subspaces $\{V_k\}$ in L^2, satisfying the following six properties:

1. *Ladder property.* $\cdots V_{-2} \subset V_{-1} \subset V_0 \subset V_1 \subset V_2 \cdots$.

2. $\displaystyle\bigcap_{k=-\infty}^{\infty} V_k = \{0\}$.

3. Closure of $\displaystyle\bigcup_{k=-\infty}^{\infty} V_k$ is equal to L^2.

4. *Scaling property.* $x(t) \in V_k$ iff $x(2t) \in V_{k+1}$. Because this implies "$x(t) \in V_0$ iff $x(2^k t) \in V_k$," all the spaces V_k are scaled versions of the space V_0. For $k > 0$, V_k is a *finer space* than V_0.

5. *Translation invariance.* If $x(t) \in V_0$, then $x(t - n) \in V_0$; that is, the space V_0 is invariant to translations by integers. By the previous property this means that V_k is invariant to translations by $2^{-k}n$.

6. *Special orthonormal basis.* A function $\phi(t) \in V_0$ exists such that the integer shifted versions $\{\phi(t - n)\}$ form an orthonormal basis for V_0. Employing property 4 this means that $\{2^{k/2}\phi(2^k t - n)\}$ is an orthonormal basis for V_k. The function $\phi(t)$ is called the **scaling function** of multiresolution analysis.

Comments on the Definition

Notice that the scaling function $\phi(t)$ determines V_0, hence all V_k. We say that $\phi(t)$ generates the entire multiresolution analysis $\{V_k\}$. The sequence $\{V_k\}$ is said to be a **ladder of subspaces** because of the inclusion property $V_k \subset V_{k+1}$. The technical terms **closed** and **closure**, which originate from metric space theory, have simple meanings in our context because L^2 is a Hilbert space. Thus, the subspace V_k is "closed" if the following is true: whenever a sequence of functions $\{f_n(t)\} \in V_k$ converges to a limit $f(t) \in L^2$ (i.e., $\|f(t) - f_n(t)\| \to 0$ as $n \to \infty$), the limit $f(t)$ is in V_k itself. In general, an infinite union of closed sets is not closed, which is why we need to take "closure" in the third property above. The third property simply means that any element $x(t) \in L^2$ can be approximated arbitrary closely (in the L^2 norm sense) by an element in $\bigcup_{k=-\infty}^{\infty} V_k$.

General meaning of W_k

In the general setting of the above definition, the subspace W_k is defined as the orthogonal complement of V_k with respect to V_{k+1}. Thus, the relation $V_{k+1} = V_k \oplus W_k$, which was valid in the ideal bandpass case (Fig. 6.34), continues to hold.

Haar Multiresolution

A simple example of multiresolution in which $\Phi(\omega)$ is not ideal low-pass is the Haar multiresolution, generated by the function $\phi(t)$ in Fig. 6.35(a). Here, V_0 is the space of all functions that are piecewise constants on intervals of the form $[n, n + 1]$. We will see later that the function $\psi(t)$ associated with this example is as in Fig. 6.35(b)—the space W_0 is spanned by $\{\psi(t - n)\}$. The space V_k contains functions which are constants in $[2^{-k}n, 2^{-k}(n+1)]$. Figure 6.35(c) and (d) show examples of functions belonging to V_0 and V_1. For this example, the six properties in the definition of multiresolution are particularly clear (except perhaps property 3, which also can be proved).

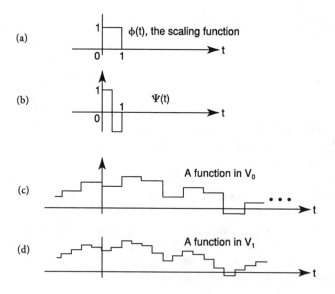

6.35 The Haar multiresolution example. (a) The scaling function $\phi(t)$ that generates multiresolution, (b) the function $\psi(t)$ which generates W_0, (c) example of a member of V_0, and (d) example of a member of V_1.

The multiresolution analysis generated by the ideal bandpass filters (Fig. 6.33 and 6.34) is another simple example, in which $\phi(t)$ is the sinc function. We see that the two elementary orthonormal wavelet examples (Haar wavelet and the ideal bandpass wavelet) also generate a corresponding multiresolution analysis. The connection between wavelets and multiresolution is deeper than this, and is elaborated in the following section.

Derivation of the Dilation Equation

Because $\{\sqrt{2}\phi(2t - n)\}$ is an orthonormal basis for V_1 (see property 6), and because $\phi(t) \in V_0 \subset V_1$, $\phi(t)$ can be expressed as a linear combination of the functions $\{\sqrt{2}\phi(2t - n)\}$:

$$\phi(t) = 2 \sum_{n=-\infty}^{\infty} g_s(n)\phi(2t - n) \qquad \text{(dilation equation)} \qquad (6.88)$$

Thus, the dilation equation arises naturally out of the multiresolution condition. For example, the Haar scaling function $\phi(t)$ satisfies the dilation equation

$$\phi(t) = \phi(2t) + \phi(2t - 1) \qquad (6.89)$$

The notation $g_s(n)$ and the factor 2 in the dilation equation might appear arbitrary now, but are convenient for future use. Orthonormality of $\{\phi(t-n)\}$ implies that $\|\phi(t)\| = 1$, and that $\{\sqrt{2}\phi(2t - n)\}$ are orthonormal. Therefore, $\sum_n |g_s(n)|^2 = 0.5$ from (6.88).

EXAMPLE 6.9: Nonorthonormal Multiresolution.

Consider the triangular function shown in Fig. 6.36(a). This has $\|\phi(t)\| = 1$ and satisfies the dilation equation

$$\phi(t) = \phi(2t) + 0.5\phi(2t - 1) + 0.5\phi(2t + 1) \qquad (6.90)$$

as demonstrated in Fig. 6.36(b). With V_k denoting the closure of the span of $\{2^{k/2}\phi(2^k t - n)\}$ it can be shown that the spaces $\{V_k\}$ satisfy all the conditions in the multiresolution definition, except one. Namely, $\{\phi(t - n)\}$ does not form an orthonormal basis [for example, compare $\phi(t)$ and $\phi(t - 1)$]. We will see later (Example 6.10) that it does form a Riesz basis and that it can be converted into an orthonormal basis by orthonormalization. This example is a special case of a family of scaling functions called **spline functions** [15].

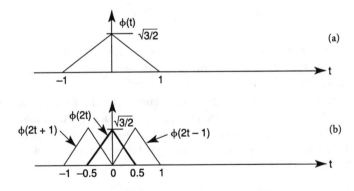

6.36   Example of a scaling function $\phi(t)$ generating nonorthogonal multiresolution. (a) The scaling function, and (b) demonstrating the dilation equation.

We will see below that starting from an orthonormal multiresolution system [in particular from the function $\phi(t)$] one can generate an orthonormal wavelet basis for L^2. The wavelet bases generated from splines $\phi(t)$ after orthonormalization are called **spline wavelets** [15]. These are also called the Battle-Lemarié family of wavelets. The link between multiresolution analysis and wavelets is explained quantitatively in the following section, "Relation between Multiresolution and Wavelets."

Multiresolution Approximation of L^2 Functions

Given a multiresolution analysis, we know that $\bigcap_{k=-\infty}^{\infty} V_k = \{0\}$ and that the closure of $\bigcup_{k=-\infty}^{\infty} V_k = L^2$. From this it can be shown that the W_k make up the entire L^2 space, that is

$$L^2 = \bigoplus_{k=-\infty}^{\infty} W_k \tag{6.91a}$$

We can approximate an arbitrary L^2 function $x(t)$ to a certain degree of accuracy by projecting it onto V_k for appropriate k. Thus, let $x_k(t)$ be this orthogonal projection (see Section 6.2). Suppose we increase k to $k+1$. Because $V_{k+1} = V_k \oplus W_k$ and W_k is orthogonal to V_k, we see that the new approximation $x_{k+1}(t)$ (projection onto the finer space V_{k+1}) is given by $x_{k+1}(t) = x_k(t) + y_k(t)$, where $y_k(t)$ is in W_k.

Thus, when we go from scale k to scale $k+1$ we go to a larger space $V_{k+1} \supset V_k$, which permits a finer approximation. This is nicely demonstrated in the two extreme examples mentioned previously. For the example with ideal filters (Figs. 6.33, 6.34), the process of passing from scale k to $k+1$ is like admitting higher frequency components, which are orthogonal to the existing low-pass components. For the Haar example (Fig. 6.35) where $\psi(t)$ and $\phi(t)$ are square pulses, when we pass from k to $k+1$ we permit finer pulses (i.e., highly localized finer variations in the time domain). For this example, Figs. 6.35(c) and (d) demonstrate the projections $x_k(t)$ and $x_{k+1}(t)$ at two successive resolutions. The projections are piecewise-constant approximations of an L^2 signal $x(t)$.

By repeated application of $V_{k+1} = V_k \oplus W_k$ we can express V_0 as

$$V_0 = \bigoplus_{k=-\infty}^{-1} W_k \tag{6.91b}$$

which, together with (6.91a), yields

$$L^2 = V_0 \oplus W_0 \oplus W_1 \oplus W_2 \oplus \cdots \tag{6.91c}$$

This has a nice interpretation based on Fig. 6.34. The L^2 signal $x(t)$ has been decomposed into orthogonal components belonging to V_0 (low-pass component), W_0 (bandpass component), W_1 (bandpass with higher bandwidth and center frequency), etc.

We can find an infinite number of multiresolution examples by choosing $\phi(t)$ appropriately. It is more important now to obtain systematic techniques for constructing such examples. The quality of the example is governed by the quality of $\psi(t)$ and $\phi(t)$—the time localization and frequency resolution they can provide, the smoothness (regularity) of these functions, and the ease with which we can implement these approximations.

Relation between Multiresolution and Wavelets

Suppose $\phi(t) \in L^2$ generates an orthonormal multiresolution $\{V_k\}$, as defined in the previous section. We know $\phi(t) \in V_0$ and that $\{\phi(t-n)\}$ is an orthonormal basis for V_0. Moreover, $\phi(t)$ satisfies the dilation equation (6.88), and the sequence $\{g_s(n)\} \in \ell^2$ defines the filter $G_s(e^{j\omega})$.

Now consider the finer space $V_1 = V_0 \oplus W_0$, where W_0 is orthonormal to V_0. If $f(t) \in W_0$ then $f(t) \in V_1$, so it is a linear combination of $\sqrt{2}\phi(2t-n)$ (property 6; see definitions). Using this and the fact that W_0 is orthogonal to V_0 we can show that $F(\omega)$ [the L^2-FT of $f(t)$] has a special form.

This is given by

$$F(\omega) = e^{j\omega/2} G_s^*(-e^{j\omega/2}) \Phi(\omega/2) H(e^{j\omega})$$

where $H(e^{j\omega})$ is 2π-periodic. The special case of this with $H(e^{j\omega}) = 1$ is denoted $\Psi(\omega)$; that is,

$$\Psi(\omega) = e^{j\omega/2} G_s^*(-e^{j\omega/2}) \Phi(\omega/2). \qquad (6.92)$$

The above definition of $\Psi(\omega)$ is equivalent to

$$\psi(t) = 2 \sum_{n=-\infty}^{\infty} (-1)^{n+1} g_s^*(-n-1) \phi(2t-n) \quad \text{[dilation equation for } \psi(t)] \qquad (6.93)$$

The function $\psi(t)$ satisfying this equation has some useful properties. First, it is in L^2. This follows from Theorem 6.2 (Riesz-Fischer Theorem), because $\sum_n |g_s(n)|^2$ is finite. It can be shown that $\psi(t-n) \in W_0$ and that $\{\psi(t-n)\}$ is an orthonormal basis for W_0. This implies that $\{2^{k/2}\psi(2^k t-n)\}$ is an orthonormal basis for W_k because $f(t) \in W_0$ iff $f(2^k t) \in W_k$, which is a property induced by the scaling property (property 4 in the definition of multiresolution). In view of (6.91a and b) we conclude that the sequence $\{2^{k/2}\psi(2^k t - n)\}$, with k and n varying over all integers, forms a basis for L^2. Summarizing we have the following result:

THEOREM 6.6 *(Multiresolution and Wavelets) Let $\phi(t) \in L^2$ generate an orthonormal multiresolution, i.e., a ladder of spaces $\{V_k\}$ satisfying the six properties in the definition of multiresolution; $\{\phi(t-n)\}$ is an orthonormal basis for V_0. Then $\phi(t)$ satisfies the dilation equation (6.88) for some $g_s(n)$ with $\sum_n |g_s(n)|^2 = 0.5$. Define the function $\psi(t)$ according to the dilation equation (6.93). Then, $\psi(t) \in W_0 \subset L^2$, and $\{\psi(t-n)\}$ is an orthonormal basis for W_0. Therefore, $\{2^{k/2}\psi(2^k t-n)\}$ is an orthonormal basis for W_k, just as $\{2^{k/2}\phi(2^k t - n)\}$ is an orthonormal basis for V_k (for fixed k). Moreover, with k and n varying over all integers, the doubly indexed sequence $\{2^{k/2}\psi(2^k t - n)\}$ is an orthonormal wavelet basis for L^2.*

Thus, to construct a wavelet basis for L^2 we have only to construct an orthonormal basis $\{\phi(t-n)\}$ for V_0. Everything else follows from that. All proofs can be found in [5, 11], and [15].

Relation between Multiresolution Analysis and Paraunitary Filter Banks

Denoting

$$h_s(n) = (-1)^{n+1} g_s^*(-1-n), \quad \text{i.e.,} \quad H_s(e^{j\omega}) = e^{j\omega} G_s^*(-e^{j\omega})$$

we see that $\phi(t)$ and $\psi(t)$ satisfy the two dilation equations in (6.84). By construction $\psi(t) \in W_0$ and $\phi(t) \in V_0$. The fact that W_0 and V_0 are mutually orthogonal subspaces can be used to show that $H_s(e^{j\omega})$ and $G_s(e^{j\omega})$ satisfy

$$G_s^*(e^{j\omega}) H_s(e^{j\omega}) + G_s^*(-e^{j\omega}) H_s(-e^{j\omega}) = 0 \qquad (6.94)$$

Also, orthonormality of $\{\phi(t-n)\}$ leads to the power complementary property

$$|G_s(e^{j\omega})|^2 + |G_s(-e^{j\omega})|^2 = 1 \qquad (6.95)$$

In other words, $G_s(e^{j\omega})$ is a power symmetric filter. That is, the filter $|G_s(e^{j\omega})|^2$ is a half-band filter. Using $H_s(e^{j\omega}) = e^{j\omega} G_s^*(-e^{j\omega})$, we also have

$$|H_s(e^{j\omega})|^2 + |H_s(-e^{j\omega})|^2 = 1 \qquad (6.96)$$

A compact way to express the above three equations is by defining the matrix

$$\mathbf{G}_s(e^{j\omega}) = \begin{bmatrix} G_s(e^{j\omega}) & H_s(e^{j\omega}) \\ G_s(-e^{j\omega}) & H_s(-e^{j\omega}) \end{bmatrix}$$

The three properties (6.94) to (6.96) are equivalent to $\mathbf{G}_s^\dagger(e^{j\omega})\mathbf{G}_s(e^{j\omega}) = \mathbf{I}$; i.e., the matrix $\mathbf{G}_s(e^{j\omega})$ is unitary for all ω. This matrix was defined in Section 6.4 in the context of paraunitary digital filter banks. Thus, the filters $G_s(e^{j\omega})$ and $H_s(e^{j\omega})$ constructed from a multiresolution setup constitute a paraunitary (CQF) synthesis bank.

Thus, *orthonormal multiresolution automatically gives rise to paraunitary filter banks.* Starting from a multiresolution analysis we obtained two functions $\phi(t)$ and $\psi(t)$. These functions generate orthonormal bases $\{\phi(t-n)\}$ and $\{\psi(t-n)\}$ for the orthogonal subspaces V_0 and W_0. The functions $\phi(t)$ and $\psi(t)$ generated in this way satisfy the dilation equation (6.84). Defining the filters $G_s(z)$ and $H_s(z)$ from the coefficients $g_s(n)$ and $h_s(n)$ in an obvious way, we find that these filters form a paraunitary pair.

This raises the following fundamental question: If we start from a paraunitary pair $\{G_s(z), H_s(z)\}$ and define the functions $\phi(t)$ and $\psi(t)$ by (successfully) solving the dilation equations, do we obtain an orthonormal basis $\{\phi(t-n)\}$ for multiresolution, and a wavelet basis $\{2^{k/2}\psi(2^k t-n)\}$ for the space of L^2 functions? The answer, fortunately, is in the affirmative, subject to some minor requirements which can be trivially satisfied in practice.

Generating Wavelet and Multiresolution Coefficients from Paraunitary Filter Banks

Recall that the subspaces V_0 and W_0 have the orthonormal bases $\{\phi(t-n)\}$ and $\{\psi(t-n)\}$, respectively. By the scaling property, the subspace V_k has the orthonormal basis $\{\phi_{kn}(t)\}$, and similarly the subspace W_k has the orthonormal basis $\{\psi_{kn}(t)\}$, where, as usual, $\phi_{kn}(t) = 2^{k/2}\phi(2^k t - n)$ and $\psi_{kn}(t) = 2^{k/2}\psi(2^k t - n)$. The orthogonal projections of a signal $x(t) \in L^2$ onto V_k and W_k are given, respectively, by

$$P_k[x(t)] = \sum_{n=-\infty}^{\infty} \langle x(t), \phi_{kn}(t) \rangle \phi_{kn}(t) \quad \text{and}$$
$$Q_k[x(t)] = \sum_{n=-\infty}^{\infty} \langle x(t), \psi_{kn}(t) \rangle \psi_{kn}(t) \tag{6.97}$$

(see Section 6.2). Denote the scale-k projection coefficients as $d_k(n) = \langle x(t), \phi_{kn}(t) \rangle$ and $c_k(n) = \langle x(t), \psi_{kn}(t) \rangle$ for simplicity. (The notation c_{kn} was used in earlier sections, but $c_k(n)$ is convenient for the present discussion.) We say that $d_k(n)$ are the **multiresolution coefficients** at scale k and $c_k(n)$ are the **wavelet coefficients** at scale k.

Assume that the projection coefficients $d_k(n)$ are known for some scale, e.g., $k = 0$. We will show that $d_k(n)$ and $c_k(n)$ for the coarser scales, i.e., $k = -1, -2, \ldots$ can be generated by using a paraunitary analysis filter bank $\{G_a(e^{j\omega}), H_a(e^{j\omega})\}$, corresponding to the synthesis bank $\{G_s(e^{j\omega}), H_s(e^{j\omega})\}$ (Section 6.4). We know $\phi(t)$ and $\psi(t)$ satisfy the dilation equations (6.84). By substituting the dilation equations into the right-hand sides of $\phi_{kn}(t) = 2^{k/2}\phi(2^k t - n)$ and $\psi_{kn}(t) = 2^{k/2}\psi(2^k t - n)$, we obtain

$$\phi_{kn}(t) = \sqrt{2} \sum_{m=-\infty}^{\infty} g_s(m - 2n)\phi_{k+1,m}(t) \quad \text{and}$$
$$\psi_{kn}(t) = \sqrt{2} \sum_{m=-\infty}^{\infty} h_s(m - 2n)\phi_{k+1,m}(t) \tag{6.98}$$

A computation of the inner products $d_k(n) = \langle x(t), \phi_{kn}(t) \rangle$ and $c_k(n) = \langle x(t), \psi_{kn}(t) \rangle$ yields

$$
\begin{aligned}
d_k(n) &= \sum_{m=-\infty}^{\infty} \sqrt{2} g_a(2n - m) d_{k+1}(m) \\
c_k(n) &= \sum_{m=-\infty}^{\infty} \sqrt{2} h_a(2n - m) d_{k+1}(m)
\end{aligned}
\tag{6.99}
$$

where $g_a(n) = g_s^*(-n)$ and $h_a(n) = h_s^*(-n)$ are the analysis filters in the paraunitary filter bank.

The beauty of these equations is that they look like **discrete time convolutions**. Thus, if $d_{k+1}(n)$ is convolved with the impulse response $\sqrt{2} g_a(n)$ and the output decimated by 2, the result is the sequence $d_k(n)$. A similar statement follows for $c_k(n)$. The above computation can therefore be interpreted in filter bank form as in Fig. 6.37. Because of the PR property of the two-channel system (Fig. 6.22), it follows that we can reconstruct the projection coefficients $d_{k+1}(n)$ from the projection coefficients $d_k(n)$ and $c_k(n)$.

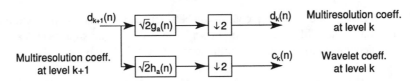

6.37   Generating the wavelet and multiresolution coefficients at level k from level $k + 1$.

Fast Wavelet Transform

Repeated application of this idea results in Fig. 6.38, which is a tree-structured paraunitary filter bank (Section 6.4) with analysis filters $\sqrt{2} g_a(n)$ and $\sqrt{2} h_a(n)$ at each stage. Thus, given the projection coefficients $d_0(n)$ for V_0, we can compute the projection coefficients $d_k(n)$ and $c_k(n)$ for the coarser spaces $V_{-1}, W_{-1}, V_{-2}, W_{-2}, \ldots$ This scheme is sometimes referred to as the **fast wavelet transform (FWT)**. Figure 6.39 shows a schematic of the computation. In this figure each node (heavy dot) represents a decimated paraunitary analysis bank $\{\sqrt{2} g_a(n), \sqrt{2} h_a(n)\}$. The subspaces W_m and V_m are indicated in the Figure rather than the projection coefficients.

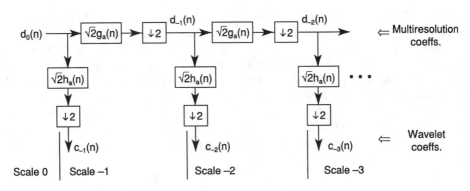

6.38   Tree-structured analysis bank generating wavelet coefficients $c_k(n)$ and multiresolution coefficients $d_k(n)$ recursively.

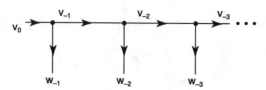

6.39   A schematic of the tree-structured filter bank which generates the coefficients of the projections onto V_k and W_k.

Computation of the Initial Projection Coefficient.   Everything depends on the computation of $d_0(n)$. Note that $d_0(n) = \langle x(t), \phi(t - n) \rangle$, which can be written as the integral $d_0(n) = \int x(t)\phi^*(t-n)\,dt$. An elaborate computation of this integral is avoided in practice. If the scale $k = 0$ is fine enough—if $x(t)$ does not change much within the duration where $\phi(t)$ is significant—we can approximate this integral with the sample value $x(n)$; i.e., $d_0(n) \approx x(n)$. Improved approximations of $d_0(n)$ have been suggested by other authors, but are not reviewed here.

Continuous Time Filter Banks and Multiresolution

The preceding discussions show the deep connection between orthonormal multiresolution analysis and discrete time paraunitary filter banks. As shown by (6.91c), any L^2 signal $x(t)$ can be written as a sum of its projections onto the mutually orthogonal spaces V_0, W_0, W_1, etc.:

$$x(t) = \sum_n d_0(n)\phi(t - n) + \sum_{k=0}^{\infty}\sum_n c_k(n)2^{k/2}\psi(2^k t - n)$$

This decomposition itself can be given a simple filter bank interpretation, with *continuous time filters and samplers.* For this, first note that the V_0 component $\sum_n d_0(n)\phi(t - n)$ can be regarded as the output of a filter with impulse response $\phi(t)$, with the input chosen as the impulse train $\sum_n d_0(n)\delta_a(t - n)$. Similarly, the W_k component $\sum_n c_k(n)2^{k/2}\psi(2^k t - n)$ is the output of a filter with impulse response $f_k(t) = 2^{k/2}\psi(2^k t)$, in response to the input $\sum_n c_k(n)\delta_a(t - 2^{-k}n)$. This interpretation is shown by the synthesis bank of Fig. 6.40(a).

The projection coefficients $d_0(n)$ and $c_k(n)$ can also be interpreted nicely. For example, we have $d_0(n) = \langle x(t), \phi(t - n) \rangle$ by orthonormality. This inner product can be explicitly written out as

$$d_0(n) = \int x(t)\phi^*(t - n)\,dt$$

The integral can be interpreted as a convolution of $x(t)$ with $\phi^*(-t)$. Consider the output of the filter with impulse response $\phi^*(-t)$, with the input chosen as $x(t)$. This output, sampled at time n, gives $d_0(n)$. Similarly, $c_k(n)$ can be interpreted as the output of the filter $h_k(t) = 2^{k/2}\psi^*(-2^k t)$, sampled at the time $2^{-k}n$. The analysis bank of Fig. 6.40(a) shows this interpretation. Thus, the projection coefficients $d_0(n)$ and $c_k(n)$ are the sampled versions of the outputs of an analysis filter bank.

Notice that all the filters in the filter bank are determined by the scaling function $\phi(t)$ and the wavelet function $\psi(t)$. Every synthesis filter $f_k(t)$ is the time-reversed conjugate of the corresponding analysis filter $h_k(t)$, that is, $f_k(t) = h_k^*(-t)$ (a consequence of orthonormality). In terms of frequency responses this means $F_k(\omega) = H_k^*(\omega)$. For completeness of the picture, Fig. 6.40(b) shows typical frequency response magnitudes of these filters.

Further Manifestations of Orthonormality

The orthonormality of the basis functions $\{\phi(t - n)\}$ and $\{\psi(t - n)\}$ have further consequences, summarized below. A knowledge of these will be useful when we generate the scaling function $\phi(t)$

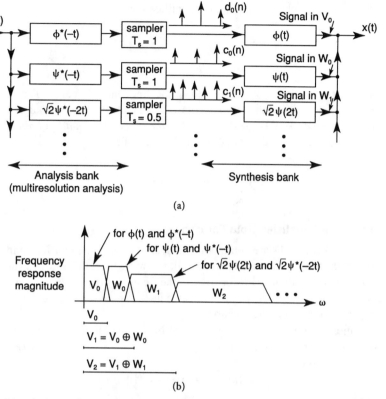

(a)

(b)

6.40 (a) The multiresolution analysis and resynthesis in filter bank form, and (b) typical frequency responses.

and the wavelet function $\psi(t)$ systematically in Section 6.11 from paraunitary filter banks.

Nyquist Property and Orthonormality

With $\phi(t) \in L^2$, the autocorrelation function $R(\tau) = \int \phi(t)\phi^*(t-\tau)dt$ exists for all τ because this is simply an inner product of two elements in L^2. Clearly, $R(0) = \|\phi(t)\|^2 = 1$. Further, the orthonormality property $\langle \phi(t), \phi(t - n) \rangle = \delta(n)$ can be rewritten as $R(n) = \delta(n)$. Thus, $R(\tau)$ has periodic zero crossings at nonzero integer values of τ (Fig. 6.41). This is precisely the *Nyquist property* familiar to communication engineers. The autocorrelation of the scaling function $\phi(t)$ is a Nyquist function. The same holds for the wavelet function $\psi(t)$.

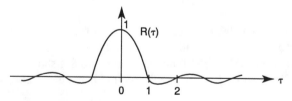

6.41 Example of an autocorrelation of the scaling function $\phi(t)$.

Next, using Parseval's identity for L^2-FTs, we obtain $\langle \phi(t), \phi(t-n) \rangle = \int \Phi(\omega)\Phi^*(\omega)e^{j\omega n}d\omega/2\pi = \delta(n)$. If we decompose the integral into a sum of integrals over intervals of length 2π and use the

2π-periodicity of $e^{j\omega n}$ we obtain, after some simplification:

$$\sum_{k=-\infty}^{\infty} |\Phi(\omega + 2\pi k)|^2 = 1 \quad \text{a.e.} \tag{6.100}$$

This is the Nyquist condition, now expressed in the frequency domain. The term **a.e.**, **almost everywhere**, arises from the fact that we have drawn a conclusion about an integrand from the value of the integral. Thus, $\{\phi(t - n)\}$ *is orthonormal iff the preceding equation holds.* A similar result follows for $\Psi(\omega)$, so orthonormality of $\{\psi(t - n)\}$ is equivalent to

$$\sum_{k=-\infty}^{\infty} |\Psi(\omega + 2\pi k)|^2 = 1 \quad \text{a.e.} \tag{6.101}$$

Case in Which Equalities Hold Pointwise

If we assume that all FTs are continuous, then equalities in the Fourier domain actually hold pointwise. This is the most common situation; in all examples here, the following are true: the filters $G_s(e^{j\omega})$ and $H_s(e^{j\omega})$ are rational (FIR or IIR), so the frequency responses are continuous functions of ω, and $\phi(t)$ and $\psi(t)$ are not only in L^2, but also in L^1; i.e., $\phi(t), \psi(t) \in L^1 \cap L^2$. Thus, $\Phi(\omega)$ and $\Psi(\omega)$ are continuous functions (Section 6.6).

With the dilation equation $\Phi(\omega) = G_s(e^{j\omega/2})\Phi(\omega/2)$ holding pointwise, we have $\Phi(0) = G_s(e^{j0})\Phi(0)$. In all our applications $\Phi(0) \neq 0$ (it is a low-pass filter), so $G_s(e^{j0}) = 1$. The power symmetry property

$$|G_s(e^{j\omega})|^2 + |G_s(-e^{j\omega})|^2 = 1$$

then implies $G_s(e^{j\pi}) = 0$. Because the high-pass synthesis filter is $H_s(e^{j\omega}) = e^{j\omega} G_s^*(-e^{j\omega})$ we conclude $H_s(e^{j0}) = 0$ and $H_s(e^{j\pi}) = -1$. Thus,

$$G_s(e^{j0}) = 1 \quad G_s(e^{j\pi}) = 0 \quad H_s(e^{j0}) = 0 \quad H_s(e^{j\pi}) = -1 \tag{6.102}$$

In particular, the low-pass impulse response $g_s(n)$ satisfies $\sum_n g_s(n) = 1$. Because we already have $\sum_n |g_s(n)|^2 = 0.5$ (Theorem 6.6), we have both of the following:

$$\sum_{n=-\infty}^{\infty} g_s(n) = 1 \quad \text{and} \quad \sum_{n=-\infty}^{\infty} |g_s(n)|^2 = 0.5 \tag{6.103}$$

From the dilation equation $\Phi(\omega) = G_s(e^{j\omega/2})\Phi(\omega/2)$ we obtain $\Phi(2\pi k) = G_s(e^{j\pi k})\Phi(\pi k)$. By using the fact that $G_s(e^{j\pi}) = 0$, and after elementary manipulations we can show that

$$\Phi(2\pi k) = 0 \quad k \neq 0 \tag{6.104}$$

In other words, $\Phi(\omega)$ is itself a Nyquist function of ω. If (6.100) is *assumed to hold pointwise,* the above implies that $|\Phi(0)| = 1$. Without loss of generality we will let $\Phi(0) = 1$, i.e., $\int \phi(t)dt = 1$. The dilation equation for the wavelet function $\Psi(\omega)$ in (6.85) shows that $\Psi(0) = 0$ [because $H_s(e^{j0}) = 0$ by (6.102)]. That is, $\int \psi(t) dt = 0$. Summarizing, the scaling and wavelet functions satisfy

$$\int_{-\infty}^{\infty} \phi(t) dt = 1 \quad \int_{-\infty}^{\infty} \psi(t) dt = 0 \quad \text{and}$$
$$\int_{-\infty}^{\infty} |\phi(t)|^2 dt = \int_{-\infty}^{\infty} |\psi(t)|^2 dt = 1 \tag{6.105}$$

where property 3 follows from orthonormality. These integrals make sense because of the assumption $\phi(t) \in L^1 \cap L^2$. Another result that follows from $\Phi(2\pi k) = \delta(k)$ is that

$$\sum_{n=-\infty}^{\infty} \phi(t-n) = 1 \quad \text{a.e.} \tag{6.106}$$

Thus, the basis functions of the subspace V_0 themselves add up to unity. Return to the Haar basis and notice how beautifully everything fits together.

Generating Wavelet and Multiresolution Basis by Design of $\phi(t)$

Most of the well-known wavelet basis families of recent times were generated by first finding a scaling function $\phi(t)$ such that it is a valid generator of multiresolution, and then generating $\psi(t)$ from $\phi(t)$. The first step, therefore, is to identify the conditions under which a function $\phi(t)$ will be a valid scaling function (i.e., it will generate a multiresolution). Once this is done and we successfully identify the coefficients $g_s(n)$ in the dilation equation for $\phi(t)$, we can identify the wavelet function $\psi(t)$ using the second dilation equation in (6.84). From Theorem 6.6 we know that if $\psi(t)$ is computed in this way, then $\{2^{k/2}\psi(2^k t - n)\}$ is an orthonormal wavelet basis for L^2. The following results can be deduced from the many detailed results presented in [5].

THEOREM 6.7 *(Orthonormal Multiresolution) Let $\phi(t)$ satisfy the following conditions: $\phi(t) \in L^1 \cap L^2$, $\int \phi(t)\,dt \neq 0$ (i.e., $\Phi(0) \neq 0$), $\phi(t) = 2\sum_n g_s(n)\phi(2t-n)$ for some $\{g_s(n)\}$, and $\{\phi(t-n)\}$ is an orthonormal sequence. Then the following are true.*

1. *$\phi(t)$ generates a multiresolution. That is, if we define the space V_k to be the closure of the span of $\{2^{k/2}\phi(2^k t - n)\}$, then the set of spaces $\{V_k\}$ satisfies the six conditions in the definition of multiresolution.*

2. *Define $\psi(t) = 2\sum_n (-1)^{n+1} g_s^*(-n-1)\phi(2t-n)$. Then $\psi(t)$ generates an orthonormal wavelet basis for L^2; that is, $\{2^{k/2}\psi(2^k t - n)\}$, with k and n varying over all integers, is an orthonormal basis for L^2. In fact, for fixed k, the functions $\{2^{k/2}\psi(2^k t - n)\}$ form an orthonormal basis for the subspace W_k (defined following the definition of multiresolution analysis).*

Comments. In many examples $\phi(t) \in L^2$, and it is compactly supported. Then it is naturally in L^1 as well, so the assumption $\phi(t) \in L^2 \cap L^1$ is not too restrictive. Because $L^1 \cap L^2$ is dense in L^2, the above construction still gives a wavelet basis for L^2. Notice also that the orthonormality of $\{\phi(t-n)\}$ implies orthonormality of $\{\sqrt{2}\phi(2t-n)\}$. The recursion $\phi(t) = 2\sum_n g_s(n)\phi(2t-n)$, therefore, is a Fourier series for $\phi(t)$ in L^2. Thus the condition $\sum_n |g_s(n)|^2 = 0.5$ is automatically implied. This is not explicitly stated as part of the conditions in the theorem.

Orthonormalization

We know that orthonormality of $\{\phi(t-n)\}$ is equivalent to

$$\sum_{k=-\infty}^{\infty} |\Phi(\omega + 2\pi k)|^2 = 1 \tag{6.107}$$

Suppose now that this is not satisfied, but the weaker condition

$$a \le \sum_{k=-\infty}^{\infty} |\Phi(\omega + 2\pi k)|^2 \le b \tag{6.108}$$

holds for some $a > 0$ and $b < \infty$. Then, it can be shown that, we can at least obtain a Riesz basis (Section 6.7) of the form $\{\phi(t - n)\}$ for V_0. We can also normalize it to obtain an orthonormal sequence $\{\widehat{\phi}(t - n)\}$ from which an orthonormal wavelet basis can be generated in the usual way. The following theorem summarizes the main results.

THEOREM 6.8 Let $\phi(t) \in L^1 \cap L^2$, $\int \phi(t)dt \neq 0$ (i.e., $\Phi(0) \neq 0$), and $\phi(t) = 2\sum_n g_s(n)\phi(2t - n)$ with $\sum_n |g_s(n)|^2 < \infty$. Instead of the orthonormality condition (6.107), let (6.108) hold for some $a > 0$ and $b < \infty$. Then the following are true:

1. $\{\phi(t - n)\}$ is a Riesz basis for the closure V_0 of its span.
2. $\phi(t)$ generates a multiresolution. That is, if we define the space V_k to be the closure of the span of $\{2^{k/2}\phi(2^k t - n)\}$, the set of spaces $\{V_k\}$ satisfies the six conditions in the definition of multiresolution.

If we define a new function $\widehat{\phi}(t)$ in terms of its FT as

$$\widehat{\Phi}(\omega) = \frac{\Phi(\omega)}{\left(\sum_k |\Phi(\omega + 2\pi k)|^2\right)^{0.5}}, \tag{6.109}$$

then $\widehat{\phi}(t)$ generates an orthonormal multiresolution, and satisfies a dilation equation similar to (6.84). Using this we can define a corresponding wavelet function $\widehat{\psi}(t)$ in the usual way. That is, if $\widehat{\phi}(t) = 2\sum_n g_s(n)\widehat{\phi}(2t - n)$, choose $\widehat{\psi}(t) = 2\sum_n h_s(n)\widehat{\phi}(2t - n)$, where $h_s(n) = (-1)^{n+1}g_s^*(-n - 1)$. This wavelet $\widehat{\psi}(t)$ generates an orthonormal wavelet basis for L^2. Note that the basis is not necessarily compactly supported if we start with compactly supported $\phi(t)$. An example is seen in Fig. 6.46(b) later.

EXAMPLE 6.10: **Battle-Lemarié Orthonormal Wavelets from Splines.**

In Example 6.9 we considered a triangular $\phi(t)$ (Fig. 6.36), which generates a nonorthonormal multiresolution. In this example we have

$$\Phi(\omega) = \sqrt{\frac{3}{2}}\left(\frac{\sin(\omega/2)}{(\omega/2)}\right)^2 \tag{6.110}$$

and it can be shown that

$$\sum_{k=-\infty}^{\infty} |\Phi(\omega + 2\pi k)|^2 = \frac{2 + \cos\omega}{2} \tag{6.111}$$

The inequality (6.108) is satisfied with $a = 1/2$ and $b = 3/2$. Thus, we have a Riesz basis $\{\phi(t - n)\}$ for V_0. From this scaling function we can obtain the normalized function $\widehat{\Phi}(\omega)$ as above and then generate the wavelet function $\widehat{\psi}(t)$ as explained earlier. This gives an orthonormal wavelet basis for L^2. $\widehat{\phi}(t)$ does not, however, have compact support [unlike $\phi(t)$]. Thus, the wavelet function $\widehat{\psi}(t)$ generating the orthonormal wavelet basis is not compactly supported either.

6.11   Orthonormal Wavelet Basis from Paraunitary Filter Banks

The wisdom gained from the multiresolution viewpoint (Section 6.10) tells us a close connection exists between wavelet bases and two-channel digital filter banks. In fact, we obtained the equations of a paraunitary filter bank just by imposing the orthonormality condition on the multiresolution basis functions $\{\phi(t-n)\}$. This section presents the complete story. Suppose we start from a two-channel digital filter bank with the paraunitary property. Can we derive an orthonormal wavelet basis from this? To be more specific, return to the dilation equations (6.84) or equivalently (6.85). Here, $g_s(n)$ and $h_s(n)$ are the impulse response coefficients of the two synthesis filters $G_s(e^{j\omega})$ and $H_s(e^{j\omega})$ in the digital filter bank. Given these two filters, can we "solve" for $\phi(t)$ and $\psi(t)$? If so, does this $\psi(t)$ generate an orthonormal basis for L^2 space? This section answers some of these questions. Unlike any other section, we also indicate a sketch of the proof for each major result, in view of the importance of these in modern signal processing theory.

Recall first that under some mild conditions (Section 6.10) we can prove that the filters must satisfy (6.102) and (6.103), if we need to generate wavelet and multiresolution bases successfully. We impose these at the outset. By repeated application of the dilation equation we obtain $\Phi(\omega) = G_s(e^{j\omega/2})G_s(e^{j\omega/4})\Phi(\omega/4)$. Further indefinite repetition yields an infinite product. Using the condition $\Phi(0) = 1$, which we justified earlier, we obtain the infinite products

$$\Phi(\omega) = \prod_{k=1}^{\infty} G_s(e^{j\omega/2^k}) = G_s(e^{j\omega/2})\prod_{k=2}^{\infty} G_s(e^{j\omega/2^k}) \qquad (6.112a)$$

$$\Psi(\omega) = H_s(e^{j\omega/2})\prod_{k=2}^{\infty} G_s(e^{j\omega/2^k}) \qquad (6.112b)$$

The first issue to be addressed is the convergence of the infinite products above. For this we need to review some preliminaries on infinite products [22, 23].

Ideal Bandpass Wavelet Rederived from the Digital Filter Bank. Before we address the mathematical details, let us consider a simple example. Suppose the pair of filters $G_s(e^{j\omega})$ and $H_s(e^{j\omega})$ are ideal brickwall low-pass and high-pass filters as in Fig. 6.28(a). Then we can verify, by making simple sketches of a few terms in (6.112a and b), that the above infinite products yield the functions $\Phi(\omega)$ and $\Psi(\omega)$ shown in Fig. 6.33. That is, the ideal bandpass wavelet is indeed related to the ideal paraunitary filter bank by means of the above infinite product.

Convergence of Infinite Products

To define convergence of a product of the form $\prod_{k=1}^{\infty} a_k$, consider the sequence $\{p_n\}$ of partial products $p_n = \prod_{k=1}^{n} a_k$. If this converges to a (complex) number A with $0 < |A| < \infty$ we say that the infinite product converges to A. Convergence to zero should be defined more carefully to avoid degenerate situations (e.g., if $a_1 = 0$, then $p_n = 0$ for all n regardless of the remaining terms $a_k, k > 1$). We use the definition in [22]. The infinite product is said to converge to zero iff $a_k = 0$ for a finite nonzero number of values of k, and if the product with these a_k deleted converges to a nonzero value.

Useful Facts About Infinite Products

1. Whenever $\prod_{k=1}^{\infty} a_k$ converges, it can be shown that $a_k \to 1$ as $k \to \infty$. For this reason it is convenient to write $a_k = 1 + b_k$.
2. We say that $\prod_{k=1}^{\infty}(1 + b_k)$ converges absolutely if $\prod_{k=1}^{\infty}(1 + |b_k|)$ converges. Absolute convergence of $\prod_{k=1}^{\infty}(1 + b_k)$ implies its convergence.
3. It can be shown that the product $\prod_{k=1}^{\infty}(1 + |b_k|)$ converges iff the sum $\sum_{k=1}^{\infty} |b_k|$ converges. That is, $\prod_{k=1}^{\infty}(1 + b_k)$ converges absolutely iff $\sum_{k=1}^{\infty} b_k$ converges absolutely.

EXAMPLE 6.11:

The product $\prod_{k=1}^{\infty}(1+k^{-2})$ converges because $\sum_{k=1}^{\infty} 1/k^2$ converges. Similarly, $\prod_{k=1}^{\infty}(1-k^{-2})$ converges because it converges absolutely, by the preceding example. The product $\prod_{k=1}^{\infty}(1+k^{-1})$ does not converge because $\sum_{k=1}^{\infty} 1/k$ diverges. Products such as $\prod_{k=1}^{\infty}(1/k^2)$ do not converge because the terms do not approach unity as $k \to \infty$.

Uniform Convergence

A sequence $\{p_n(z)\}$ of functions of the complex variable z converges *uniformly* to a function $p(z)$ on a set S in the complex plane if the convergence rate is the same everywhere in S. More precisely, if we are given $\epsilon > 0$, we can find N such that $|p_n(z) - p(z)| < \epsilon$ for every $z \in S$, as long as $n \geq N$. The crucial thing is that N depends only on ϵ and not on z, as long as $z \in S$. A similar definition applies for functions of real variables.

We say that an infinite product of functions $\prod_{k=1}^{\infty} a_k(z)$ converges at a point z if the sequence of partial products $p_n(z) = \prod_{k=1}^{n} a_k(z)$ converges as described previously. If this convergence of $p_n(z)$ is uniform in a set S, we say that the infinite product converges *uniformly* on S. Uniform convergence has similar advantages, as in the case of infinite summations. For example, if each of the functions $a_k(\omega)$ is continuous on the real interval $[\omega_1, \omega_2]$, then uniform convergence of the infinite product $A(\omega) = \prod_{k=1}^{\infty} a_k(\omega)$ on $[\omega_1, \omega_2]$ implies that the limit $A(\omega)$ is continuous on $[\omega_1, \omega_2]$. We saw above that convergence of infinite products can be related to that of infinite summations. The following theorem [23] makes the connection between uniform convergence of summations and uniform convergence of products.

THEOREM 6.9   *Let $b_k(z), k \geq 1$ be a sequence of bounded functions of the complex variable z, such that $\sum_{k=1}^{\infty} |b_k(z)|$ converges uniformly on a compact set[14] S in the complex z plane. Then the infinite product $\prod_{k=1}^{\infty}(1 + b_k(z))$ converges uniformly on S. This product is zero for some z_0 iff $1 + b_k(z_0) = 0$ for some k.*

Uniform convergence and analyticity.   We know that if a sequence of continuous functions converges uniformly to a function, then the limit is also continuous. A similar result is true for analytic functions. If a sequence $\{f_n(s)\}$ of analytic functions converges uniformly to a function $f(s)$, then $f(s)$ is analytic as well. For a more precise statement of this result see Theorem 10.28 in [23].

Infinite Product Defining the Scaling Function

Return now to the infinite product (6.112a). As justified in Section 6.10, we assume $G_s(e^{j\omega})$ to be continuous, $G_s(e^{j0}) = 1$, and $\Phi(0) \neq 0$. Note that $G_s(e^{j0}) = 1$ is necessary for the infinite product to converge (because convergence of $\prod_k a_k$ implies that $a_k \to 1$; apply this for $\omega = 0$). The following convergence result is fundamental.

THEOREM 6.10   *(Convergence of the Infinite Product) Let $G_s(e^{j\omega}) = \sum_{n=-\infty}^{\infty} g_s(n)e^{-j\omega n}$. Assume that $G_s(e^{j0}) = 1$, and $\sum_n |n g_s(n)| < \infty$. Then,*

---

[14]For us, a compact set means any closed bounded set in the complex plane or on the real line. Examples are all points on and inside a circle in the complex plane, and the closed interval $[a, b]$ on the real line.

1. *The infinite product (6.112a) converges pointwise for all ω. In fact, it converges absolutely for all ω, and uniformly on compact sets (i.e., closed bounded sets, such as sets of the form $[\omega_1, \omega_2]$).*

2. *The quantity $G_s(e^{j\omega})$ as well as the limit $\Phi(\omega)$ of the infinite product (6.112a) are continuous functions of ω.*

3. *$G_s(e^{j\omega})$ is in L^2.*

Because the condition $\sum_n |ng_s(n)| < \infty$ implies $\sum_n |g_s(n)| < \infty$, the filter $G_s(e^{j\omega})$ is restricted to be stable, but the above result holds whether $g_s(n)$ is FIR or IIR.

Sketch of Proof. Theorem 6.9 allows us to reduce the convergence of the product to the convergence of an infinite sum. For this we must write $G_s(e^{j\omega})$ in the form $1 - F(e^{j\omega})$ and then consider the summation $\sum_{k=1}^{\infty} |F(e^{j\omega/2^k})|$. Because $G_s(e^{j0}) = 1 = \sum_n g_s(n)$, we can write $G_s(e^{j\omega}) = 1 - \left(1 - G_s(e^{j\omega})\right) = 1 - \sum_n g_s(n)\left(1 - e^{-j\omega n}\right)$. However, $|\sum_n g_s(n)(1 - e^{-j\omega n})| \leq 2\sum_n |g_s(n) \sin(\omega n/2)| \leq |\omega| \sum_n |ng_s(n)|$ (use $|\sin x/x| \leq 1$). $\sum_n |ng_s(n)|$ is assumed to converge, thus we have $|\sum_n g_s(n)\left(1 - e^{-j\omega n}\right)| \leq c|\omega|$. Using this and the fact that $\sum_{k=1}^{\infty} 2^{-k}$ converges, we can complete the proof of part 1 (by applying Theorem 6.9). $\sum_n |ng_s(n)| < \infty$ implies in particular that $g_s(n) \in \ell^1$, therefore, its ℓ^1-FT $G_s(e^{j\omega})$ is continuous (see Section 6.6). The continuity of $G_s(e^{j\omega})$, together with uniform convergence of the infinite product, implies that the pointwise limit $\Phi(\omega)$ is also continuous. Finally, because $\ell^1 \subset \ell^2$ (Section 6.6), we have $g_s(n) \in \ell^2$, that is, $G_s(e^{j\omega}) \in L^2[0, 2\pi]$ as well.

Orthonormal Wavelet Basis from Paraunitary Filter Bank

We now consider the behavior of the infinite product $\prod_{k=1}^{\infty} G_s(e^{j\omega/2^k})$, when $G_s(e^{j\omega})$ comes from a paraunitary filter bank. The paraunitary property implies that $G_s(e^{j\omega})$ is power symmetric. If we impose some further mild conditions on $G_s(e^{j\omega})$, the scaling function $\phi(t)$ generates an orthonormal multiresolution basis $\{\phi(t - n)\}$. We can then obtain an orthonormal wavelet basis $\{\psi_{kn}(t)\}$ (Theorems 6.6 and 6.7). The main results are given in Theorems 6.11 to 6.15.

First, we define the truncated partial products $P_n(\omega)$. Because $G_s(e^{j\omega})$ has period 2π, the term $G_s(e^{j\omega/2^k})$ has period $2^{k+1}\pi$. For this reason the partial product $\prod_{k=1}^{n} G_s(e^{j\omega/2^k})$ has period $2^{n+1}\pi$, and we can regard the region $[-2^n\pi, 2^n\pi]$ to be the fundamental period. Let us truncate the partial product to this region, and define

$$P_n(\omega) = \begin{cases} \prod_{k=1}^{n} G_s(e^{j\omega/2^k}), & \text{for } -2^n\pi \leq \omega \leq 2^n\pi, \\ 0, & \text{otherwise} \end{cases} \qquad (6.113)$$

This quantity will be useful later. We will see that this is in $L^2(R)$, and we can discuss $p_n(t)$, its inverse L^2-FT.

THEOREM 6.11 *Let $G_s(e^{j\omega})$ be as in Theorem 6.10. In addition let it be power symmetric; in other words, $|G_s(e^{j\omega})|^2 + |G_s(-e^{j\omega})|^2 = 1$. [Notice in particular that this implies $G_s(e^{j\pi}) = 0$, because $G_s(e^{j0}) = 1$]. Then the following are true.*

1. *$\int_0^{2\pi} |G_s(e^{j\omega})|^2 d\omega/2\pi = 0.5$.*

2. *The truncated partial product $P_n(\omega)$ is in L^2, and $\int_{-\infty}^{\infty} |P_n(\omega)|^2 d\omega/2\pi = 1$ for all n. Further, the inverse L^2-FT, denoted as $p_n(t)$, gives rise to an orthonormal sequence $\{p_n(t - k)\}$, i.e., $\langle p_n(t - k), p_n(t - i) \rangle = \delta(k - i)$ for any $n \geq 1$.*

3. The limit $\Phi(\omega)$ of the infinite product (6.112a) is in L^2, hence it has an inverse L^2-FT, $\phi(t) \in L^2$. Moreover, $\|\phi(t)\|_2 \leq 1$.

Sketch of Proof.   Part 1 follows by integrating both sides of $|G_s(e^{j\omega})|^2 + |G_s(-e^{j\omega})|^2 = 1$. The integral in part 2 is $\int_0^{2^{n+1}\pi} \prod_{k=1}^n |G_s(e^{j\omega/2^k})|^2 d\omega/2\pi$, which we can split into two terms such as $\int_0^{2^n\pi} + \int_{2^n\pi}^{2^{n+1}\pi}$. Using the 2π-periodicity and the power symmetric property of $G_s(e^{j\omega})$, we obtain $\int |P_n|^2 d\omega = \int |P_{n-1}|^2 d\omega$. Repeated application of this, together with part 1, yields $\int_{-\infty}^{\infty} |P_n(\omega)|^2 d\omega/2\pi = 1$. The proof of orthonormality of $\{p_n(t-k)\}$ follows essentially similarly by working with the modified integral $\int_{-\infty}^{\infty} |P_n(\omega)|^2 e^{j\omega(k-i)} d\omega/2\pi$, and using the half-band property of $|G_s(e^{j\omega})|^2$.

The third part is the most subtle, and uses Fatou's lemma for Lebesgue integrals (Section 6.6). For this, define $g_n(\omega) = |P_n(\omega)|^2$. Then $\{g_n(\omega)\}$ is a sequence of non-negative integrable functions such that $g_n(\omega) \to |\Phi(\omega)|^2$ pointwise for each ω. Because $\int g_n(\omega) d\omega = 2\pi$ (from part 2), Fatou's lemma assures us that $|\Phi(\omega)|^2$ is integrable with integral $\leq 2\pi$. This proves part 3.

It is most interesting that the truncated partial products $P_n(\omega)$ give rise to orthonormal sequences $\{p_n(t-k)\}$. This orthonormality is induced by the paraunitary property, more precisely the power symmetry property of $G_s(e^{j\omega})$. This is consistent with the fact that the *filter bank type of basis* introduced in Section 6.4 is an orthonormal basis for ℓ^2 whenever the filter bank is paraunitary.

As the scaling function $\Phi(\omega)$ is the pointwise limit of $\{P_n(\omega)\}$ as $n \to \infty$, this leads to the hope that $\{\phi(t-k)\}$ is also an orthonormal sequence, so that we can generate a multiresolution and then a wavelet basis as in Theorems 6.6 and 6.7). This, however, is not always true. The crux of the reason is that $\Phi(\omega)$ is only the pointwise limit of $\{P_n(\omega)\}$, and not necessarily the L^2 limit. The distinction is subtle (see below). The pointwise limit property means that for any fixed ω, the function $P_n(\omega)$ approaches $\Phi(\omega)$. The L^2 limit property means that $\int |P_n(\omega) - \Phi(\omega)|^2 d\omega \to 0$. Neither of these limit properties implies the other; neither is stronger than the other. It can be shown that it is the L^2 limit which propagates the orthonormality property, and this is what we want.

THEOREM 6.12   Let $\{p_n(t-k)\}$ be an orthonormal sequence for each n. That is, $\langle p_n(t-k), p_n(t-i)\rangle = \delta(k-i)$. Suppose $p_n(t) \to \phi(t)$ in the L^2 sense. Then $\{\phi(t-k)\}$ is an orthonormal sequence.

PROOF 6.1   If we take limits as $n \to \infty$, we can write

$$\lim_{n \to \infty} \langle p_n(t-k), p_n(t-i)\rangle = \left\langle \lim_{n \to \infty} p_n(t-k), \lim_{n \to \infty} p_n(t-i)\right\rangle \qquad (6.114)$$

This movement of the "limit" sign past the inner product sign is allowed (by continuity of inner products, Section 6.6), provided the limits in the second expression are L^2 limits. By the conditions of the theorem, the left side of the above equation is $\delta(k-i)$, whereas the right side is $\langle \phi(t-k), \phi(t-i)\rangle$. So the result follows. $\square$

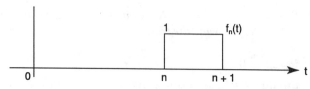

6.42   A sequence $\{f_n(t)\}$ whose pointwise limit is not a limit in the L^2 sense.

L^2 Convergence vs. Pointwise Convergence

The fact that L^2 limits are not necessarily pointwise limits is obvious from the fact that differences at a countable set of points do not affect integrals. The fact that pointwise limits are not necessarily L^2 limits is demonstrated by the sequence of L^2 functions $\{f_n(t)\}$, with $f_n(t)$ as in Fig. 6.42. Note that $f_n(t) \to 0$ pointwise for each t, that is, the pointwise limit is $f(t) \equiv 0$. Hence, $\|f_n(t) - f(t)\| = \|f_n(t)\| = 1$ for all n, so $\|f_n(t) - f(t)\|$ does not go to zero as $n \to \infty$, and thus, $f(t)$ is not the L^2 limit of $f_n(t)$. Notice in this example that $1 = \lim_{n\to\infty} \int |f_n(t)|^2 dt \neq \int \lim_{n\to\infty} |f_n(t)|^2 dt = 0$. This is consistent with the fact that the Lebesgue dominated convergence theorem cannot be applied here—no integrable function dominates $|f_n(t)|^2$ for all n. In this example, the sequence $\{f_n(t)\}$ does not converge in the L^2 sense. In fact, $\|f_n(t) - f_m(t)\|^2 = 2$ for $n \neq m$. Thus, $\{f_n\}$ is not a Cauchy sequence [22] in L^2.

Some facts pertaining to pointwise and L^2 convergences: It can be shown that if $f_n(t) \to f(t)$ in L^2 sense and $f_n(t) \to g(t) \in L^2$ pointwise as well, then $f(t) = g(t)$ a.e. In particular, $\|f(t) - g(t)\| = 0$ and $\|f(t)\| = \|g(t)\|$. It also can be shown that if $f_n(t) \to f(t)$ in L^2 sense, then $\|f_n(t)\| \to \|f(t)\|$. Finally, if $f_n(t) \to f(t) \in L^2$ pointwise a.e., and $\|f_n(t)\| \to \|f(t)\|$ then $f_n(t) \to f(t)$ in L^2 sense as well [23].

THEOREM 6.13 *(Orthonormal Wavelet Basis) Let the filter $G_s(e^{j\omega}) = \sum_{n=-\infty}^{\infty} g_s(n)e^{-j\omega n}$ satisfy the following properties:*

1. $G_s(e^{j0}) = 1$
2. $\sum_n |ng_s(n)| < \infty$
3. $|G_s(e^{j\omega})|^2 + |G_s(-e^{j\omega})|^2 = 1$ *(power symmetry)*
4. $G_s(e^{j\omega}) \neq 0$ *for $\omega \in [-0.5\pi, 0.5\pi]$*

Then the infinite product (6.112a) converges to a limit $\Phi(\omega) \in L^2$, and its inverse FT $\phi(t)$ is such that $\{\phi(t - n)\}$ is an orthonormal sequence. Defining the wavelet function $\psi(t)$ as usual, i.e., as in (6.93), the sequence $\{2^{k/2}\psi(2^k t - n)\}$ (with k and n varying over all integers) forms an orthonormal wavelet basis for L^2.

Sketch of Proof. We will show that the sequence $\{P_n(\omega)\}$ of partial products converges to $\Phi(\omega)$ in the L^2 sense, i.e., $\int |P_n(\omega) - \Phi(\omega)|^2 d\omega \to 0$, so that $p_n(t) \to \phi(t)$ in L^2 sense. The desired result then follows in view of Theorems 6.11 and 6.12. The key tool in the proof is the dominated convergence theorem for Lebesgue integrals (Section 6.6). First, the condition $G(e^{j\omega}) \neq 0$ in $[-0.5\pi, 0.5\pi]$ implies that $\Phi(\omega) \neq 0$ in $[-\pi, \pi]$. Because $|\Phi(\omega)|^2$ is continuous (Theorem 6.10) it has a minimum value $c^2 > 0$ in $[-\pi, \pi]$. Now the truncated partial product $P_n(\omega)$ can always be written as $P_n(\omega) = \Phi(\omega)/\Phi(\omega/2^n)$ in its region of support. Because $|\Phi(\omega/2^n)|^2 \geq c^2$ in $[-2^n\pi, 2^n\pi]$, we have $|P_n(\omega)|^2 \leq |\Phi(\omega)|^2/c^2$ for all ω. Define $Q_n(\omega) = |P_n(\omega) - \Phi(\omega)|^2$. Then using $|P_n(\omega)|^2 \leq |\Phi(\omega)|^2/c^2$ we can show that $Q_n(\omega) \leq \alpha |\Phi(\omega)|^2$ for some constant α. Because the right-hand side is integrable, and because $Q_n(\omega) \to 0$ pointwise (Theorem 6.10) we can use the dominated convergence theorem (Section 6.6) to conclude that $\lim_n \int Q_n(\omega) \, d\omega = \int \lim_n Q_n(\omega) \, d\omega = 0$. This completes the proof.

Computing the Scaling and Wavelet Functions

Given the coefficients $g_s(n)$ of the filter $G_s(e^{j\omega})$, how do we compute the scaling function $\phi(t)$ and the wavelet function $\psi(t)$? Because we can compute $\psi(t)$ using $\psi(t) = 2\sum_{n=-\infty}^{\infty}(-1)^{n+1}g_s^*(-n-1)\phi(2t - n)$, the key issue is the computation of $\phi(t)$. In the preceding theorems $\phi(t)$ was defined only as an inverse L^2-FT of the infinite product $\Phi(\omega)$ given in (6.112a). Because an L^2 function is determined only in the a.e. sense, this way of defining $\phi(t)$ itself does not fully determine $\phi(t)$.

Recall, however, that the infinite product for $\Phi(\omega)$ was only a consequence of the more fundamental equation, the dilation equation $\phi(t) = 2\sum_{n=-\infty}^{\infty} g_s(n)\phi(2t-n)$. In practice $\phi(t)$ is computed using this equation, which is often a finite sum (see Section 6.12). The procedure is recursive; we assume an initial solution for the function $\phi(t)$, substitute it into the right-hand side of the dilation equation, thereby recomputing $\phi(t)$, and then repeat the process. Details of this and discussions on convergence of this procedure can be found in [5, 15], and [26]

Lawton's Eigenfunction Condition for Orthonormality [5]

Equation (6.100) is equivalent to the orthonormality of $\{\phi(t-n)\}$. Let $S(e^{j\omega})$ denote the left-hand side of (6.100), which evidently has period 2π in ω. Using the frequency domain version of the dilation equation (6.85), it can be shown that the scaling function $\phi(t)$ generated from $G_s(e^{j\omega})$ is such that

$$|G_s(e^{j\omega})|^2 S(e^{j\omega})\Big|_{\downarrow 2} = 0.5 S(e^{j\omega}) \tag{6.115}$$

where the notation $\downarrow 2$ indicates decimation (Section 6.4). Thus, the function $S(e^{j\omega})$ can be regarded as an eigenfunction (with eigenvalue = 0.5) of the operator $\mathcal{F}$, which performs filtering by $|G_s(e^{j\omega})|^2$ followed by decimation.

Now consider the case in which the digital filter bank is paraunitary, so that $G_s(e^{j\omega})$ is power symmetric [i.e., satisfies (6.95)]. The power symmetric condition can be rewritten in the form $|G_s(e^{j\omega})|^2|_{\downarrow 2} = 0.5$. Thus, in the power symmetric case the identity function is an eigenfunction of the operator $\mathcal{F}$. If the only eigenfunction of the operator $\mathcal{F}$ is the identity function, it then follows that $S(e^{j\omega}) = 1$; i.e., (6.100) holds and $\{\phi(t-n)\}$ is orthonormal.

The FIR Case. Section 6.12 shows that restricting $G_s(z)$ to be FIR ensures that $\phi(t)$ has finite duration. For the FIR case, Lawton and Cohen independently showed that the above eigenfunction condition also works in the other direction. That is, if $\{\phi(t-n)\}$ has to be orthonormal, then the trignometric polynomial $S(e^{j\omega})$ satisfying (6.115) must be unique up to a scale factor.[15] Details can be found in [5].

Examples and Counter-Examples

We already indicated after the introduction of (6.112a and b) that the example of the ideal bandpass wavelet can be generated formally by starting from the ideal brickwall paraunitary filter bank. We now discuss some other examples.

EXAMPLE 6.12:   Haar Basis from Filter Banks.

A filter bank of the form Fig. 6.22(a) with filters

$$G_a(z) = \frac{1+z^{-1}}{2} \quad H_a(z) = \frac{z^{-1}-1}{2} \quad G_s(z) = \frac{1+z^{-1}}{2} \quad H_s(z) = \frac{1-z^{-1}}{2}$$

is paraunitary. The magnitude responses of the synthesis filters, $|G_s(e^{j\omega})| = |\cos(\omega/2)|$ and $|H_s(e^{j\omega})| = |\sin(\omega/2)|$, are shown in Fig. 6.43(a). $G_s(z)$ satisfies all the conditions of Theorem 6.13. In this case we can evaluate the infinite products for $\Phi(\omega)$ and $\Psi(\omega)$ explicitly by using the identity $\prod_{m=1}^{\infty} \cos(2^{-m}\omega) = \sin\omega/\omega$. The resulting $\phi(t)$ and $\psi(t)$ are as shown in Fig. 6.43(b) and (c). These are precisely the functions that generate the Haar orthonormal basis.

---

[15] A finite sum of the form $\sum_{n=N_1}^{N_2} p_n e^{j\omega n}$ is said to be a trigonometric polynomial. If $G_s(e^{j\omega})$ is FIR, it can be shown that the left-hand side of (6.100) is not only periodic in ω, but is in fact a trigonometric polynomial.

EXAMPLE 6.13: Paraunitary Filter Bank which Does Not Give Orthonormal Wavelets.

Consider the filter bank with analysis filters $G_a(z) = (1 + z^{-3})/2$, $H_a(z) = -(1 - z^{-3})/2$, and synthesis filters $G_s(z) = (1 + z^{-3})/2$, $H_s(z) = (1 - z^{-3})/2$. As this is obtained from the preceding example by the substitution $z \rightarrow z^3$, it remains paraunitary and satisfies the PR property. $G_s(z)$ satisfies all the properties of Theorem 6.13, except the fourth condition. With $\phi(t)$ and $\psi(t)$ obtained from $G_s(e^{j\omega})$ using the usual dilation equations, the functions $\{\phi(t - n)\}$ are *not orthonormal*. In addition, the wavelet functions $\{2^{k/2}\psi(2^k t - n)\}$ do not form an orthonormal basis. These statements can be verified from the sketches of the functions $\phi(t)$ and $\psi(t)$ shown in Fig. 6.44. Clearly, $\phi(t)$ and $\phi(t-1)$ are not orthogonal, and $\psi(t)$ and $\psi(t-2)$ are not orthogonal. In this example $\|P_n(\omega)\| = 1$ for all n, whereas $\|\Phi(\omega)\| = 1/\sqrt{3}$. The limit of $\|P_n(\omega)\|$ does not agree with $\|\Phi(\omega)\|$, and our conclusion is that $\Phi(\omega)$ is not the L^2 limit of $P_n(\omega)$. The L_2 limit of $P_n(\omega)$ does not exist in this example.

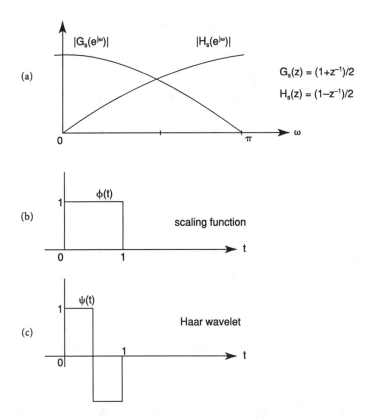

6.43   Haar basis generated from paraunitary filter bank. (a) The synthesis filters in the paraunitary filter bank, (b) the scaling function, and (c) the wavelet function generated using dilation equations.

Thus, a paraunitary filter bank may not generate an orthonormal wavelet basis if the fourth condition in Theorem 6.13 is violated. However, this is hardly of concern in practice, because any reasonable low-pass filter designed for a two-channel filter bank will be free from zeroes in the region $[-0.5\pi, 0.5\pi]$. In fact, a stronger result was proved by Cohen, who derived necessary and sufficient conditions for an FIR paraunitary filter bank to generate an orthonormal wavelet basis. One outcome of Cohen's analysis is that the fourth condition in Theorem 6.13 can be replaced by the even milder condition that $G_s(e^{j\omega})$ not be zero in $[-\pi/3, \pi/3]$. In this sense the condition for obtaining an orthonormal

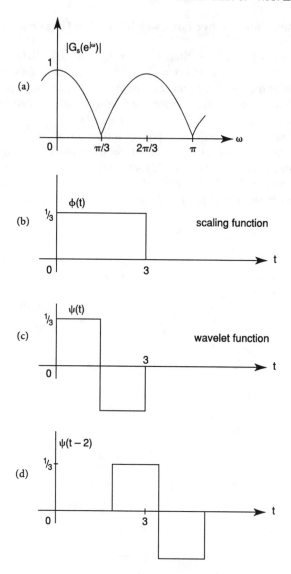

6.44 A paraunitary filter bank generating nonorthonormal $\{\phi(t-n)\}$. (a) The synthesis filter response, (b) the scaling function, (c) the wavelet function, and (d) a shifted version.

wavelet basis is trivially satisfied in practice. The case in which the fourth condition fails is primarily of theoretical interest; an attractive result in this context is Lawton's tight frame theorem.

Wavelet Tight Frames

Although the wavelet functions $\{2^{k/2}\psi(2^kt-n)\}$ generated from a paraunitary filter bank may not form an orthonormal basis when the fourth condition of Theorem 6.13 is violated, the functions always form a tight frame for L^2. Thus, any L^2 function can be expressed as an infinite linear combination of the functions $\{2^{k/2}\psi(2^kt-n)\}$. More precisely, we have the following result due to Lawton [5].

THEOREM 6.14 *(Tight Frames from Paraunitary Filter Banks) Let* $G_s(e^{j\omega}) = \sum_{n=0}^{N} g_s(n)e^{-j\omega n}$
be a filter satisfying the following properties:

1. $G_s(e^{j0}) = 1$.
2. $|G_s(e^{j\omega})|^2 + |G_s(-e^{j\omega})|^2 = 1$ *(power symmetry)*.

Then, $\phi(t) \in L^2$. *Defining the wavelet function* $\psi(t)$ *as in (6.93), the sequence* $\{2^{k/2}\psi(2^k t - n)\}$
(with k and n varying over all integers) forms a tight frame for L^2, *with frame bound unity (i.e.,*
$A = B = 1$; *see Section 6.8).*

Thus, the functions $\psi_{kn}(t)$ in Example 6.13 constitute a tight frame for L^2. From Section 6.8 we
know that this tight frame property means that any $x(t) \in L^2$ can be expressed as

$$x(t) = \sum_{k=-\infty}^{\infty} \sum_{n=-\infty}^{\infty} \langle x(t), \psi_{kn}(t) \rangle \psi_{kn}(t) \qquad (6.116)$$

where $\psi_{kn}(t) = 2^{k/2}\psi(2^k t - n)$. This expression is pretty much like an expansion into an orthonormal
basis. We can find the wavelet coefficients $c_{kn} = \langle x(t), \psi_{kn}(t) \rangle$ exactly as in the orthonormal case.
We also know that frames offer stability of reconstruction. Thus, in every respect this resembles
an orthonormal basis, the only difference being that the functions are not linearly independent
(redundancy exists in the wavelet tight frame $\{\psi_{kn}(t)\}$).

6.12 Compactly Supported Orthonormal Wavelets

Section 6.11 showed how to construct an orthonormal wavelet basis for L^2 space by starting from a
paraunitary filter bank. Essentially, we defined two infinite products $\Phi(\omega)$ and $\Psi(\omega)$ starting from
the digital low-pass filter $G_s(e^{j\omega})$. Under some mild conditions on $G_s(e^{j\omega})$, the products converge
(Theorem 6.10). Under the further condition that $G_s(e^{j\omega})$ be power symmetric and nonzero in
$[-0.5\pi, 0.5\pi]$, we saw that $\{\phi(t-k\}$ forms an orthonormal set, and the corresponding $\{2^{k/2}\psi(2^k t -
n)\}$ forms an orthonormal wavelet basis for L^2 (Theorem 6.13). If we further constrain $G_s(e^{j\omega})$ to
be FIR, that is, $G_s(z) = \sum_{n=0}^{N} g_s(n)z^{-n}$, then the scaling function $\phi(t)$ and the wavelet function
$\psi(t)$ have finite duration [6, 5].

THEOREM 6.15 *Let* $G_s(z) = \sum_{n=0}^{N} g_s(n)z^{-n}$, *with* $G_s(e^{j0}) = 1$ *and* $H_s(e^{j\omega}) = e^{j\omega} G_s^*(-e^{j\omega})$.
Define the infinite products as in (6.112a) and (b), and assume that the limits $\Phi(\omega)$ *and* $\Psi(\omega)$ *are* L^2
functions, for example, by imposing power symmetry condition on $G_s(z)$ *as in Theorem 6.11. Then*
$\phi(t)$ *and* $\psi(t)$ *(the inverse* L^2*-FTs) are compactly supported, with support in* $[0, N]$.

The time decay of the wavelet $\psi(t)$ is therefore excellent. In particular, all the basis functions
$2^{k/2}\psi(2^k t - n)$ are compactly supported. By further restricting the low-pass filter $G_s(z)$ to have a
sufficient number of zeroes at $\omega = \pi$, we also ensure (Section 6.13) that the FT $\Psi(\omega)$ has excellent
decay (equivalently $\psi(t)$ is regular or smooth in the sense to be quantified in Section 6.13).

The rest of this section is devoted to the technical details of the above result. The reader not interested
in these details can move to Section 6.13 without loss of continuity. The theorem might seem
"obvious" at first sight, and indeed a simple engineering argument based on Dirac delta functions
can be given (p. 521 of [7]). However, the correct mathematical justification relies on a number of
deep results in function theory. One of these is the celebrated Paley-Wiener theorem for band-limited
functions.

Paley-Wiener Theorem. A beautiful result in the theory of signals is that if an L^2 function $f(t)$ is band-limited, that is, $F(\omega) = 0$, $|\omega| \geq \sigma$, then $f(t)$ is the "real-axis restriction of an entire function." We say that a function $f(s)$ of the complex variable s is entire if it is analytic for all s. Examples are polynomials in s, exponentials such as e^s, and simple combinations of these. The function $f(t)$ obtained from $f(s)$ for real values of s ($s = t$) is the real-axis restriction of $f(s)$.

Thus, if $f(t)$ is a band-limited signal then an entire function $f(s)$ exists such that its real-axis restriction is $f(t)$. In particular, therefore, a band-limited function $f(t)$ is continuous and infinitely differentiable with respect to the time variable t. The entire function $f(s)$ associated with the band-limited function has the further property $|f(s)| \leq ce^{\sigma|s|}$ for some $c > 0$. We express this by saying that $f(s)$ is **exponentially bounded** or of the exponential type. What is even more interesting is that the converse of this result is true: if $f(s)$ is an entire function of the exponential type, and the real-axis restriction $f(t)$ is in L^2, then $f(t)$ is band-limited. By interchanging the time and frequency variables we can obtain similar conclusions for time-limited signals; this is what we need in the discussion of time-limited (compactly supported) wavelets.

THEOREM 6.16 *(Paley-Wiener) Let $W(s)$ be an entire function such that for all s, we have $|W(s)| \leq c \exp(A|s|)$ for some c, $A > 0$, and the real-axis restriction $W(\omega)$ is in L^2. Then, there exists a function $w(t)$ in L^2 such that $W(s) = \int_{-A}^{A} w(t)e^{-jts}\,dt$.*

A proof can be found in [23]. Thus, $w(t)$ can be regarded as a compactly supported function with support in $[-A, A]$. Recall (6.48) that $L^2[-A, A] \subset L^1[-A, A]$, so $w(t)$ is in $L^1[-A, A]$ and $L^2[-A, A]$. Therefore, $W(\omega)$ is the L^1-FT of $w(t)$, and agrees with the L^2-FT a.e.

Our aim is to show that the infinite product for $\Phi(\omega)$ satisfies the conditions of the Paley-Wiener theorem, and therefore that $\phi(t)$ is compactly supported. A modified version of the above result is more convenient for this. The modification allows the support to be more general, namely $[-A_1, A_2]$, and permits us to work with the imaginary part of s rather than the absolute value.

THEOREM 6.17 *(Paley-Wiener, Modified) Let $W(s)$ be an entire function such that*

$$|W(s)| \leq \begin{cases} c_1 \exp(A_1|\operatorname{Im} s|), & \operatorname{Im} s \geq 0 \\ c_2 \exp(A_2|\operatorname{Im} s|), & \operatorname{Im} s \leq 0 \end{cases} \tag{6.117}$$

for some c_1, c_2, A_1, $A_2 > 0$, and such that the real-axis restriction $W(\omega)$ is in L^2. Then a function $w(t)$ exists in L^2 such that $W(s) = \int_{-A_2}^{A_1} w(t)e^{-jts}\,dt$. We can regard $W(\omega)$ as the FT of the function $w(t)$ supported in $[-A_2, A_1]$.

This result can be made more general; the condition (6.117) can be replaced with one in which the right-hand sides have the form $P_i(s) \exp(A_i|\operatorname{Im} s|)$, where $P_i(s)$ are polynomials. We are now ready to sketch the proof that $\phi(t)$ and $\psi(t)$ have the compact support $[0, N]$.

1. Using the fact that $G_s(z)$ is FIR and that $G_s(e^{j0}) = 1$, show that the product $\prod_{k=1}^{\infty} G_s(e^{js/2^k})$ converges uniformly on any compact set of the complex s-plane. (For real s, namely $s = \omega$, this holds even for the IIR case as long as $\sum_n |ng_s(n)|$ converges. This was shown in Theorem 6.10.)

2. Uniformity of convergence of the product guarantees that its limit $\Phi(s)$ is an entire function of the complex variable s (Theorem 10.28, [23]).

3. The FIR nature of $G_s(z)$ allows us to establish the exponential bound (6.117) for $\Phi(s)$ with $A_2 = 0$ and $A_1 = N$. This shows that $\phi(t)$ is compactly supported in $[0, N]$.

Because $\psi(t)$ is obtained from the dilation equation (6.93), the same result follows for $\psi(t)$ as well.

6.13   Wavelet Regularity

From the preceding section we know that if we construct the power symmetric FIR filter $G_s(z)$ properly, we can get an orthonormal multiresolution basis $\{\phi(t-n)\}$, and an orthonormal wavelet basis $\{2^{k/2}\psi(2^k t - n)\}$ for L^2. Both of these bases are compactly supported. These are solutions to the two-scale dilation equations

$$\phi(t) = 2\sum_{n=0}^{N} g_s(n)\phi(2t - n) \qquad (6.118)$$

$$\psi(t) = 2\sum_{n=0}^{N} h_s(n)\phi(2t - n) \qquad (6.119)$$

where $h_s(n) = (-1)^{n+1} g_s^*(-n-1)$. In the frequency domain we have the explicit infinite product expressions (6.112a and b) connecting the filters $G_s(z)$ and $H_s(z)$ to the L^2-FTs $\Phi(\omega)$ and $\Psi(\omega)$.

Figure 6.45(a) shows two cases of a ninth-order FIR filter $G_s(e^{j\omega})$ used to generate the compactly supported wavelet. The resulting wavelets are shown in Figs. 6.45(b) and (c). In both cases all conditions of Theorem 6.13 are satisfied so we obtain orthonormal wavelet bases for L^2. The filter $G_s(e^{j\omega})$ has more zeroes at π for case 2 than for case 1. The corresponding wavelet looks much smoother or "regular;" this is an example of a Daubechies wavelet. By designing $G_s(z)$ to have a sufficient number of zeroes at π we can make the wavelet "as regular as we please." A quantitative discussion of the connection between the number of zeroes at π and the smoothness of $\psi(t)$ is given in the following discussions.

Qualitatively, the idea is that if $G_s(e^{j\omega})$ has a large number of zeroes at π, the function $\Phi(\omega)$ given by the infinite product (6.112a) decays "fast," as $\omega \to \infty$. This fast asymptotic decay in the frequency domain implies that the time function $\phi(t)$ is "smooth." Because $\psi(t)$ is derived from $\phi(t)$ using a finite sum (6.119), the smoothness of $\phi(t)$ is transmitted to $\psi(t)$. We will make the ideas more quantitative in the next few sections.

Why Regularity?

The point made above was that if we design an FIR paraunitary filter bank with the additional constraint that the low-pass filter $G_s(e^{j\omega})$ have a sufficient number of zeroes at π, the wavelet basis functions $\psi_{kn}(t)$ are sufficiently smooth. The smoothness requirement is perhaps the main new component brought into the filter bank theory from the wavelet theory. Its importance can be understood in a number of ways.

Consider the expansion $x(t) = \sum_{k,n} c_{k,n}\psi_{kn}(t)$. Suppose we truncate this to a finite number of terms, as is often done in practice. If the basis functions are not smooth, the error can produce perceptually annoying effects in applications such as audio and image coding, even though the L^2 norm of the error may be small.

Next, consider a tree-structured filter bank. An example is shown in Fig. 6.26. In the synthesis bank, the first path can be regarded as an effective interpolation filter, or an expander [e.g., $\uparrow 8$ in Fig. 6.26(b)] followed by a filter of the form $G_s(e^{j\omega})G_s(e^{2j\omega})G_s(e^{4j\omega})\cdots G_s(e^{2^L j\omega})$. The same finite product can be obtained by truncating to $L+1$ terms the infinite product defining $\Phi(\omega)$ (6.112a and b), and making a change of variables. Similarly, the remaining paths can be related to interpolation filters which are various truncated versions of the infinite product defining

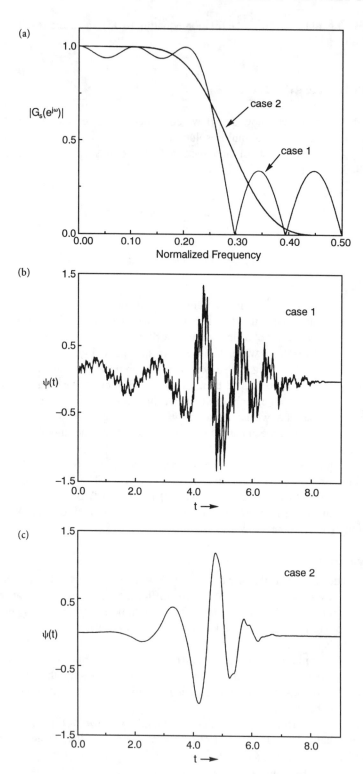

6.45   Demonstrating the importance of zeroes at π. (a) The response of the FIR filter $G_s(z)$ for two cases, and (b) and (c) the corresponding wavelet functions.

$\Psi(\omega)$ in (6.112a and b). Imagine that we use the tree-structured system in subband coding. The quantization error in each subband is filtered through an interpolation filter. If the impulse response of the interpolation filter is not smooth enough [e.g., if it resembles Fig. 6.45(b)], the filtered noise tends to show severe perceptual effects, for example, in image reconstruction. This explains, qualitatively, the importance of having "smooth impulse responses" for the synthesis filters.

Smoothness and Hölder Regularity Index

We are familiar with the notion of continuous functions. We say that $f(t)$ is continuous at t_0 if, for any $\epsilon > 0$, we can find a $\delta > 0$ such that $|f(t) - f(t_0)| < \epsilon$ for all t satisfying $|t - t_0| < \delta$. A stronger type of continuity, called Hölder continuity, is defined as follows: $f(t)$ is Hölder continuous in a region S if $|f(t_0) - f(t_1)| \leq c|t_0 - t_1|^\beta$ for some $c, \beta > 0$, for all $t_0, t_1 \in S$. This implies, in particular, continuity in the ordinary sense. If $\beta > 1$ the above would imply that $f(t)$ is constant on S. For this reason, we have the restriction $0 < \beta \leq 1$. As β increases from 0 to 1, the function becomes increasingly "smoother." The constant β is called the **Lipschitz constant** of the function $f(t)$.

Suppose the function $f(t)$ is n times differentiable in some region S and the nth derivative $f^{(n)}(t)$ is Hölder continuous with Lipschitz constant β. Define $\alpha = n + \beta$. We say that $f(t)$ belongs to the class C^α. The coefficient α is called the **Hölder regularity index** of $f(t)$. For example, $C^{3.4}$ is the class of functions that are three times differentiable and the third derivatives are Hölder continuous with Lipschitz constant equal to 0.4.

The Hölder regularity index α is taken as a quantitative measure of regularity or smoothness of the function $\psi(t)$. We sometimes say $\psi(t)$ has regularity α. Qualitatively speaking, a function with a large Hölder index is regarded as more "smooth" or "well-behaved." Because the dilation equations in the FIR case are finite summations, the Hölder indices of $\phi(t)$ and $\psi(t)$ are identical.

Some functions are differentiable an infinite number of times. That is, they belong to C^∞. Examples are e^t, $\sin t$, and polynomials. C^∞ functions even exist that are compactly supported (i.e., have finite duration; they will not be discussed here).

Frequency-Domain Decay and Time-Domain Smoothness

We can obtain time-domain smoothness of a certain degree by imposing certain conditions on the FT $\Psi(\omega)$. This is made possible by the fact that the rate of decay of $\Psi(\omega)$ as $\omega \to \infty$ (i.e., the asymptotic decay) governs the Hölder regularity index α of $\psi(t)$. Suppose $\Psi(\omega)$ decays faster than $(1 + |\omega|)^{-(1+\alpha)}$:

$$|\Psi(\omega)| \leq \frac{c}{(1 + |\omega|)^{1+\alpha+\epsilon}} \quad \text{for all } \omega \tag{6.120}$$

for some $c > 0$, $\epsilon > 0$. Then, $\Psi(\omega)(1 + |\omega|)^\alpha$ is bounded by the integrable function $c/(1 + |\omega|)^{1+\epsilon}$, and is therefore (Lebesgue) integrable. Using standard Fourier theory it can be shown that this implies $\psi(t) \in C^\alpha$. In the wavelet construction of Section 6.11, which begins with a digital filter bank, the above decay of $\Psi(\omega)$ can be accomplished by designing the digital filter $G_s(e^{j\omega})$ such that it has a sufficient number of zeroes at $\omega = \pi$.

Thus, the decay in the frequency domain translates into regularity in the time domain. Similarly, one can regard time-domain decay as an indication of smoothness in frequency. When comparing two kinds of wavelets, we can usually compare them in terms of time domain regularity (frequency domain decay) and time domain decay (frequency domain smoothness). An extreme example is one in which $\psi(t)$ is band-limited. This means that $\Psi(\omega)$ is zero outside the passband, and so the "decay" is the best possible. Correspondingly, the smoothness of $\psi(t)$ is excellent; in fact, $\psi(t) \in C^\infty$. However, the *decay* of $\psi(t)$ may not be excellent (certainly it cannot be time-limited if it is band-limited).

Return to the two familiar wavelet examples, the Haar wavelet (Fig. 6.12) and the bandpass wavelet (Figs. 6.9 and 6.11). We see that the Haar wavelet has poor decay in the frequency domain because $\Psi(\omega)$ decays only as ω^{-1}. Correspondingly, the time-domain signal $\psi(t)$ is not even continuous, hence, not differentiable.[16] The bandpass wavelet, on the other hand, is band-limited, so the decay in frequency is excellent. Thus, $\psi(t) \in C^{\infty}$, but it decays slowly, behaving similarly to t^{-1} for large t. These two examples represent two extremes of orthonormal wavelet bases for L^2.

The game, therefore, is to construct wavelets that have good **decay in time** as well as good **regularity in time.** An extreme hope is where $\psi(t) \in C^{\infty}$, and has compact support as well. It can be shown that such $\psi(t)$ can never give rise to an orthonormal basis, so we must strike a compromise between *regularity in time* and *decay in time.*

Regularity and Decay in Early Wavelet Constructions

In 1982 Stromberg showed how to construct wavelets in such a way that $\psi(t)$ has exponential decay, and at the same time has arbitrary regularity (i.e., $\psi(t) \in C^k$ for any chosen integer k). In 1985 Meyer constructed wavelets with band-limited $\psi(t)$ [so $\psi(t) \in C^{\infty}$ as for the bandpass wavelet], but he also showed how to design this $\psi(t)$ to decay faster than any chosen inverse polynomial, as $t \to \infty$. Figure 6.46(a) shows an example of a Meyer wavelet; a detailed description of this wavelet can be found in [5]. In both of the above constructions the wavelets gave rise to orthonormal bases for L^2.

In 1987 and 1988 Battle and Lemarié independently constructed wavelets with similar properties as Stromberg's wavelets, namely $\psi(t) \in C^k$ for arbitrary k, and $\psi(t)$ decays exponentially. Their construction is based on spline functions and an orthonormalization step, as described in Section 6.10. The resulting wavelets, while not compactly supported, decay exponentially and generate orthonormal bases. Figure 6.46(b) shows an example of the Battle-Lemarié wavelet.

Table 6.1 gives a summary of the main features of these early wavelet constructions (first three entries). When these examples were constructed, the relation between wavelets and digital filter banks was not known. The constructions were not systematic or unified by a central theory. Moreover, it was not clear whether one could get a compactly supported (i.e., finite duration) wavelet $\psi(t)$ which at the same time had arbitrary regularity (i.e., $\psi(t) \in C^k$ for any chosen k), *and generated an orthonormal wavelet basis.* This was made possible for the first time when the relation between wavelets and digital filter banks was observed by Daubechies in [6]. Simultaneously and independently Mallat invented the multiresolution framework and observed the relation between his framework, wavelets, and paraunitary digital filter banks (the CQF bank, Section 6.4). These discoveries have made the wavelet construction easy and systematic, as described in Sections 6.11 and 6.12. The way to obtain arbitrary wavelet regularity with this scheme is described next.

Time-Domain Decay and Time-Domain Regularity

We now state a fundamental limitation which arises when trying to impose regularity and decay simultaneously [5].

THEOREM 6.18   *(Vanishing Moments) Let* $\{2^{k/2}\psi(2^k t - n)\}$, $-\infty \le k, n \le \infty$ *be an orthonormal set in* L^2. *Suppose the wavelet* $\psi(t)$ *satisfies the following properties:*

---

[16]It is true that $\psi(t)$ is differentiable almost everywhere, but the discontinuities at the points $t = 0, 0.5, 1.0$ will be very noticeable if we take linear combinations such as $\sum_{k,n} c_{kn} \psi_{kn}(t)$.

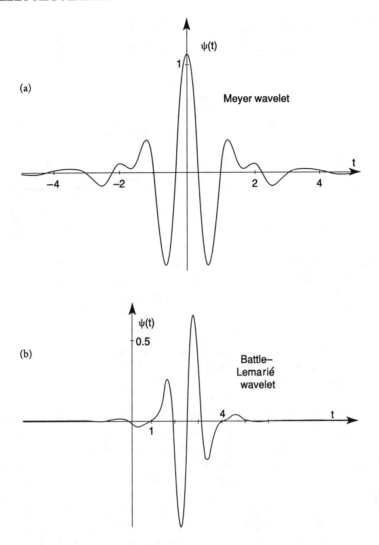

6.46 (a) An example of the Meyer wavelet, and (b) an example of the Battle-Lemarié wavelet.

1. $|\psi(t)| \leq c(1 + |t|)^{-(m+1+\epsilon)}$ *for some integer m and some* $\epsilon > 0$; *that is, the wavelet decays faster than* $(1 + |t|)^{-(m+1)}$.

2. $\psi(t) \in C^m$ *[i.e.,* $\psi(t)$ *differentiable m times], and the m derivatives are bounded.*

Then the first m moments of $\psi(t)$ *are zero, that is,* $\int t^i \psi(t) dt = 0$ *for* $0 \leq i \leq m$.

Impossibility of Compact Support, Infinite Differentiability, and Orthonormality. Suppose we have an orthonormal wavelet basis such that $\psi(t)$ is compactly supported, and infinitely differentiable [i.e., $\psi(t) \in C^\infty$]. Then all the conditions of Theorem 6.18 are satisfied. So the moments of $\psi(t)$ are zero, and therefore $\psi(t) = 0$ for all t violating the unit-norm property of $\psi(t)$. Thus, we cannot design compactly supported orthonormal wavelets which are infinitely differentiable; only a finite Hölder index can be accomplished. A similar observation can be made even when $\psi(t)$ is

TABLE 6.1   Summary of Several Types of Wavelet Bases for $L^2(R)$

Type of Wavelet	Decay of $\psi(t)$ in Time	Regularity of $\psi(t)$ in Time	Type of Wavelet Basis
Stromberg, 1982	Exponential	$\psi(t) \in C^k$; k can be chosen arbitrarily large	Orthonormal
Meyer, 1985	Faster than any chosen inverse polnomial	$\psi(t) \in C^\infty$ (band-limited)	Orthonormal
Battle-Lemarié, 1987, 1988 (Splines)	Expotential	$\psi(t) \in C^k$; k can be chosen arbitrarily large	Orthonormal
Daubechies, 1988	Compactly supported	$\psi(t) \in C^\alpha$; α can be chosen as large as we please	Orthonormal

not compactly supported as long as it decays faster than any inverse polynomial (e.g., exponential decay).

The vanishing moment condition $\int t^i \psi(t)dt = 0, 0 \le i \le m$ implies that the L^2-FT $\Psi(\omega)$ has $m+1$ zeroes at $\omega = 0$. This follows by using standard theorems on the L^1-FT [23].[17] Thus, the first m derivatives of $\Psi(\omega)$ vanish at $\omega = 0$. This implies a certain degree of flatness at $\omega = 0$. Summarizing, we have the following result.

THEOREM 6.19   (Flatness in Frequency and Regularity in Time) *Suppose we have a compactly supported $\psi(t)$ generating an orthonormal wavelet basis $\{2^{k/2}\psi(2^k t - n)\}$, and let $\psi(t) \in C^m$, with m derivatives bounded. Then $\Psi(\omega)$ has $m+1$ zeroes at $\omega = 0$.*

Return now to the wavelet construction technique described in Section 6.11. We started from a paraunitary FIR filter bank (Fig. 6.22(a)) and obtained the scaling function $\phi(t)$ and wavelet function $\psi(t)$ as in (6.118) and (6.119). The FIR nature implies that $\psi(t)$ has compact support (Section 6.12). With the mild conditions of Theorem 6.13 satisfied, we have an orthonormal wavelet basis for L^2. We see that if the wavelet $\psi(t)$ has Hölder index α, it satisfies all the conditions of Theorem 6.19, where m is the integer part of α. Thus, $\Psi(\omega)$ has $m+1$ zeroes at $\omega = 0$, but because $\Phi(0) \ne 0$ (Section 6.10), we conclude from the dilation equation $\Psi(\omega) = H_s(e^{j\omega/2})\Phi(\omega/2)$ that the high-pass FIR filter $H_s(z)$ has $m+1$ zeroes at $\omega = 0$ (i.e., at $z = 1$). Using the relation $H_s(e^{j\omega}) = e^{j\omega}G_s^*(-e^{j\omega})$ we conclude that $G_s(e^{j\omega})$ has $m+1$ zeroes at $\omega = \pi$; that is, the low-pass FIR filter $G_s(z)$ has the form $G_s(z) = (1 + z^{-1})^{m+1}F(z)$, where $F(z)$ is FIR. Summarizing, we have the theorem below.

THEOREM 6.20   (Zeroes at π and Regularity) *Suppose we wish to design a compactly supported orthonormal wavelet basis for L^2 by designing an FIR filter $G_s(z)$ satisfying the conditions of Theorem 6.13. If $\psi(t)$ must have the Hölder regularity index α then it is necessary that $G_s(z)$ have the form $G_s(z) = (1 + z^{-1})^{m+1}F(z)$, where $F(z)$ is FIR, and m is the integer part of α.*

One zero at π is essential. From Theorem 6.10 we know that we must have $G_s(e^{j0}) = 1$ for the infinite product (6.112a) to converge. Theorem 6.13 imposes further conditions that enable us to obtain an orthonormal wavelet basis for L^2. One of these conditions is the power symmetric property $|G_s(e^{j\omega})|^2 + |G_s(-e^{j\omega})|^2 = 1$. Together with $G_s(e^{j0}) = 1$, this implies $G_s(e^{j\pi}) = 0$. Thus, it is necessary to have at least one zero of $G_s(e^{j\omega})$ at π. The filter that generates the Haar

---

[17]Because $\psi(t) \in L^2$ and has compact support, $\psi(t) \in L^1$ as well.

basis (Example 6.12) has exactly one zero at π, but the Haar wavelet $\psi(t)$ is not even continuous. If we desire increased regularity (continuity, differentiability, etc.), we need to put additional zeroes at π, as the above theorem shows.

Design techniques for paraunitary filter banks do not automatically yield filters which have zeroes at π. This condition must be incorporated separately. The maximally flat filter bank solution (Section 6.4) does satisfy this property, and in fact even allows us to specify the number of zeroes at π.

Wavelets with Specified Regularity

The fundamental connection between digital filter banks and continuous time wavelets, elaborated in the preceding sections, allows us to construct the scaling function $\phi(t)$ and the wavelet function $\psi(t)$ with specified regularity index α. If $G_s(z)$ has a certain number of zeroes at π, this translates into the Hölder regularity index α. What really matters is not only the number of zeroes at π, but also the order of the FIR filter $G_s(z)$.

For a given order N of the filter $G_s(z)$, suppose we wish to put as many of its zeroes as possible at π. Let this number be K. What is the largest possible K? Not all N zeroes can be at π because we have imposed the power symmetric condition on $G_s(z)$. The best we can do is to put all the unit circle zeroes at π. The power symmetric condition says that $G(z) \stackrel{\triangle}{=} \widetilde{G}_s(z)G_s(z)$ is a half-band filter. This filter has order $2N$, with $2K$ zeroes at π. Because we wish to maximize K for fixed N, the solution for $G(z)$ is the maximally flat FIR filter (Fig. 6.25), given in (6.45). As the filter in (6.45) has $2K$ zeroes at π and order $2N = 4K - 2$, we conclude that $K = (N + 1)/2$. For example, if $G_s(z)$ is a fifth-order power symmetric filter it can have at most three zeroes at π.

The 20% Regularity Rule

Suppose $G_s(z)$ has been designed to be FIR power symmetric of order N, with the number K of zeroes at π adjusted to be maximum (i.e., $K = (N + 1)/2$). It can be shown that the corresponding scaling and wavelet functions have a Hölder regularity index $\alpha \approx 0.2\,K$. This estimate is poor for small K, but improves as K grows. Thus, every additional zero at π contributes to $\approx 20\%$ improvement in regularity.

For $K = 4$ (i.e., seventh-order $G_s(z)$) we have $\alpha = 1.275$, which means that the wavelet $\psi(t)$ is once differentiable and the derivative is Hölder continuous with Lipschitz constant 0.275. For $K = 10$ [19th-order $G_s(z)$] we have $\alpha = 2.9$, so the wavelet $\psi(t)$ is twice differentiable and the second derivative has Hölder regularity index 0.9.

Design Procedure. The design procedure is therefore very simple. For a specified regularity index α, we can estimate K and hence $N = 2K - 1$. For this K, we compute the coefficients of the FIR half-band maximally flat filter $G(z)$ using (6.45). From this we compute a spectral factor $G_s(z)$ of the filter $G(z)$. Tables of the filter coefficients $g_s(n)$ for various values of N can be found in [5]. From the coefficients $g_s(n)$ of the FIR filter $G_s(z)$, the compactly supported scaling and wavelet functions are fully determined via the dilation equations. These wavelets are called Daubechies wavelets and were first generated in [6]. Figure 6.45(c) is an example, generated with a ninth-order FIR filter $G_s(z)$, whose response is shown as case 6.6 in Fig. 6.45(a).

The above regularity estimates, based on frequency domain behavior, give a single number α, which represents the regularity of $\psi(t)$ for all t. It is also possible to define **pointwise** or **local regularity** of the function $\psi(t)$ so that its smoothness can be estimated as a function of time t. These estimation methods, based on time domain iterations, are more sophisticated, but give a detailed view of the behavior of $\psi(t)$. Detailed discussions on obtaining various kinds of estimates for regularity can be found in [5] and [26].

6.14   Concluding Remarks

We introduced the WT, and studied its connection to filter banks and STFTs. A number of mathematical concepts such as frames and Riesz bases were reviewed and used later for a more careful study of wavelets. We introduced the idea of multiresolution analysis, and explained the connections both to filter banks and wavelets. This connection was then used to generate orthonormal wavelet bases from paraunitary filter banks. Such wavelets have compact support when the filter bank is FIR. The regularity or smoothness of the wavelet was quantified in terms of the Hölder exponent. We showed that we can achieve any specified Hölder exponent for compactly supported wavelets by restricting the low-pass filter of the FIR paraunitary filter bank to be a maximally flat power symmetric filter, with a sufficient number of zeroes at π.

Why Wavelets?

Discussions comparing wavelets with other types of time-frequency transforms appear at several places in this chapter. Here is a list of these discussions:

1. Section 6.2 discusses basic properties of wavelets, and gives an elementary comparison of wavelet basis with the Fourier basis.

2. Section 6.3 compares the WT to the STFT, and shows the time-frequency tilings for both cases (e.g., see Figs. 6.18 and 6.20).

3. Section 6.9 gives a deeper comparison with the STFT in terms of stability properties of the inverse, existence of frames, etc.

4. Section 6.13 shows a comparison to the traditional filter bank design approach. In traditional designs the appearance of zero(es) at π is not considered important. At the beginning of Section 6.13 (under "Why Regularity?"), we discuss the importance of these zeroes in wavelets as well as in tree-structured filter banks.

Further Reading

The literature on wavelet theory and applications is enormous. This chapter is only a brief introduction, concentrating on one-dimensional orthonormal wavelets. Many results can be found on the topics of multidimensional wavelets, biorthogonal wavelets, and wavelets based on IIR filter banks. Two special issues of the *IEEE Transactions* have appeared on the topic thus far [27, 28]. Multidimensional wavelets are treated by several authors in the edited volume of [15], and the filter bank perspective can be found in the work by Kovačević and Vetterli [27]. Advanced results on multidimensional wavelets can be found in [29]. Advanced results on wavelets constructed from M-channel filter banks can be found in the chapter by Gopinath and Burrus in the edited volume of [15], and in the work by Steffen et al. [28]. The reader can also refer to the collections of chapters in [15] and [16], and the many references therein.

Acknowledgment

The authors are grateful to Dr. Ingrid Daubechies, Princeton University, Princeton, NJ, for many useful e-mail discussions on wavelets.

This work was supported in parts by Office of Naval Research grant N00014-93-1-0231, Rockwell International, and Tektronix, Inc.

References

[1] Van Valkenburg, M.E., *Introduction to Modern Network Synthesis*, New York, John Wiley & Sons, 1960.

[2] Oppenheim, A.V. and Schafer, R.W., *Discrete-Time Signal Processing*, Englewood Cliffs, NJ, Prentice-Hall, 1989.

[3] Grossman, A. and Morlet, J., "Decomposition of Hardy functions into square integrable wavelets of constant shape," *SIAM J. Math. Anal.*, 15, 723–736, 1984.

[4] Meyer, Y., *Wavelets and Operators*, Cambridge, Cambridge University Press, 1992.

[5] Daubechies, I., *Ten Lectures on Wavelets*, SIAM, CBMS Series, April 1992.

[6] Daubechies, I., "Orthonormal bases of compactly supported wavelets," *Commun. Pure Appl. Math.*, 41, 909–996, November 1988.

[7] Vaidyanathan, P.P., *Multirate Systems and Filter Banks*, Englewood Cliffs, NJ, Prentice-Hall, 1993.

[8] Vetterli, M., "A theory of multirate filter banks," *IEEE Trans. Acoust. Speech Signal Process.*, ASSP-35, 356–372, March 1987.

[9] Akansu, A.N. and Haddad, R.A., *Multiresolution Signal Decomposition: Transforms, Subbands, and Wavelets*, Orlando, FL, Academic Press, 1992.

[10] Malvar, H.S., *Signal Processing with Lapped Transforms*, Norwood, MA, Artech House, 1992.

[11] Mallat, S., "Multiresolution approximations and wavelet orthonormal bases of $L^2(R)$," *Trans. Am. Math. Soc.*, 315, 69–87, September 1989.

[12] Heil, C.E. and Walnut, D.F., "Continuous and discrete wavelet transforms," *SIAM rev.*, 31, 628–666, December 1989.

[13] Vetterli, M. and Herley, C., "Wavelets and filter banks," *IEEE Trans. Signal Process.*, SP-40, 1992.

[14] Gopinath, R.A. and Burrus, C.S., "A tutorial overview of filter banks, wavelets, and interrelations," *Proc. IEEE Int. Symp. Circuits Syst.*, 104–107, May 1993.

[15] Chui, C.K., Vol. 1, *An Introduction to Wavelets*, and Vol. 2 (edited), *Wavelets: A Tutorial in Theory and Applications*, Orlando, FL, Academic Press, 1992.

[16] Benedetto, J.J. and Frazier, M.W., *Wavelets: Mathematics and Applications*, Boca Raton, FL, CRC Press, 1994.

[17] Allen, J.B. and Rabiner, L.R., "A unified theory of short-time spectrum analysis and synthesis," *Proc. IEEE*, 65, 1558–1564, November 1977.

[18] Smith, M.J.T. and Barnwell, III, T.P., "A procedure for designing exact reconstruction filter banks for tree structured subband coders," *Proc. IEEE Int. Conf. Acoust. Speech, Signal Proc.*, 27.1.1–27.1.4, San Diego, CA, March 1984.

[19] Mintzer, F., "Filters for distortion-free two-band multirate filter banks," *IEEE Trans. Acoust., Speech Signal Process.*, ASSP-33, 626–630, June 1985.

[20] Vaidyanathan, P.P., "Theory and design of M-channel maximally decimated quadrature mirror filters with arbitrary M, having perfect reconstruction property," *IEEE Trans. Acoustics, Speech Signal Process.*, ASSP-35, 476–492, April 1987.

[21] Belevitch, V., *Classical Network Theory*, San Francisco, Holden Day, 1968.

[22] Apostol, T.M., *Mathematical Analysis*, Reading, MA, Addison-Wesley, 1974.

[23] Rudin, W., *Real and Complex Analysis*, New York, McGraw-Hill, 1966.

[24] Young, R.M., *An Introduction to Nonharmonic Fourier Series*, New York, Academic Press, 1980.

[25] Duffin, R.J. and Schaeffer, A.C., "A class of nonharmonic Fourier series," *Trans. Am. Math. Soc.*, 72, 341–366, 1952.

[26] Rioul, O., "Simple regularity criteria for subdivision schemes," *SIAM J. Math. Anal.*, 23, 1544–1576, November 1992.

[27] Special issue on wavelet transforms and multiresolution signal analysis, *IEEE Trans. Info. Theory*, 38, March 1992.

[28] Special issue on wavelets and signal processing, *IEEE Trans. Signal Process*, 41, December 1993.

[29] Cohen, A. and Daubechies, I., "Non-separable bidimensional wavelet bases," *Rev. Mat. IberoAm.*, 9, 51–137, 1993.

7

Graph Theory

Krishnaiyan Thulasiraman
University of Oklahoma

7.1 Introduction

Graph theory had its beginning in Euler's solution of what is known as the Konigsberg Bridge problem. Kirchhoff developed the theory of trees in 1847 as a tool in the study of electrical networks. This was the first application of graph theory to a problem in physical science. Electrical network theorists have since played a major role in the phenomenal advances of graph theory that have taken place in this century. A comprehensive treatment of these developments may be found in [1]. In this chapter we develop most of those results which form the foundation of graph theoretic study of electrical networks.

Our development of graph theory is self-contained, except for the definitions of standard set-theoretic operations and elementary results from matrix theory. We wish to note that the **ring sum** of two sets S_1 and S_2 refers to the set consisting of all those elements which are in S_1 or in S_2 but not in S_1 and S_2.

7.2 Basic Concepts

A graph $G = (V, E)$ consists of two sets: a finite set $V = (v_1, v_2, \ldots, v_n)$ of elements called **vertices** and a finite set $E = (e_1, e_2, \ldots, e_m)$ of elements called **edges**. Each edge is identified with a pair of vertices. If the edges of G are identified with ordered pairs of vertices, then G is called a **directed** or an **oriented graph**. Otherwise G is called an **undirected** or a **nonoriented graph**. Graphs are amenable for pictorial representations. In a pictorial representation each vertex is represented by a dot and each edge is represented by a line segment joining the dots associated with the edge. In directed graphs we assign an orientation or direction to each edge. If the edge is associated with the ordered pair (v_i, v_j), then this edge is oriented from v_i to v_j. If an edge e connects vertices v_i and v_j then it is denoted by $e = (v_i, v_j)$. In a directed graph (v_i, v_j) refers to an edge directed from v_i to v_j. A graph and a directed graph are shown in Fig. 7.1. Unless explicitly stated, the term "graph" may refer to an undirected graph or to a directed graph.

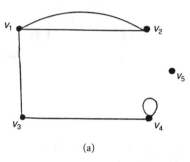
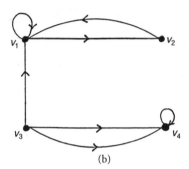

        (a)   (b)

7.1   (a) An undirected graph; (b) a directed graph.

The vertices v_i and v_j associated with an edge are called the **end vertices** of the edge. All edges having the same pair of end vertices are called **parallel edges**. In a directed graph parallel edges refer to edges connecting the same pair of vertices v_i and v_j and oriented in the same way from v_i to v_j or from v_j to v_i. For instance, in the graph of Fig. 7.1(a), the edges connecting v_1 and v_2 are parallel edges. In the directed graph of Fig. 7.1(b) the edges connecting v_3 and v_4 are parallel edges. However, the edges connecting v_1 and v_2 are not parallel edges because they are not oriented the same way. If the end vertices of an edge are not distinct, then the edge is called a **self-loop**. The graph of Fig. 7.1(a) has one self-loop and the graph of Fig. 7.1(b) has two self-loops.

An edge is said to be **incident on** its end vertices. In a directed graph the edge (v_i, v_j) is said to be **incident out** of v_i and is said to be **incident into** v_j. Vertices v_i and v_j are adjacent if an edge connects v_i and v_j.

The number of edges incident on a vertex v_i is called the **degree** of v_i and is denoted by $d(v_i)$. In a directed graph $d_{in}(v_i)$ refers to the number of edges incident into vertex v_i, and it is called the **in-degree** of v_i. $d_{out}(v_i)$ refers to the number of edges incident out of vertex v_i and it is called the **out-degree** of v_i. If $d(v_i) = 0$, then v_i is called an **isolated vertex**. If $d(v_i) = 1$, then v_i is called a **pendant vertex**. A self-loop at a vertex v_i is counted twice while computing $d(v_i)$. As an example, in the graph of Fig. 7.1(a), $d(v_1) = 3$, $d(v_4) = 3$, and v_5 is an isolated vertex. In the directed graph of Fig. 7.1(b) $d_{in}(v_1) = 3$, $d_{out}(v_1) = 2$.

Note that in a directed graph, for every vertex v_i,

$$d(v_i) = d_{in}(v_i) + d_{out}(v_i)$$

THEOREM 7.1

1. *The sum of the degrees of the vertices of a graph G is equal to 2m, where m is the number of edges of G.*

2. *In a directed graph with m edges, the sum of the in-degrees and the sum of the out-degrees are both equal to m.*

PROOF 7.1

1. Because each edge is incident on two vertices, it contributes 2 to the sum of the degrees

of G. Hence, all edges together contribute $2m$ to the sum of the degrees.

2. Proof follows if we note that each edge is incident out of exactly one vertex and incident into exactly one vertex.

□

THEOREM 7.2 *The number of vertices of odd degree in any graph is even.*

PROOF 7.2 By Theorem 7.1, the sum of the degrees of the vertices is even. Thus, the sum of the odd degrees must be even. This is possible only if the number of vertices of odd degree is even. □

Consider a graph $G = (V, E)$. The graph $G' = (V', E')$ is a subgraph of G if $V' \subseteq V$ and $E' \subseteq E$. If every vertex in V' is an end vertex of an edge in E', then G' is called the **induced subgraph** of G on E'. As an example, a graph G and two subgraphs of G are shown in Fig. 7.2.

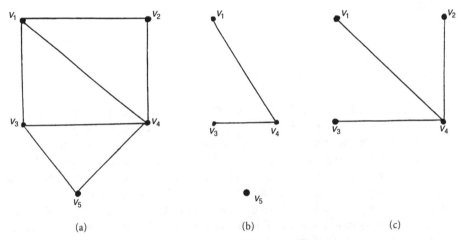

7.2 (a) Graph G; (b) subgraph of G; (c) an edge-induced subgraph of G.

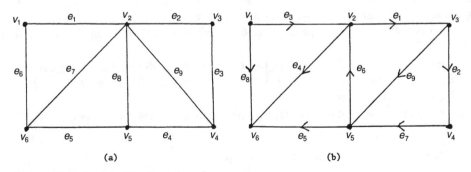

7.3 (a) An undirected graph; (b) a directed graph.

In a graph G a **path** P connecting vertices v_i and v_j is an alternating sequence of vertices and edges starting at v_i and ending at v_j, with all vertices except v_i and v_j being distinct. In a directed graph a *path* P connecting vertices v_i and v_j is called a **directed path** from v_i to v_j if all the edges in P are oriented in the same direction as we traverse P from v_i toward v_j. If a path starts and ends at the same vertex, it is called a

circuit.[1] In a directed graph, a circuit in which all the edges are oriented in the same direction is called a **directed circuit**. It is often convenient to represent paths and circuits by the sequences of edges representing them.

For example, in the undirected graph of Fig. 7.3(a) P: e_1, e_2, e_3, e_4 is a path connecting v_1 and v_5 and C: $e_1, e_2, e_3, e_4, e_5, e_6$ is a circuit. In the directed graph of Fig. 7.3(b) $P : e_1, e_2, e_7, e_5$ is a directed path and $C : e_1, e_2, e_7, e_6$ is a directed circuit. Note that e_7, e_5, e_4, e_1, e_2 is a circuit in this directed graph, although it is not a directed circuit.

Two vertices v_i and v_j are said to be **connected** in a graph G if a path in G connects v_i and v_j. A graph G is **connected** if every pair of vertices in G is connected; otherwise it is a **disconnected graph**. For example, the graph G in Fig. 7.4(a) is connected, but the graph in Fig. 7.4(b) is not connected.

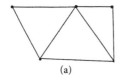

(a)   (b)

7.4   (a) A connected graph; (b) a disconnected graph.

A connected subgraph $G' = (V', E')$ of a graph $G = (V, E)$ is a **component** of G if adding to G' an edge $e \in E - E'$ results in a disconnected graph. Thus, a connected graph has exactly one component. For example, the graph in Fig. 7.4(b) is not connected and has two components.

A **tree** is a graph that is connected and has no circuits. Consider a connected graph G. A subgraph of G is a **spanning tree**[2] of G if the subgraph is a tree and contains all the vertices of G. A tree and a spanning tree of the graph of Fig. 7.5(a) are shown in Fig. 7.5(b) and (c), respectively.

The edges of a spanning tree T are called the **branches** of T. Given a spanning tree of a connected graph G, the **cospanning tree**[2] relative to T is the subgraph of G induced by the edges that are not present in T. For example, the cospanning tree relative to the spanning tree T of Fig. 7.5(c) consists of the edges e_3, e_6, and e_7. The edges of a cospanning tree are called **chords**.

A subgraph of a graph G is a **k-tree** of G if the subgraph has exactly k components and has no circuits. For example, a 2-tree of the graph of Fig. 7.5(a) is shown in Fig. 7.5(d). If a graph has k components, then a **forest** of G is a spanning subgraph that has k components and no circuits. Thus, each component of the forest is a spanning tree of a component of G. A graph G and a forest of G are shown in Fig. 7.6.

Consider a directed graph G. A spanning tree T of G is called a **directed spanning tree** with root v_i if T is a spanning tree of G, and $d_{in}(v_i) = 0$ and $d_{in}(v_j) = 1$ for all $v_j \neq v_i$. A directed graph G and a directed spanning tree with root v_1 are shown in Fig. 7.7.

It can easily be verified that in a tree exactly one path connects any two vertices.

THEOREM 7.3   *A tree on n vertices has $n - 1$ edges.*

---

[1] In electrical network theory literature the term **loop** is also used to refer to a circuit.

[2] In electrical network theory literature the terms **tree** and **cotree** are usually used to mean spanning tree and cospanning tree, respectively.

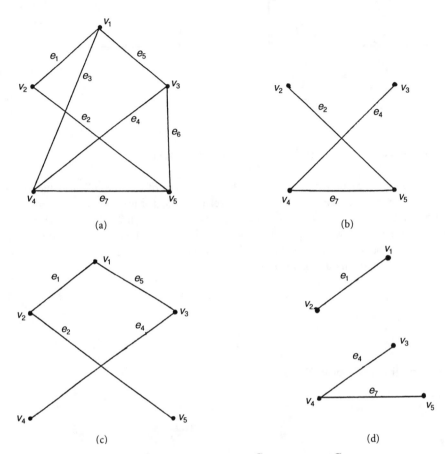

7.5 (a) Graph G; (b) a tree of graph G; (c) a spanning tree of G; (d) a 2-tree of G.

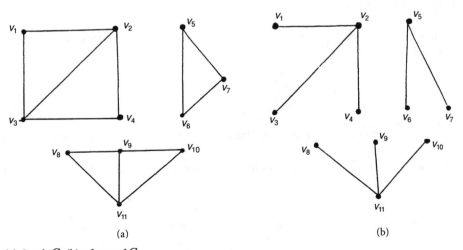

7.6 (a) Graph G; (b) a forest of G.

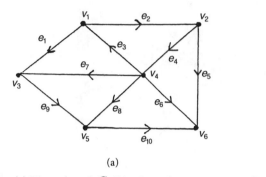 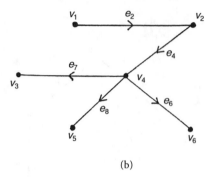

(a)   (b)

7.7   (a) Directed graph G; (b) a directed spanning tree of G with root v_1.

PROOF 7.3   Proof is by induction on the number of vertices of the tree. Clearly, the result is true if a tree has one or two vertices. Assume that the result is true for trees on $n \geq 2$ or fewer vertices. Consider now a tree T on $n+1$ vertices. Pick an edge $e = (v_i, v_j)$ in T. Removing e from T would disconnect it into exactly two components T_1 and T_2. Both T_1 and T_2 are themselves trees. Let n_1 and m_1 be the number of vertices and the number of edges in T_1, respectively. Similarly n_2 and m_2 are defined. Then, by the induction hypothesis

$$m_1 = n_1 - 1$$

and

$$m_2 = n_2 - 1$$

Thus, the number m of edges in T is given by

$$
\begin{aligned}
m &= m_1 + m_2 + 1 \\
  &= (n_1 - 1) + (n_2 - 1) + 1 \\
  &= n_1 + n_2 - 1 \\
  &= n - 1
\end{aligned}
$$

This completes the proof of the theorem.   □

If a connected graph G has n vertices, m edges, and k components, then the **rank** ρ and **nullity** μ of G are defined as follows:

$$\rho(G) = n - k \tag{7.1}$$
$$\mu(G) = m - n + k \tag{7.2}$$

Clearly, if G is connected, then any spanning tree of G has $\rho = n - 1$ branches and $\mu = m - n + 1$ chords.

We conclude this subsection with the following theorems. Proofs of these theorems may be found in [2].

THEOREM 7.4   *A tree on $n \geq 2$ vertices has at least two pendant vertices.*

THEOREM 7.5   *A subgraph of an n-vertex connected graph G is a spanning tree of G if and only if the subgraph has no circuits and has $n - 1$ edges.*

THEOREM 7.6   *If a subgraph G' of a connected graph G has no circuits then there exists a spanning tree of G that contains G'.*

7.3   Cuts, Circuits, and Orthogonality

We introduce here the notions of a cut and a cutset and develop certain results which bring out the dual nature of circuits and cutsets.

Consider a connected graph $G = (V, E)$ with n vertices and m edges. Let V_1 and V_2 be two mutually disjoint nonempty subsets of V such that $V = V_1 \cup V_2$. Thus, $V_2 = \overline{V_1}$, the complement of V_1 in V. V_1 and V_2 are also said to form a partition of V. Then the set of all those edges which have one end vertex in V_1 and the other in V_2 is called a **cut** of G and is denoted by $< V_1, V_2 >$. As an example, a graph G and a cut $< V_1, V_2 >$ of G are shown in Fig. 7.8.

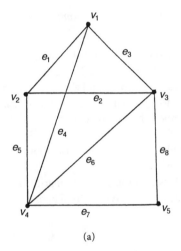 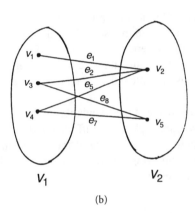

(a)	(b)

7.8   (a) Graph G; (b) cut $\langle v_1, v_2 \rangle$ of G.

The graph G' which results after removing the edges in a cut will have at least two components and so will not be connected. G' may have more than two components. A **cutset** S of a connected graph G is a minimal set of edges of G such that removal of S disconnects G into exactly two components. Thus, a cutset is also a cut. Note that the minimality property of a cutset implies that no proper subset of a cutset is a cutset.

Consider a spanning tree T of a connected graph G. Let b be a branch of T. Removal of the branch b disconnects T into exactly two components, T_1 and T_2. Let V_1 and V_2 denote the vertex sets of T_1 and T_2, respectively. Note that V_1 and V_2 together contain all the vertices of G. We can verify that the cut $< V_1, V_2 >$ is a cutset of G and is called the **fundamental cutset** of G with respect to branch b of T. Thus, for a given connected graph G and a spanning tree T of G, we can construct $n - 1$ fundamental cutsets, one for each branch of T. As an example, for the graph shown in Fig. 7.8, the fundamental cutsets with respect to the spanning tree $T = [e_1, e_2, e_6, e_8]$ are

$$
\begin{array}{ll}
\text{Branch } e_1: & (e_1, e_3, e_4) \\
\text{Branch } e_2: & (e_2, e_3, e_4, e_5) \\
\text{Branch } e_6: & (e_6, e_4, e_5, e_7) \\
\text{Branch } e_8: & (e_8, e_7)
\end{array}
$$

Note that the fundamental cutset with respect to branch b contains b. Furthermore, the branch b is not present in any other fundamental cutset with respect to T.

Next we identify a special class of circuits of a connected graph G. Again, let T be a spanning tree of G. Because exactly one path exists between any two vertices of T, adding a chord c to T produces a unique circuit. This circuit is called the **fundamental circuit** of G with respect to chord c of T. Note again that the fundamental circuit with respect to chord c contains c, and the chord c is not present in any other fundamental circuit with respect to T. As an example, the set of fundamental circuits with respect to the spanning tree $T = (e_1, e_2, e_6, e_8)$ of the graph shown in Fig. 7.8 is

$$
\begin{array}{ll}
\text{Chord } e_3 : & (e_3, e_1, e_2) \\
\text{Chord } e_4 : & (e_4, e_1, e_2, e_6) \\
\text{Chord } e_5 : & (e_5, e_2, e_6) \\
\text{Chord } e_7 : & (e_7, e_8, e_6)
\end{array}
$$

We now present a result which is the basis of what is known as the orthogonality relationship.

THEOREM 7.7   *A circuit and a cutset of a connected graph have an even number of common edges.*

PROOF 7.4   Consider a circuit C and a cutset $S = \langle V_1, V_2 \rangle$ of G. The result is true if C and S have no common edges. Suppose that C and S possess some common edges. Let us traverse the circuit C starting from a vertex, e.g., v_1 in V_1. Because the traversing should end at v_1, it is necessary that every time we encounter an edge of S leading us from V_1 to V_2 an edge of S must lead from V_2 back to V_1. This is possible only if S and C have an even number of common edges.   □

The above result is the foundation of the theory of duality in graphs. Several applications of this simple result are explored in different parts of this chapter.

A comprehensive treatment of the duality theory and its relationship to planarity may be found in [2]. The following theorem establishes a close relationship between fundamental circuits and fundamental cutsets.

THEOREM 7.8

1. *The fundamental circuit with respect to a chord of a spanning tree T of a connected graph consists of exactly those branches of T whose fundamental cutsets contain the chord.*
2. *The fundamental cutset with respect to a branch of a spanning tree T of a connected graph consists of exactly those chords of T whose fundamental circuits contain the branch.*

PROOF 7.5   Let C be the fundamental circuit of a connected graph G with respect to a chord c of a spanning tree T of G. Let C contain, in addition to the chord c, the branches $b_1, b_2, \ldots, b_k$ of T. Let S_i be the fundamental cutset with respect to branch b_i.

We first show that each S_i, $1 \le i \le k$ contains c. Note that b_i is the only branch common to S_i and C, and c is the only chord in C. Because by Theorem 7.7, S_i and C must have an even number of common edges, it is necessary that S_i contains c.

Next we show that no other fundamental cutset of T contains c. Suppose the fundamental cutset S_{k+1} with respect to some branch b_{k+1} of T contains c. Then c will be the only edge common to S_{k+1} and C, contradicting Theorem 7.7. Thus the chord c is present only in those cutsets defined by the branches $b_1, b_2, \ldots, b_k$.

The proof for item 2 of the theorem is similar to that of item 1.   □

7.4   Incidence, Circuit, and Cut Matrices of a Graph

The incidence, circuit, and cut matrices are coefficient matrices of Kirchhoff's equations which describe an electrical network. We develop several properties of these matrices which have proved useful in the study of electrical networks. Our discussions are mainly in the context of directed graphs. The results become valid in the case of undirected graphs if addition and multiplication are in $GF(2)$, the field of integers modulo 2. (Note that $1 + 1 = 0$ in this field.)

Incidence Matrix

Consider a connected directed graph G with n vertices and m edges and having no self-loops. The **all-vertex incidence matrix** $A_c = [a_{ij}]$ of G has n rows, one for each vertex, and m columns, one for each edge. The element a_{ij} of A_c is defined as follows:

$$a_{ij} = \begin{cases} 1, & \text{if the } j\text{th edge is incident out of the } i\text{th vertex,} \\ -1, & \text{if the } j\text{th edge is incident into the } i\text{th vertex,} \\ 0, & \text{if the } j\text{th edge is not incident on the } i\text{th vertex} \end{cases}$$

A row of A_c will be referred to as an **incidence vector**. As an example, for the directed graph shown in Fig. 7.9, the matrix A_c is given below.

$$A_c = \begin{array}{c} \\ v_1 \\ v_2 \\ v_3 \\ v_4 \\ v_5 \end{array} \begin{array}{c} \begin{array}{ccccccc} e_1 & e_2 & e_3 & e_4 & e_5 & e_6 & e_7 \end{array} \\ \left[\begin{array}{ccccccc} 1 & 0 & 0 & 0 & 0 & 1 & -1 \\ -1 & 1 & 0 & 0 & 0 & 0 & 0 \\ 0 & -1 & 1 & 0 & 1 & 0 & 0 \\ 0 & 0 & -1 & -1 & 0 & -1 & 0 \\ 0 & 0 & 0 & 1 & -1 & 0 & 1 \end{array} \right] \end{array}$$

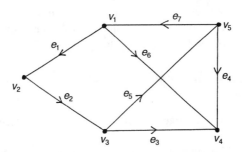

7.9   A directed graph.

From the definition of A_c it should be clear that each column of this matrix has exactly two nonzero entries, one $+1$ and one -1, and therefore, we can obtain any row of A_c from the remaining rows. Thus,

$$\text{rank}(A_c) \leq n - 1 \tag{7.3}$$

An $(n - 1)$ rowed submatrix of A_c is referred to as an **incidence matrix** of G. The vertex which corresponds to the row of A_c that is not in A is called the **reference vertex** of A.

THEOREM 7.9   *The determinant of an incidence matrix of a tree is ± 1.*

PROOF 7.6   Proof is by induction on the number m of edges in the tree. We can easily verify that the result is true for any tree with $m \leq 2$ edges. Assume that the result is true for all trees having $m \geq 2$ or fewer edges. Consider a tree T with $m + 1$ edges. Let A be the incidence matrix of T with reference vertex v_r. Because, by Theorem 7.4, T has at least two pendant vertices, we can find a pendant vertex $v_i \neq v_r$. Let (v_i, v_j) be the only edge incident on v_i. Then the remaining edges form a tree T_1. Let A_1 be the incidence matrix of T_1 with vertex v_r as reference. Now let us rearrange the rows and columns of A so that the first $n - 1$ rows correspond to the vertices in T_1, (except v_r) and the first $n - 1$ columns correspond to the edges of T_1. Then we have

$$A = \begin{bmatrix} A_1 & A_3 \\ 0 & \pm 1 \end{bmatrix}$$

So

$$\det A = \pm(\det A_1) \tag{7.4}$$

A_1 is the incidence matrix of T_1 and T_1 has m edges, it follows from the induction hypothesis that $\det A_1 = \pm 1$. Hence the theorem.   □

Because a connected graph has at least one spanning tree, it follows from the above theorem that any incidence matrix A of a connected graph has a nonsingular submatrix of order $n - 1$. Therefore,

$$\text{rank}(A_c) \geq n - 1 \tag{7.5}$$

Combining (7.3) and (7.5) yields the following theorem.

THEOREM 7.10   *The rank of any incidence matrix of a connected directed graph G is equal to $n - 1$, the rank of G.*

Cut Matrix

Consider a cut $\langle V_a, \overline{V_a} \rangle$ in a connected directed graph G with n vertices and m edges. Recall that $\langle V_a, \overline{V_a} \rangle$ consists of all those edges connecting vertices in V_a to those in $\overline{V_a}$. This cut may be assigned an orientation from V_a to $\overline{V_a}$ or from $\overline{V_a}$ to V_a. Suppose the orientation of $\langle V_a, \overline{V_a} \rangle$ is from V_a to $\overline{V_a}$. Then the orientation of an edge (v_i, v_j) is said to agree with the cut orientation if $v_i \in V_a$, and $v_j \in \overline{V_a}$.

The **cut matrix** $Q_c = [q_{ij}]$ of G has m columns, one for each edge, and has one row for each cut. The element q_{ij} is defined as follows:

$$q_{ij} = \begin{cases} 1, & \text{if the } j\text{th edge is in the } i\text{th cut and its orientation agrees with} \\ & \text{the cut orientation,} \\ -1, & \text{if the } j\text{th edge is in the } i\text{th cut and its orientation does not} \\ & \text{agree with the cut orientation,} \\ 0, & \text{if the } j\text{th edge is not in the } i\text{th cut} \end{cases}$$

Each row of Q_c is called a **cut vector**.

The edges incident on a vertex form a cut. Thus it follows that the matrix A_c is a submatrix of Q_c. Next we identify another important submatrix of Q_c.

Recall that each branch of a spanning tree T of a connected graph G defines a fundamental cutset. The submatrix of Q_c corresponding to the $n - 1$ fundamental cutsets defined by T is called the **fundamental cutset matrix** Q_f of G with respect to T.

Let $b_1, b_2, \ldots, b_{n-1}$ denote the branches of T. Let us assume that the orientation of a fundamental cutset is chosen so as to agree with that of the defining branch. Suppose we arrange the rows and the columns of Q_f so that the ith column corresponds to branch b_i, and the ith row corresponds to the fundamental cutset defined by b_i. Then the matrix Q_f can be displayed in a convenient form as follows:

$$Q_f = [U \mid Q_{fc}] \tag{7.6}$$

where U is the unit matrix of order $n - 1$ and its columns correspond to the branches of T.

As an example, the fundamental cutset matrix of the graph in Fig. 7.9 with respect to the spanning tree $T = (e_1, e_2, e_5, e_6)$ is given below:

$$Q_f = \begin{array}{c} \\ e_1 \\ e_2 \\ e_5 \\ e_6 \end{array} \begin{array}{c} \begin{array}{ccccccc} e_1 & e_2 & e_5 & e_6 & e_3 & e_4 & e_7 \end{array} \\ \left[\begin{array}{ccccccc} 1 & 0 & 0 & 0 & -1 & -1 & -1 \\ 0 & 1 & 0 & 0 & -1 & -1 & -1 \\ 0 & 0 & 1 & 0 & 0 & -1 & -1 \\ 0 & 0 & 0 & 1 & 1 & 1 & 0 \end{array} \right] \end{array}$$

It is clear from (7.6) that the rank of Q_f is $n - 1$. Hence,

$$\text{rank}(Q_c) \geq n - 1 \tag{7.7}$$

Circuit Matrix

Consider a circuit C in a connected directed graph G with n vertices and m edges. This circuit can be traversed in one of two directions, clockwise or counterclockwise. The direction we choose for traversing C is called the orientation of C. If an edge $e = (v_i, v_j)$ directed from v_i to v_j is in C, and if v_i appears before v_j as we traverse C in the direction specified by its orientation, then we say that the orientation of e agrees with the orientation of C.

The **circuit matrix** $B_c = [b_{ij}]$ of G has m columns, one for each edge, and has one row for each circuit in G. The element b_{ij} is defined as follows:

$$b_{ij} = \begin{cases} 1, & \text{if the } j\text{th edge is in the } i\text{th circuit, and its orientation agrees} \\ & \text{with the circuit orientation,} \\ -1, & \text{if the } j\text{th edge is in the } i\text{th circuit, and its orientation does not} \\ & \text{agree with the circuit orientation,} \\ 0, & \text{if the } j\text{th edge is not in the } i\text{th circuit} \end{cases}$$

Each row of B_c is called a **circuit vector**.

The submatrix of B_c corresponding to the fundamental circuits defined by the chords of a spanning tree T is called the **fundamental circuit matrix** B_f of G with respect to the spanning tree T.

Let $c_1, c_2, \ldots, c_{m-n+1}$ denote the chords of T. Suppose we arrange the columns and the rows of B_f so that the ith row corresponds to the fundamental circuit defined by the chord c_i, and the ith column corresponds to the chord c_i.

If, in addition, we choose the orientation of a fundamental circuit to agree with the orientation of the defining chord, we can write B_f as

$$B_f = [U \mid B_{ft}] \tag{7.8}$$

where U is the unit matrix of order $m - n + 1$, and its columns correspond to the chords of T.

As an example, the fundamental circuit matrix of the graph shown in Fig. 7.9 with respect to the tree $T = (e_1, e_2, e_5, e_6)$ is given below:

$$B_f = \begin{array}{c} \\ e_3 \\ e_4 \\ e_7 \end{array} \begin{array}{c} \begin{array}{ccccccc} e_3 & e_4 & e_7 & e_1 & e_2 & e_5 & e_6 \end{array} \\ \left[\begin{array}{ccccccc} 1 & 0 & 0 & 1 & 1 & 0 & -1 \\ 0 & 1 & 0 & 1 & 1 & 1 & -1 \\ 0 & 0 & 1 & 1 & 1 & 1 & 0 \end{array} \right] \end{array}$$

It is clear from (7.8) that the rank of B_f is $m - n + 1$. Hence,

$$\text{rank}(B_c) \geq m - n + 1 \tag{7.9}$$

7.5   Orthogonality Relation and Ranks of Circuit and Cut Matrices

THEOREM 7.11   *If a cut and a circuit in a directed graph have $2k$ edges in common, then k of these edges have the same relative orientation in the cut and in the circuit, and the remaining k edges have one orientation in the cut and the opposite orientation in the circuit.*

PROOF 7.7   Consider a cut $\langle V_a, \overline{V_a} \rangle$ and a circuit C in a directed graph. Suppose we traverse C starting from a vertex in V_a. Then, for every edge e_1 that leads from V_a to $\overline{V_a}$, an edge e_2 leads from $\overline{V_a}$ to V_a. Suppose the orientation of e_1 agrees with the orientation of the cut and that of the circuit. Then we can easily verify that e_2 has one orientation in the cut and the opposite orientation in the circuit (see Fig. 7.10). On the other hand, we can also verify that if e_1 has one orientation in the cut and the opposite orientation in the circuit, the e_2 will have the same relative orientation in the circuit and in the cut. This proves the theorem.   □

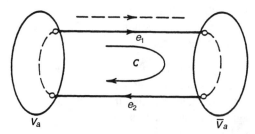

7.10   Relative orientations of an edge in a cut and a circuit.

Next we prove the orthogonality relation.

THEOREM 7.12   *If the columns of the circuit matrix B_c and the columns of the cut matrix Q_c are arranged in the same edge order, then*

$$B_c Q_c^t = 0 \tag{7.10}$$

PROOF 7.8   Each entry of the matrix $B_c Q_c^t$ is the inner product of a circuit vector and a cut vector. Suppose a circuit and a cut have $2k$ edges in common. The inner product of the corresponding vectors is zero, because by Theorem 7.11, this product is the sum of k 1's and $k - 1$'s.   □

The orthogonality relation is a profound result with interesting applications in electrical network theory. Consider a connected graph G with m edges and n vertices. Let Q_f be the fundamental cutset matrix and B_f be the fundamental circuit matrix of G with respect to a spanning tree T. If we write Q_f and B_f as in (7.6) and (7.8), then using the orthogonality relation we get

$$B_f Q_f^t = 0$$

that is,

$$[B_{ft} U]\begin{bmatrix} U \\ Q^t_{fc} \end{bmatrix} = 0$$

that is,

$$B_{ft} = -Q^t_{fc} \tag{7.11}$$

Using (7.11) each circuit vector can now be expressed as a linear combination of the fundamental circuit vectors. Consider a circuit vector $\beta = [\beta_1, \beta_2, \ldots, \beta_\rho | \beta_{\rho+1} \cdots \beta_m]$ of G where $\rho = n - 1$, is the rank of G. Then, again by the orthogonality relation we have

$$\beta Q^t_f = [\beta_1, \beta_2, \ldots, \beta_\rho | \beta_{\rho+1} \cdots \beta_m]\begin{bmatrix} U \\ Q^t_{fc} \end{bmatrix} = 0 \tag{7.12}$$

Therefore,

$$\begin{aligned}
[\beta_1, \beta_2, \ldots, \beta_\rho] &= -[\beta_{\rho+1}, \beta_{\rho+2} \cdots \beta_m]Q^t_{fc} \\
&= [\beta_{\rho+1}, \beta_{\rho+2} \cdots \beta_m]B_{ft}
\end{aligned}$$

So,

$$\begin{aligned}
[\beta_1, \beta_2, \ldots, \beta_m] &= [\beta_{\rho+1}, \beta_{\rho+2} \cdots \beta_m][B_{ft} \quad U] \\
&= [\beta_{\rho+1}, \beta_{\rho+2} \cdots \beta_m]B_f
\end{aligned} \tag{7.13}$$

Thus, any circuit vector can be expressed as a linear combination of the fundamental circuit vectors. So

$$\text{rank}(B_c) \leq \text{rank}(B_f) = m - n + 1$$

Combining the above with (7.9) we obtain

$$\text{rank}(B_c) = m - n + 1 \tag{7.14}$$

Starting from a cut vector and using the orthogonality relation we can prove in an exactly similar manner that

$$\text{rank}(Q_c) \leq \text{rank}(Q_f) = n - 1$$

Combining the above with (7.7) we get

$$\text{rank}(Q_c) = n - 1$$

Summarizing, we have the following theorem.

THEOREM 7.13 *For a connected graph G with m edges and n vertices*

$$\text{rank}(B_c) = m - n + 1$$
$$\text{rank}(Q_c) = n - 1$$

We wish to note from (7.12) that the vector corresponding to a circuit C can be expressed as an appropriate linear combination of the circuit vectors corresponding to the chords present in C. Similarly, the vector corresponding to a cut can be expressed as an appropriate linear combination of the cut vectors corresponding to the branches present in the cut. Because modulo 2 addition of two vectors corresponds to the ring sum of the corresponding subgraphs we have the following results for undirected graphs.

THEOREM 7.14 *Let G be a connected undirected graph.*

1. *Every circuit can be expressed as a ring sum of the fundamental circuits with respect to a spanning tree.*

2. *Every cut can be expressed as a ring sum of the fundamental cutsets with respect to a spanning tree.*

We can easily verify the following consequences of the orthogonality relation:

1. A linear relationship exists among the columns of the cut matrix (also of the incidence matrix) which correspond to the edges of a circuit.

2. A linear relationship exists among the columns of the circuit matrix which correspond to the edges of a cut.

The following theorem characterizes the submatrices of A_c, Q_c and B_c which correspond to spanning trees and cospanning trees. Proof follows from the above results and may be found in [2].

THEOREM 7.15   *Let G be a connected graph G with n vertices, and m edges.*

1. *A square submatrix of order $n - 1$ of Q_c (also of A_c) is nonsingular iff the edges corresponding to the columns of this submatrix form a spanning tree of G.*

2. *A square submatrix of order $m - n + 1$ of B_c is nonsingular iff the edges corresponding to the columns of this submatrix form a cospanning tree of G.*

7.6   Spanning Tree Enumeration

Here we first establish a formula for counting the number of spanning tress of an undirected graph. We then state a generalization of this result for the case of a directed graph. These formulas have played key roles in the development of topological formulas for electrical network functions. A detailed development of topological formulas for network functions may be found in Swamy and Thulasiraman [1].

The formula for counting the number of spanning trees of a graph is based on Theorem 7.9 and a result in matrix theory, known as the Binet-Cauchy theorem.

A **major** of a matrix is a determinant of maximum order. Consider a matrix P of order $p \times q$ and a matrix Q of order $q \times p$, with $p \leq q$. The majors of P and Q are of order p. If a major of P consists of columns $i_1, i_2, \ldots, i_p$ the corresponding major of Q is formed by rows $i_1, i_2, \ldots, i_p$ of Q. For example, if

$$P = \begin{bmatrix} 1 & -2 & -2 & 4 \\ 2 & 3 & -1 & 2 \end{bmatrix} \quad \text{and} \quad Q = \begin{bmatrix} -5 & 0 \\ 2 & 1 \\ -2 & 2 \\ 3 & 1 \end{bmatrix}$$

Then for the major

$$\begin{vmatrix} -2 & 4 \\ 3 & 2 \end{vmatrix}$$

of P

$$\begin{vmatrix} 2 & 1 \\ 3 & 1 \end{vmatrix}$$

is the corresponding major of Q.

The **Binet-Cauchy theorem** is stated next. Proof of this theorem may be found in Hohn [3].

THEOREM 7.16   *If P is a $p \times q$ matrix and Q is a $q \times p$ matrix, with $p \leq q$, then*

$$\det(PQ) = \sum(\text{product of the corresponding majors of } P \text{ and } Q).$$

THEOREM 7.17   *Let G be a connected undirected graph and A an incidence matrix of a directed graph obtained by assigning orientations to the edges of G. Then*

$$\tau(G) = \det(AA^t) \tag{7.15}$$

where $\tau(G)$ is the number of spanning trees of G.

PROOF 7.9   By the Binet-Cauchy theorem we have

$$\det(AA^t) = \sum(\text{product of the corresponding majors of } A \text{ and } A^t) \tag{7.16}$$

Recall from Theorem 7.15 that a major of A is nonzero iff the edges corresponding to the columns of the major form a spanning tree of G. Also, the corresponding majors of A and A^t have the same value equal to 0, 1, or -1 (Theorem 7.9). Thus, each nonzero term in the sum on the right-hand side of (7.16) has the value 1, and it corresponds to a spanning tree and vice versa. Hence the theorem.
□

For example, consider the undirected graph G shown in Fig. 7.11(a). Assigning arbitrary orientations to the edges of G, we obtain the directed graph in Fig. 7.11(b). If A is the incidence matrix of this directed graph with vertex v_4 as reference vertex then it can be verified that

$$AA^t = \begin{bmatrix} 3 & -1 & -1 \\ -1 & 2 & 0 \\ -1 & 0 & 2 \end{bmatrix}$$

and $\det(AA^t) = 8$. Thus, G has eight spanning trees.

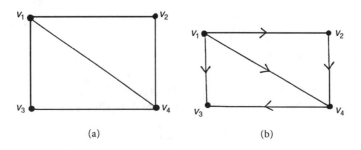

(a)   (b)

7.11   (a) An undirected graph G; (b) directed graph obtained after assigning arbitrary orientations to the edges of G.

An interesting and useful interpretation of the matrix AA^t now follows. Let $v_1, v_2, \dots, v_n$ be the vertices of an undirected graph G. The **degree matrix** $K = [k_{ij}]$ of G is an $n \times n$ matrix defined as follows.

$$k_{ij} = \begin{cases} -p, & \text{if } i \neq j \text{ and } p \text{ parallel edges connect } v_i \text{ and } v_j \\ d(v_i), & \text{if } i = j \end{cases}$$

We may easily verify that $K = A_c A_c^t$, and that it is independent of the choice of orientations for the edges of G. Also, if v_i is the reference vertex, then AA^t is obtained by removing row i and column i of K. In other words, $\det(AA^t)$ is the (i, i) cofactor of K. It then follows from Theorem 7.17 that all the cofactors of K are equal to the number of spanning trees of G. Thus, Theorem 7.17 may be stated in the following form originally presented by Kirchhoff [4].

THEOREM 7.18   *All the cofactors of the degree matrix of an undirected graph G have the same value equal to the number of spanning trees of G.*

Consider a connected undirected graph G. Let A be the incidence matrix of G with reference vertex v_n. Let $\tau_{i,n}$ denote the number of spanning 2-trees of G such that the vertices v_i and v_n are in different components of these spanning 2-trees. Also, let $\tau_{ij,n}$ denote the number of spanning 2-trees such that vertices v_i and v_j are in the same component, and vertex v_n is in a different component of these spanning 2-trees. If Δ_{ij} denotes the (i, j) cofactor of (AA^t), then we have the following result, proof of which may be found in [2].

THEOREM 7.19   *For a connected graph G,*

$$\tau_{i,n} = \Delta_{ii} \qquad (7.17)$$
$$\tau_{ij,n} = \Delta_{ij} \qquad (7.18)$$

Consider next a directed graph $G = (V, E)$ without self-loops and with $V = (v_1, v_2, \ldots, v_n)$. The **in-degree matrix** $K = [k_{ij}]$ of G is an $(n \times n)$ matrix defined as follows:

$$k_{ij} = \begin{cases} -p, & \text{if } i \neq j \text{ and } p \text{ parallel edges are directed from } v_i \text{ to } v_j \\ d_{in}(v_i), & \text{if } i = j \end{cases}$$

The following result is due to Tutte [5]. Proof of this result may also be found in [2].

THEOREM 7.20   *Let K be the in-degree matrix of a directed graph G without self-loops. Let the ith row of K correspond to vertex v_i. Then the number τ_d of directed spanning trees of G having v_r as root is given by*

$$\tau_d = \Delta_{rr} \qquad (7.19)$$

where Δ_{rr} is the (r, r) co-factor of K.

Note the similarity between Theorem 7.18 and Theorem 7.20.

To illustrate Theorem 7.20, consider the directed graph G shown in Fig. 7.12. The in-degree matrix K of G is

$$K = \begin{bmatrix} 1 & -1 & -2 \\ -1 & 2 & -1 \\ 0 & -1 & 3 \end{bmatrix}$$

Then

$$\Delta_{11} = \begin{bmatrix} 2 & -1 \\ -1 & 3 \end{bmatrix} = 5$$

The five directed spanning trees of G with vertex v_1 as root are (e_1, e_5), $(e_1, e_6, (e_1, e_3), (e_4, e_5)$, and (e_4, e_6)

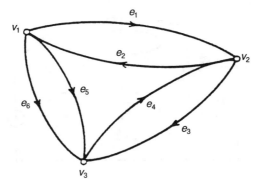

7.12 A directed graph G.

7.7 Graphs and Electrical Networks

An electrical network is an interconnection of electrical network elements such as resistances, capacitances, inductances, voltage and current sources, etc. Each network element is associated with two variables, the voltage variable $v(t)$ and the current variable $i(t)$. We also assign reference directions to the network elements (see Fig. 7.13) so that $i(t)$ is positive whenever the current is in the direction of the arrow, and $v(t)$ is positive whenever the voltage drop in the network element is in the direction of the arrow. Replacing each element and its associated reference direction by a directed edge results in the directed graph representing the network. For example, a simple electrical network and the corresponding directed graph are shown in Fig. 7.14.

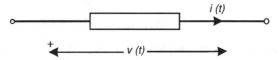

7.13 A network element with reference convention.

The physical relationship between the current and voltage variables of a network element is specified by Ohm's law. For voltage and current sources, the voltage and current variables are required to have specified values. The linear dependence among the voltage variables in the network and the linear dependence among the current variables are governed by Kirchhoff's voltage and current laws.

> **Kirchhoff's Voltage Law (KVL):** The algebraic sum of the voltages around any circuit is equal to zero.
>
> **Kirchhoff's Current Law (KCL):** The algebraic sum of the currents flowing out of a node is equal to zero.

As an example, the KVL equation for the circuit 1, 3, 5 and the KCL equation for the vertex b in the graph of Fig. 7.14 are

$$\text{Circuit: } 1, 3, 5 \qquad v_1 + v_3 + v_5 = 0$$
$$\text{Vertex } b: \qquad -i_1 + i_2 + i_3 = 0$$

It can easily be seen that KVL and KCL equations for an electrical network N can be conveniently written as

$$A_c I_e = 0 \qquad\qquad (7.20)$$

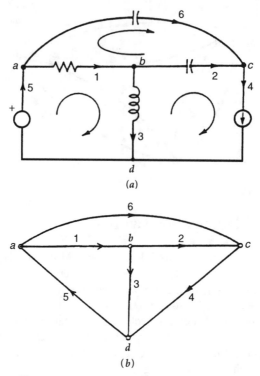

7.14   (a) An electrical network N; (b) directed graph representation of N.

and

$$B_c V_e = 0 \qquad (7.21)$$

where A_c and B_c are, respectively, the incidence and circuit matrices of the directed graph representing N, and I_e and V_e are, respectively, the column vectors of element currents and voltages in N. Because each row in the cut matrix Q_c can be expressed as a linear combination of the rows of the matrix, in (7.20) we can replace A_c by Q_c. Thus, we have:

$$\text{KCL:} \quad Q_c I_e \;=\; 0 \qquad (7.22)$$
$$\text{KVL:} \quad B_c V_e \;=\; 0 \qquad (7.23)$$

From (7.22) we can see that KCL can also be stated as: The algebraic sum of the currents in any cut of N is equal to zero.

If a network N has n vertices, m elements, and its graph is connected then there are only $(n - 1)$ linearly independent cuts and only $(m - n + 1)$ linearly independent circuits (Theorem 7.13). Thus, in writing KVL and KCL equations we need to use only B_f, a fundamental circuit matrix and Q_f, a fundamental circuit matrix, respectively. Thus, we have

$$\text{KCL:} \quad Q_f I_e \;=\; 0 \qquad (7.24)$$
$$\text{KVL:} \quad B_f V_e \;=\; 0 \qquad (7.25)$$

We note that the KCL and the KVL equations depend only on the way the network elements are interconnected and not on the nature of the network elements. Thus, several results in electrical network theory are essentially graph theoretic in nature. Some of these results and their usefulness in electrical network analysis are presented in the remainder of this chapter. In the following a network N and its directed graph are both denoted by N.

THEOREM 7.21   *Consider an electrical network N. Let T be a spanning tree of N, and let B_f and Q_f denote the fundamental circuit and the fundamental cutset matrices of N with respect to T. If I_e and V_e are the column vectors of element currents and voltages and I_c and V_t are, respectively, the column vector of currents associated with the chords of T and the column vector of voltages associated with the branches of T, then*

$$\text{Loop Transformation:} \quad I_e \;=\; B_f^t I_c \tag{7.26}$$

$$\text{Cutset Transformation:} \quad V_e \;=\; Q_f^t V_t \tag{7.27}$$

PROOF 7.10   From Kirchhoff's laws we have

$$Q_f I_e = 0 \tag{7.28}$$

and

$$B_f V_e = 0 \tag{7.29}$$

Let us partition I_e and V_e as

$$I_e = \begin{bmatrix} I_c \\ I_t \end{bmatrix}$$

and

$$V_e = \begin{bmatrix} V_c \\ V_t \end{bmatrix}$$

where the vectors which correspond to the chords and branches of T are distinguished by the subscripts c and t, respectively. Then (7.28) and (7.29) can be written as

$$\begin{bmatrix} Q_{fc} & U \end{bmatrix} \begin{bmatrix} I_c \\ I_t \end{bmatrix} = 0 \tag{7.30}$$

and

$$\begin{bmatrix} U & B_{ft} \end{bmatrix} \begin{bmatrix} V_c \\ V_t \end{bmatrix} = 0 \tag{7.31}$$

Recall (7.11) that

$$B_{ft} = -Q_{fc}^t$$

Then we get from (7.30)

$$\begin{aligned} I_t &= -Q_{fc} I_c \\ &= B_{ft}^t I_c \end{aligned}$$

Thus

$$I_e = \begin{bmatrix} U \\ B_{ft}^t \end{bmatrix} I_c = B_f^t I_c$$

This establishes the loop transformation.

Starting from (7.31) we can show in a similar manner that

$$V_e = Q_f^t V_t$$

thereby establishing the cutset transformation.   □

In the special case in which the incidence matrix A is used in place of the fundamental cutset matrix, the cutset transformation (7.27) is called the **node transformation**. The loop, cutset, and node transformations have been extensively employed to develop different methods of network analysis. The loop method of analysis develops a system of network equations which involve only the chord currents as variables. The cutset (node) method of analysis develops a system of equations involving only the branch (node) voltages as variables. Thus, the loop and cutset (node) methods result in systems of equations involving $m - n + 1$ and $n - 1$ variables, respectively. In the mixed-variable method of analysis, which is essentially a combination of both the loop and cutset methods, some of the independent variables are currents and the others are voltages. The minimum number of variables required in the mixed-variable method of analysis is determined by what is known as the principal partition of a graph introduced by Kishi and Kajitani in a classic paper [6]. Ohtsuki, Ishizaki, and Watanabe [7] discuss several issues relating to the mixed-variable method of analysis. A detailed discussion of the principal partition of a graph and the different methods of network analysis including the state-variable method may be found in [1].

7.8   Tellegen's Theorem and Network Sensitivity Computation

Here we first present a simple and elegant theorem due to Tellegen [8]. The proof of this theorem is essentially graph theoretic in nature and is based on the loop and the cutset transformations, (7.26) and (7.27), and the orthogonality relation (Theorem 7.12). Using this theorem we develop the concept of the adjoint of a network and its application in network sensitivity computations.

THEOREM 7.22   *Consider two electrical networks N and $\hat{N}$ such that the graphs associated with them are identical. Let V_e and ψ_e denote the element voltage vectors of N and $\hat{N}$, respectively, and let I_e and Λ_e be the corresponding element current vectors. Then*

$$V_e^t \Lambda_e \ = \ 0$$

$$I_e^t \psi_e \ = \ 0$$

PROOF 7.11   If B_f and Q_f are the fundamental circuit and cutset matrices of N (and hence also of $\hat{N}$), then from the loop and cutset transformations we obtain

$$V_e = Q_f^t V_t$$

and

$$\Lambda_e = B_f^t \Lambda_c$$

So

$$
\begin{aligned}
V_e^t \Lambda_e \ &= \ V_t^t \left(Q_f B_f^t \right) \Lambda_c \\
&= \ 0, \quad \text{by Theorem 12}
\end{aligned}
$$

Proof follows in a similar manner.   □

The adjoint network was introduced by Director and Rohrer [9], and our discussion is based on their work. A more detailed discussion may be found in [1].

Consider a lumped, linear time-invariant network N. We assume, without loss of generality, that N is a 2-port network. Let $\hat{N}$ be a 2-port network which is topologically equivalent to N. In other words, the graph of $\hat{N}$ is identical to that of N. The corresponding elements of N and $\hat{N}$ are denoted by the same symbol. Our goal now is to define the elements of $\hat{N}$ so that $\hat{N}$ in conjunction with N can be used in computing the sensitivities of network functions of N.

Let V_e and I_e denote, respectively, the voltage and the current associated with the element e in N, and ψ_e and λ_e denote, respectively, the voltage and the current associated with the corresponding element e in $\hat{N}$. Also, V_i and I_i, $i = 1, 2$, denote the voltage and current variables associated with the ports of N, and ψ_i and λ_i, $i = 1, 2$, denote the corresponding variables for the ports of $\hat{N}$ (see Fig. 7.15).

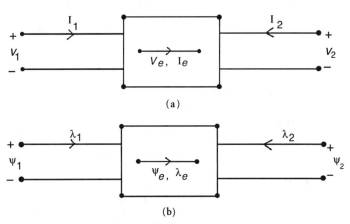

(a)

(b)

7.15 (a) A 2-port network N; (b) adjoint network $\hat{N}$ of N.

Applying Tellegen's theorem to N and $\hat{N}$ we get

$$V_1\lambda_1 + V_2\lambda_2 = \sum_e V_e\lambda_e \tag{7.32}$$

and

$$I_1\psi_1 + I_2\psi_2 = \sum_e I_e\psi_e \tag{7.33}$$

Suppose we now perturb the values of elements of N and apply Tellegen's theorem to $\hat{N}$ and the perturbed network N:

$$(V_1 + \Delta V_1)\lambda_1 + (V_2 + \Delta V_2)\lambda_2 = \sum_e (V_e + \Delta V_e)\lambda_e \tag{7.34}$$

and

$$(I_1 + \Delta I_1)\psi_1 + (I_2 + \Delta I_2)\psi_2 = \sum_e (I_e + \Delta I_e)\psi_e \tag{7.35}$$

where ΔV and ΔI represent the changes in the voltage and current which result as a consequence of the perturbation of the element values in N. Subtracting (7.32) from (7.34) and subtracting (7.33) from (7.35)

$$\Delta V_1\lambda_1 + \Delta V_2\lambda_2 = \sum_e \Delta V_e\lambda_e \tag{7.36}$$

and

$$\Delta I_1\psi_1 + \Delta I_2\psi_2 = \sum_e \Delta I_e\psi_e \tag{7.37}$$

Subtracting (7.37) from (7.36) yields

$$(\Delta V_1 \lambda_1 - \Delta I_1 \psi_1) + (\Delta V_2 \lambda_2 - \Delta I_2 \psi_2) = \sum_e (\Delta V_e \lambda_e - \Delta I_e \psi_e) \tag{7.38}$$

We wish to define the corresponding element of $\hat{N}$ for every element in N so that each term in the summation on the right-hand side of (7.38) reduces to a function of the voltage and current variables and the change in value of the corresponding network element. We illustrate this for resistance elements. Consider a resistance element R in N. For this element we have

$$V_R = R I_R \tag{7.39}$$

Suppose we change R to ΔR, then

$$(V_R + \Delta V_R) = (R + \Delta R)(I_R + \Delta I_R) \tag{7.40}$$

Neglecting second-order terms, (7.40) simplifies to

$$V_R + \Delta V_R = R I_R + R \Delta I_R + I_R \Delta R \tag{7.41}$$

Subtracting (7.39) from (7.41),

$$\Delta V_R = R \Delta I_R + I_R \Delta R \tag{7.42}$$

Now using (7.42) the terms in (7.38) corresponding to the resistance elements of N can be written as

$$\sum_R [R \lambda_R - \psi_R] \Delta I_R + I_R \lambda_R \Delta R \tag{7.43}$$

If we now choose

$$\psi_R = R \lambda_R \tag{7.44}$$

then (7.43) reduces to

$$\sum_R I_R \lambda_R \Delta R \tag{7.45}$$

which involves only the network variables in N (before perturbation) and $\hat{N}$ and the changes in resistance values. Equation (7.44) is the relation for a resistance. Therefore, the element in $\hat{N}$ corresponding to a resistance element of value R in N is also a resistance of value R.

Proceeding in a similar manner we can determine the elements of $\hat{N}$ corresponding to other types of network elements (inductance, capacitance, controlled sources, etc.) The network $\hat{N}$ so obtained is called the **adjoint** of N. A table defining adjoint elements corresponding to different types of network elements may be found in [1].

We now illustrate the application of the adjoint network in the computation of the sensitivity of a network function. Note that the sensitivity of a network function F with respect to a parameter x is a measure of the effect on F of an incremental change in x. Computing this sensitivity essentially involves determining $\partial F / \partial x$.

For the sake of simplicity consider the resistance network shown in Fig. 7.16(a). Let us assume that resistance R is perturbed from its nominal value of 3 Ω. Assume that no changes occur in the values of the other resistance elements. We wish to compute $\partial F / \partial R$ where F is the open-circuit voltage ratio, that is,

$$F = \frac{V_2}{V_1} \bigg|_{I_2 = 0}$$

In other words, to compute F, we connect a voltage source of value $V_1 = 1$ across port 1 of N and open-circuit port 2 of N (so that $I_2 = 0$). So, $\Delta V_1 = 0$ and $\Delta I_2 = 0$ and (7.38) reduces to

$$-\Delta I_1 \psi_1 + \Delta V_2 \lambda_2 = I_R \lambda_R \Delta R \qquad (7.46)$$

Now we need to determine ΔV_2 as a function of ΔR. This could be achieved if we set $\psi_1 = 0$ and $\lambda_2 = 1$ for the adjoint network $\hat{N}$. Connect a current source of value $\lambda_2 = 1$ across port 2 and short circuit port 1 of $\hat{N}$. The resulting adjoint network is shown in Fig. 7.16(b). With port variables of $\hat{N}$ defined as above, (7.46) reduces to

$$\Delta V_2 = I_R \lambda_R \Delta_R$$

Thus,

$$\partial F / \partial R = \partial V_2 / \partial R = I_R \lambda_R$$

where I_R and λ_R are the currents in the networks N and $\hat{N}$ shown in Fig. 7.16. Thus, in general, computing the sensitivity of a network function essentially reduces to the analysis of N and $\hat{N}$ under appropriate excitations at their ports. Note that we do not need to express the network function explicitly in terms of the network elements nor do we need to calculate partial derivatives.

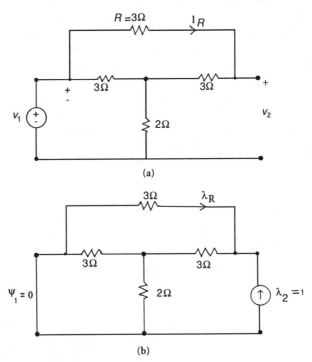

7.16 (a) A 2-port network N; (b) adjoint network $\hat{N}$.

For the example under consideration, we calculate $I_R = 1/12$ A and $\lambda_R = -7/12$ A with the result that $\partial F / \partial R = -7/144$. A further discussion of the adjoint network and related results may be found in Section 7.3.

7.9   Arc Coloring Theorem and the No-Gain Property

We now derive a profound result in graph theory, the arc coloring theorem for directed graphs, and discuss its application in establishing the no-gain property of resistance networks. In the special case of undirected graphs the arc coloring theorem reduces to the "painting" theorem. Both of these theorems (Minty [10]) are based on the notion of **painting a graph**.

Given an undirected graph with edge set E, a painting of the graph is a partitioning of E into three subsets, R, G, and B, such that $|G| = 1$. We may consider the edges in the set R as being "painted red," the edge in G as being "painted green" and the edges in B as being "painted blue."

THEOREM 7.23   *For any painting of a graph, there exists a circuit C consisting of the green edge and no blue edges, or a cutset C^* consisting of the green edge and no red edges.*

PROOF 7.12   Consider a painting of the edge set E of a graph G. Assuming that there does not exist a required circuit, we shall establish the existence of a required cutset.

Let $E' = R \cup G$ and T' denote a spanning forest of the subgraph induced by E', containing the green edge. (Note that the subgraph induced by E' may not be connected). Then construct a spanning tree T of G such that $T' \subseteq T$.

Now consider any red edge y which is not in T', and hence not in T. Because the fundamental circuit of y with respect to T is the same as the fundamental circuit of y with respect to T', this circuit consists of no blue edges. Furthermore, this circuit will not contain the green edge, for otherwise a circuit consisting of the green edge and no blue edges would exist contrary to our assumption. Thus, the fundamental circuit of a red edge with respect to T does not contain the green edge. Then it follows from Theorem 8 that the fundamental cutset of the green edge with respect to T contains no red edges. Thus, this cutset satisfies the requirements of the theorem.   □

A painting of a directed graph with edge set E is a partitioning of E into three sets R, G, and B, and the distinguishing of one element of the set G. Again, we may regard the edges of the graph as being colored red, green, or blue with exactly one edge of G being colored dark green. Note that the dark green edge is also to be treated as a green edge.

Next we state and prove Minty's arc coloring theorem.

THEOREM 7.24   *For any painting of a directed graph exactly one of the following is true:*

1. *A circuit exists containing the dark green edge, but no blue edges, in which all the green edges are similarly oriented.*

2. *A cutset exists containing the dark green edge, but no red edges, in which all the green edges are similarly oriented.*

PROOF 7.13   Proof is by induction on the number of green edges. If only one green edge exists, then the result will follow from Theorem 7.23. Assume then that the result is true when the number of green edges is $m \geq 1$. Consider a painting in which $m + 1$ edges are colored green. Pick a green edge x other than the dark green edge (see Fig. 7.17). Color the edge x red. In the resulting painting we find m green edges. If a cutset of type 2 is now found, then the theorem is proved. On the other hand if we color the edge x blue and in the resulting painting a circuit of type 1 exists, then the theorem is proved.

Suppose neither occurs. Then, using the induction hypothesis we have the following:

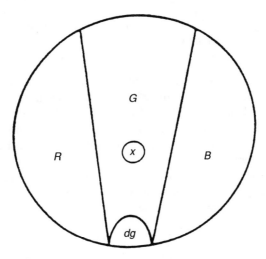

7.17 Painting of a directed graph.

1. A cutset of type 2 exists when x is colored blue.
2. A circuit of type 1 exists when x is colored red.

Now let the corresponding rows of the circuit and cutset matrices be

	dg	R	B	G	x
Cutset	$+1$	$00\ldots0\ \ 0$	$1-1\ldots01$	$111\ldots0$	?
Circuit	$+1$	$-11\ldots0-1$	$0\ 0\ldots00$	$011\ldots0$	?

Here we have assumed, without loss of generality, that +1 appears in the dark green position of both rows.

By the orthogonality relation (Theorem 7.12) the inner product of these two row vectors is zero. No contribution is made to this inner product from the red edges or from the blue edges. The contribution from the green edges is a non-negative integer p. The dark green edge contributes 1 and the edge x contributes an unknown integer q which is 0, 1, or -1. Thus, we have $1+p+q = 0$. This equation is satisfied only for $p = 0$ and $q = -1$. Therefore, in one of the rows, the question mark is $+1$ and in the other it is -1. The row in which the question mark is 1 corresponds to the required circuit or cutset. Thus, either statement 1 or 2 of the theorem occurs. Both cannot occur simultaneously because the inner product of the corresponding circuit and cutset vectors will then be nonzero. □

THEOREM 7.25 *Each edge of a directed graph belongs to either a directed circuit or to a directed cutset but no edge belongs to both. (Note: A cutset is a **directed cutset** if all its edges are similarly oriented.)*

PROOF 7.14 Proof will follow if we apply the arc coloring theorem to a painting in which all the edges are colored green and the given edge is colored dark green. □

We next present an application of the arc coloring theorem in the study of electrical networks. We prove what is known as the **no-gain property** of resistance networks. Our proof is the result of the work of Wolaver [11] and is purely graph theoretic in nature.

THEOREM 7.26 *In a network of sources and (linear/nonlinear) positive resistances the magni-*

tude of the current through any resistance with nonzero voltage is not greater than the sum of the magnitudes of the currents through the sources.

PROOF 7.15   Let us eliminate all the elements with zero voltage by considering them to be short-circuits and then assign element reference directions so that all element voltages are positive.

Consider a resistance with nonzero voltage. Thus, no directed circuit can contain this resistance, for if such a directed circuit were present, the sum of all the voltages in the circuit would be nonzero, contrary to Kirchhoff's voltage law. It then follows from Theorem 7.25 that a directed cutset contains the resistance under consideration.

Pick a directed cutset that contains the considered resistance. Let the current through this resistance be i_o. Let R be the set of all other resistances in this cutset and let S be the set of all sources. Then, applying Kirchhoff's current law to the cutset, we obtain

$$i_o + \sum_{k \in R} i_k + \sum_{s \in S} \pm i_s = 0 \qquad (7.47)$$

Because all the resistances and voltages are positive, every resistance current is positive. Therefore, we can write the above equation as

$$|i_o| + \sum_{k \in R} |i_k| + \sum_{s \in S} \pm i_s = 0 \qquad (7.48)$$

and so

$$|i_o| \leq \sum_{s \in S} \mp i_s \leq \sum_{s \in S} |i_s| \qquad (7.49)$$

Thus follows the theorem.   □

The following result is the dual of the above theorem. Proof of this theorem follows in an exactly dual manner, if we replace current with voltage, voltage with current, and circuit with cutset in the proof of the above theorem.

THEOREM 7.27   *In a network of sources and (linear/nonlinear) positive resistances, the magnitude of the voltage across any resistance is not greater than the sum of the voltages across all the sources.*

Chua and Green [12] used the arc-coloring theorem to establish several properties of nonlinear networks and nonlinear multiport resistive networks.

References

[1] Swamy, M.N.S. and Thulasiraman, K., *Graphs, Networks and Algorithms*, New York, Wiley-Interscience, 1981.

[2] Thulasiraman, K. and Swamy, M.N.S., *Graphs: Theory and Algorithms*, New York, Wiley-Interscience, 1992.

[3] Hohn, F.E., *Elementary Matrix Algebra*, New York, Macmillan, 1958.

[4] Kirchhoff, G., "Uber die Auflosung der Gleichungen auf welche mon bei der untersuchung der linearen Verteilung galvanischer strome gefuhrt wind," *Ann. Phys. Chem.*, 72, 497–508, 1847.

[5] Tutte, W.T., "The dissection of equilateral triangles into equilateral triangles," *Proc. Cambr. Philos. Soc.*, 44, 203–217, 1948.

[6] Kishi, G. and Kajitani, Y., "Maximally distant trees and principal partition of a linear graph," *IEEE Trans. Circuit Theory*, CT-15, 247–276, 1968.

[7] Ohtsuki, T., Ishizaki, Y., and Watanabe, H., "Topological degrees of freedom and mixed analysis of electrical networks," *IEEE Trans. Circuit Theory*, CT-17, 491–499, 1970.

[8] Tellegen, B.D.H., "A general network theorem with applications," Philips Res. Rep., 7, 259–269, 1952.

[9] Director, S.W. and Rohrer, R.A., "Automated network design—the frequency domain case," *IEEE Trans. Circuit Theory*, CT-16, 330–337, 1969.

[10] Minty, G.J.,"On the axiomatic foundations of the theories of directed linear graphs, electrical networks and network programming," *J. Math. Mech.*, 15, 485–520, 1966.

[11] Wolaver, D.H., "Proof in graph of the 'no-gain' property of resistor networks," *IEEE Trans. Circuit Theory*, CT-17, 436–437, 1970.

[12] Chua, L.O. and Green, D.N., "Graph-theoretic properties of dynamic nonlinear networks," *IEEE Trans. Circuit Theory*, CAS-23, 292–312, 1976.

8

Signal Flow Graphs

Krishnaiyan Thulasiraman
University of Oklahoma

8.1 Introduction

Signal flow graph theory is concerned with the development of a graph theoretic approach to solving a system of linear algebraic equations. Two closely related methods proposed by Coates [1] and Mason [2, 3] have appeared in the literature and have served as elegant aids in gaining insight into the structure and nature of solutions of systems of equations. In this chapter we develop these two methods. Our development follows closely [4].

An extensive discussion of signal flow theory may be found in [5]. Applications of signal flow theory in the analysis and synthesis electrical networks may be found in Sections 4 and 5. Coates' and Mason's methods may be viewed as generalizations of a basic theorem in graph theory due to Harary [6], which provides a formula for finding the determinant of the adjacency matrix of a directed graph. Thus, our discussion begins with the development of this theorem. For graph theoretic terminology the reader may refer to Chapter 7.

8.2 Adjacency Matrix of a Directed Graph

Consider a directed graph $G = (V, E)$ with no parallel edges. Let $V = \{v_1, \ldots, v_n\}$. The **adjacency matrix** $M = [m_{ij}]$ of G is an $n \times n$ matrix defined as follows:

$$m_{ij} = \begin{cases} 1, & \text{if } (v_i, v_j) \in E \\ 0, & \text{otherwise} \end{cases}$$

The graph shown in Fig. 8.1 has the following adjacency matrix:

$$M = \begin{array}{c} \\ v_1 \\ v_2 \\ v_3 \\ v_4 \end{array} \begin{array}{c} \begin{array}{cccc} v_1 & v_2 & v_3 & v_4 \end{array} \\ \left[\begin{array}{cccc} 1 & 1 & 1 & 0 \\ 0 & 1 & 0 & 0 \\ 1 & 0 & 0 & 1 \\ 1 & 1 & 1 & 1 \end{array} \right] \end{array}$$

In the following we shall develop a topological formula for det M. Toward this end we introduce some basic terminology. A **1-factor** of a directed graph G is a spanning subgraph of G in which the

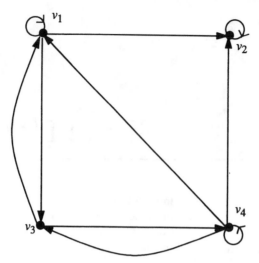

8.1   The graph G.

in-degree and the out-degree of every vertex are both equal to 1. It is easy to see that a 1-factor is a collection of vertex-disjoint directed circuits. Because a self-loop at a vertex contributes 1 to the in-degree and 1 to the out-degree of the vertex, a 1-factor may have some self-loops. As an example, the three 1-factors of the graph of Fig. 8.1 are shown in Fig. 8.2.

A **permutation** $(j_1, j_2, \ldots, j_n)$ of integers $1, 2, \ldots, n$ **is even (odd)** if an even (odd) number of interchanges are required to rearrange it as $(1, 2, \ldots, n)$. The notation

$$\begin{pmatrix} 1, 2, \ldots, n \\ j_1, j_2, \ldots, j_n \end{pmatrix}$$

is also used to represent the permutation $(j_1, j_2, \ldots, j_n)$. As an example, the permutation $(4, 3, 1, 2)$ is odd because it can be rearranged as $(1, 2, 3, 4)$ using the following sequence of interchanges:

1. Interchange 2 and 4.
2. Interchange 1 and 2.
3. Interchange 2 and 3.

For a permutation $(j) = (j_1, j_2, \ldots, j_n)$, $\varepsilon_{j_1, j_2, \ldots, j_n}$, is defined as equal to 1, if (j) is an even permutation; otherwise, $\varepsilon_{j_1, j_2, \ldots, j_n}$, is equal to -1.

Given an $n \times n$ square matrix $X = [x_{ij}]$, we note that det X is given by

$$\det X = \sum_{(j)} \varepsilon_{j_1, j_2, j_3, \ldots, j_n} \; x_{1j_1}, x_{2j_2} \cdots x_{nj_n}$$

where the summation $\sum_{(j)}$ is over all permutations of $1, 2, \ldots, n$ [7].

The following theorem is due to Harary [6].

THEOREM 8.1   *Let H_i, $i = 1, 2, \ldots, p$ be the 1-factors of an n-vertex directed graph G. Let L_i denote the number of directed circuits in H_i, and let M denote the adjacency matrix of G. Then*

$$\det M = (-1)^n \sum_{i=1}^{p} (-1)^{L_i}$$

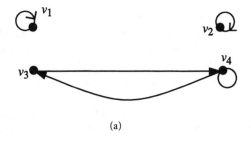

(a)

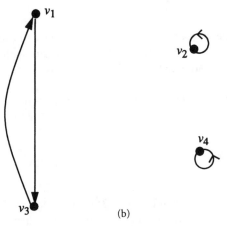

(b)

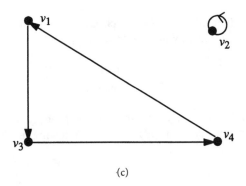

(c)

8.2 The three 1-factors of the graph of Fig. 8.1.

PROOF 8.1 From the definition of a determinant, we have

$$\det M = \sum_{(j)} \varepsilon_{j_1, j_2, \ldots, j_n} \quad m_{1j_1} \cdot m_{2j_2} \cdots m_{nj_n} \tag{8.1}$$

Proof will follow if we establish the following:

1. Each nonzero term $m_{1j_1} \cdot m_{2j_2} \cdots m_{nj_n}$ corresponds to a 1-factor of G, and conversely, each 1-factor of G corresponds to a non-zero term $m_{1j_1} \cdot m_{2j_2} \cdots m_{nj_n}$.

2. $\varepsilon_{j_1, j_2, \ldots, j_n} = (-1)^{n+L}$ if the 1-factor corresponding to a nonzero $m_{1j_1} \cdot m_{2j_2} \cdots m_{nj_n}$ has L directed circuits.

A nonzero term $m_{1j_1} m_{1j_2} \cdots m_{nj_n}$ corresponds to the set of edges $(v_1, v_{j1}), (v_2, v_{j2}) \cdots (v_n, v_{jn})$. Each vertex appears exactly twice in this set, once as an initial vertex and once as a terminal vertex of a pair of edges. Therefore, in the subgraph induced by these edges, for each vertex its in-degree and its out-degree are both equal to 1, and this subgraph is a 1-factor of G. In other words, each non-zero term in the sum in (8.1) corresponds to a 1-factor of G. The fact that each 1-factor of G corresponds to a non-zero term $m_{1j_1} \cdot m_{2j_2} \cdots m_{nj_n}$ is obvious.

As regards $\varepsilon_{j_1, j_2, \ldots, j_n}$, consider a directed circuit C in the 1-factor corresponding to $m_{1j_1} \cdot m_{2j_2} \cdots m_{nj_n}$. Without loss of generality, assume that C consists of the w edges

$$(v_1, v_2), \quad (v_2, v_3), \quad \cdots, \quad (v_w, v_1)$$

It is easy to see that the corresponding permutation $(2, 3, \ldots, w, 1)$ can be rearranged as $(1, 2, \ldots, w)$ using $w - 1$ interchanges. If the 1-factor has L directed circuits with lengths $w_1, \ldots, w_L$, the permutation $(j_1, \ldots, j_n)$ can be rearranged as $(1, 2, \ldots, n)$ using

$$(w_1 - 1) + (w_2 - 1) + \cdots + (w_L - 1) = n - L$$

interchanges. So,

$$\varepsilon_{j_1, j_2, j_n} = (-1)^{n+L}$$

$\square$

As an example, for the 1-factors (shown in Fig. 8.2) of the graph of Fig. 8.1, the corresponding L_i are $L_1 = 3$, $L_2 = 3$, and $L_3 = 2$. So, the determinant of the adjacency matrix of the graph of Fig. 8.1 is

$$(-1)^4 \left[(-1)^3 + (-1)^3 + (-1)^2 \right] = -1$$

Consider next a weighted directed graph G in which each edge (v_i, v_j) as associated with a weight w_{ij}. Then we may define the **adjacency matrix** $M = [m_{ij}]$ of G as follows:

$$m_{ij} = \begin{cases} w_{ij} & \text{if } (v_i, v_j) \in E \\ 0, & \text{otherwise} \end{cases}$$

Given a subgraph H of G, let us define weight $w(H)$ of H as the product of the weights of all edges in H. If H has no edges, then we define $w(H) = 1$. The following result is an easy generalization of Theorem 8.1.

THEOREM 8.2   *The determinant of the adjacency matrix of an n-vertex directed graph G is given by*

$$\det M = (-1)^n \sum_H (-1)^{L_H} w(H),$$

where H is a 1-factor, $w(H)$ is the weight of H, and L_H is the number of directed circuits in H.

8.3   Coates' Gain Formula

Consider a linear system described by the equation

$$AX = Bx_{n+1} \tag{8.2}$$

where A is a nonsingular $n \times n$ matrix, X is a column vector of unknown variables $x_1, x_2, \ldots, x_n$, B is a column vector of elements $b_1, b_2, \ldots, b_n$ and x_{n+1} is the input variable. It is well known that

$$\frac{x_k}{x_{n+1}} = \frac{\sum_{i=1}^{n} b_i \Delta_{ik}}{\det A} \tag{8.3}$$

where Δ_{ik} is the (i, k) cofactor of A.

To develop Coates' topological formulas for the numerator and the denominator of (8.3), let us first augment the matrix A by adding $-B$ to the right of A and adding a row of zeroes at the bottom of the resulting matrix. Let this matrix be denoted by A'. **The Coates flow graph**[1] $G_c(A')$, or simply the **Coates graph** associated with matrix A', is a weighted directed graph whose adjacency matrix is the transpose of the matrix A'. Thus, $G_c(A')$ has $n + 1$ vertices $x_1, x_2, \ldots, x_{n+1}$, and if $a_{ji} \cdots \neq 0$, then $G_c(A')$ has an edge directed from x_i to x_j with weight a_{ji}. Clearly, the Coates graph $G_c(A)$ associated with matrix A can be obtained from $G_c(A')$ by removing the vertex x_{n+1}.

As an example, for the following system of equations

$$\begin{bmatrix} 3 & -2 & 1 \\ -1 & 2 & 0 \\ 3 & -2 & 2 \end{bmatrix} \begin{bmatrix} x_1 \\ x_2 \\ x_3 \end{bmatrix} = \begin{bmatrix} 3 \\ 1 \\ -2 \end{bmatrix} x_4 \tag{8.4}$$

the matrix A' is

$$A' = \begin{bmatrix} 3 & -2 & 1 & -3 \\ -1 & 2 & 0 & -1 \\ 3 & -2 & 2 & 2 \\ 0 & 0 & 0 & 0 \end{bmatrix}$$

The Coates graphs $G_c(A')$ and $G_c(A)$ are shown in Fig. 8.3.

Because a matrix and its transpose have the same determinant value and because A is the transpose of the adjacency matrix of $G_c(A')$, we obtain the following result from Theorem 8.2.

THEOREM 8.3 *If a matrix A is nonsingular, then*

$$\det A = (-1)^n \sum_{H} (-1)^{L_H} w(H) \tag{8.5}$$

where H is a 1-factor of $G_c(A)$, $w(H)$ is the weight of H and L_H is the number of directed circuits in H.

To derive a similar expression for the sum in the numerator of (8.3), we first define the concept of a **1-factorial connection**. A 1-factorial connection H_{ij} from x_i to x_j in $G_c(A)$ is a spanning subgraph of G which contains a directed path P from x_i to x_j and a set of vertex-disjoint directed circuits which include all the vertices of $G_c(A)$ other than those which lie on P. Similarly, a 1-factorial connection of $G_c(A')$ can be defined. As an example, a 1-factorial connection from x_4 to x_3 of the graph $G_c(A')$ of Fig. 8.3(a) is shown in Fig. 8.3(c).

---

[1] In network and systems theory literature, the Coates graph is referred to as a **flow graph**.

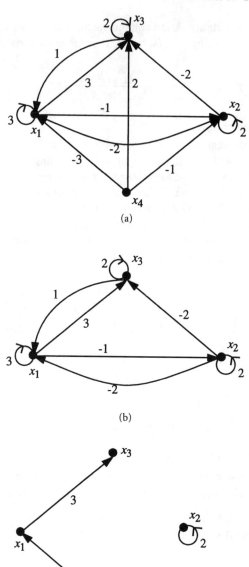

8.3   (a) The Coates graph $G_c(A')$; (b) the graph $G_c(A)$; (c) A-factorial connection $H_{4,3}$ of the graph $G_c(A')$

THEOREM 8.4   *Let $G_c(A)$ be the Coates graph associated with an $n \times n$ matrix A. Then*

$$1. \quad \Delta_{ii} \;=\; (-1)^{n-1} \sum_{H} (-1)^{L_H}\, w(H)$$

$$2. \quad \Delta_{ij} \;=\; (-1)^{n-1} \sum_{H_{ij}} (-1)^{L'_H}\, w(H_{ij}) \qquad i \neq j$$

where H is a 1-factor in the graph obtained by removing vertex x_i from $G_c(A)$, H_{ij} is a 1-factorial connection in $G_c(A)$ from vertex x_i to vertex x_j, and L_H and L'_H are the numbers of directed circuits in H and H_{ij}, respectively.

PROOF 8.2   1. Note that Δ_{ii} is the determinant of the matrix obtained from A by removing its row i and column i. Also, the Coates graph of the resulting matrix can be obtained from $G_c(A)$ by removing vertex x_i. Proof follows from these observations and Theorem 8.3.

2. Let A_α denote the matrix obtained from A by replacing its jth column by a column of zeroes, except for the element in row i, which is 1. Then it is easy to see that

$$\Delta_{ij} = \det A_\alpha$$

Now, the Coates graph $G_c(A_\alpha)$ can be obtained from $G_c(A)$ by removing all edges incident out of vertex x_j and adding an edge directed from x_j to x_i with weight 1. Then from Theorem 8.3, we get

$$
\begin{aligned}
\Delta_{ij} &= \det A_\alpha \\
&= (-1)^n \sum_{H_\alpha} (-1)^{L_\alpha} w(H_\alpha)
\end{aligned}
\tag{8.6}
$$

where H_α is a 1-factor of $G_c(A_\alpha)$ and L_α is the number of directed circuits in H_α.

Consider now a 1-factor H_α in $G_c(A_\alpha)$. Let C be the directed circuit of H_α containing x_i. Because in $G_c(A_\alpha)$, (x_j, x_i) is the only edge incident out of x_j, it follows that x_j also lies in C. If we remove the edge (x_j, x_i) from H_α we get a 1-factorial connection, H_{ij}. Furthermore, $L'_H = L_\alpha - 1$ and $w(H_{ij}) = w(H_\alpha)$ because (x_j, x_i) has weight equal to 1. Thus, each H_α corresponds to a 1-factorial connection H_{ij} of $G_c(A_\alpha)$ with $w(H_\alpha) = w(H_{ij})$ and $L'_H = L_\alpha - 1$. The converse of this is also easy to see. Thus, in (8.6) we can replace H_α by H_{ij} and L_α by $(L'_H + 1)$. Then we obtain

$$\Delta_{ij} = (-1)^{n-1} \sum_{H_{ij}} (-1)^{L'_H} w(H_{ij})$$

$\square$

Having shown that each Δ_{ij} can be expressed in terms of the weights of the 1-factorial connections H_{ij} in $G_c(A)$, we now show that $\sum b_i \Delta_{ik}$ can be expressed in terms of the weights of the 1-factorial connections $H_{n+1, k}$ in $G_c(A')$.

First, note that adding the edge (x_{n+1}, x_i) to H_{ik} results in a 1-factorial connection $H_{n+1, k}$, with $w(H_{n+1, k}) = -b_i\, w(H_{ik})$. Also, $H_{n+1, k}$ has the same number of directed circuits as H_{ik}. Conversely, from each $H_{n+1, k}$ that contains the edge (x_{n+1}, x_i) we can construct a 1-factorial connection H_{ik} satisfying $w(H_{n+1, k}) = -b_i\, w(H_{ik})$. Also, $H_{n+1, k}$ and the corresponding H_{ik} will have the same number of directed circuits. Thus, a one-to-one correspondence exists between the set of all 1-factorial connections $H_{n+1, k}$ in $G_c(A')$ and the set of all 1-factorial connections in $G_c(A)$ of the form H_{ik} such that each $H_{n+1, k}$ and the corresponding H_{ik} have the same number of directed circuits and satisfy the relation $w(H_{n+1, k}) = -b_i\, w(H_{ik})$. Combining this result with Theorem 8.4, we get

$$\sum_{i=1}^{n} b_i \Delta_{ik} = (-1)^n \sum_{H_{n+1, k}} (-1)^{L'_H} w(H_{n+1, k})
\tag{8.7}$$

where the summation is over all 1-factorial connections, $H_{n+1, k}$ in $G_c(A')$, and L'_H is the number of directed circuits in $H_{n+1, k}$. From (8.5) and (8.7) we get the following theorem.

THEOREM 8.5   *If the coefficient matrix A is nonsingular, then the solution of (8.2) is given by*

$$\frac{x_k}{x_{n+1}} = \frac{\sum_{H_{n+1,k}}(-1)^{L'_H}w(H_{n+1,k})}{\sum_H (-1)^{L_H}w(H)} \tag{8.8}$$

for $k = 1, 2, \ldots, n$, where $H_{n+1,k}$ is a 1-factorial connection of $G_c(A')$ from vertex x_{n+1} to vertex x_k, H is a 1-factor of $G_c(A)$, and L'_H and L_H are the numbers of directed circuits in $H_{n+1,k}$ and H, respectively.

Equation (8.8) is the called **Coates' gain formula**. We now illustrate Coates' method by solving the system (8.4) for x_2/x_4. First, we determine the 1-factors of the Coates' graph $G_c(A)$ shown in Fig. 8.3(b). These 1-factors, along with their weights, are listed below. The vertices enclosed within parentheses represent a directed circuit.

1-Factor H	Weight $w(H)$	L_H
$(x_1)(x_2)(x_3)$	12	3
$(x_2)(x_1, x_3)$	6	2
$(x_3)(x_1, x_2)$	4	2
(x_1, x_2, x_3)	2	1

From the above we get the denominator in (8.8) as

$$\sum_H (-1)^{L_H}w(H) = (-1)^3 \cdot 12 + (-1)^2 \cdot 6 + (-1)^2 \cdot 4 + (-1)^1 \cdot 2 = -4$$

To compute the numerator in (8.8) we need to determine the 1-factorial connections $H_{4,2}$ in the Coates graph $G_c(A')$ shown in Fig. 8.3(a). They are listed below along with their weights. The vertices in a directed path from x_4 to x_2 are given within parentheses.

1-Factorial connection $H_{4,2}$	$w(H_{4,2})$	L'_H
$(x_4, x_1, x_2)(x_3)$	6	1
$(x_4, x_2)(x_1)(x_3)$	−6	2
$(x_4, x_2)(x_1, x_3)$	−3	1
(x_4, x_3, x_1, x_2)	−2	0

From the above we get the numerator in (8.8) as

$$\sum_{H_{4,2}}(-1)^{L'_H}w(H_{4,2}) = (-1)^1 \cdot 6 + (-1)^2(-6) + (-1)^1(-3) + (-1)^0(-2) = -11$$

Thus, we get

$$\frac{x_2}{x_4} = \frac{11}{4}$$

8.4   Mason's Gain Formula

Consider again the system of equations

$$AX = Bx_{n+1}$$

We can rewrite the above as

$$x_j = (a_{jj} + 1)x_j + \sum_{\substack{k=1 \\ k \neq j}}^{n} a_{jk}x_k - b_j x_{n+1}, \quad j = 1, 2, \ldots, n, \qquad x_{n+1} = x_{n+1} \tag{8.9}$$

Letting X' denote the column vector of the variables $x_1, x_2, \ldots x_{n+1}$, and U_{n+1} denote the unit matrix of order n, we can write (8.9) in matrix form as follows:

$$(A' + U_{n+1})X' = X' \qquad (8.10)$$

where A' is the matrix defined earlier in Section 8.3.

The Coates graph $G_c(A' + U_{n+1})$ is called the **Mason's signal flow graph** or simply the **Mason graph**[2] associated with A', and it is denoted by $G_m(A')$. The Mason graph $G_m(A)$ is defined in a similar manner. The Mason graphs $G_m(A')$ and $G_m(A)$ associated with the system (8.4) are shown in Fig. 8.4. Mason's graph elegantly represents the flow of variables in a system. If we associate each vertex with a variable and if an edge is directed from x_i to x_j, then we may consider the variable x_i as contributing $(a_{ji}x_i)$ to the variable x_j. Thus, x_j is equal to the sum of the products of the weights of the edges incident into vertex x_j and the variables corresponding to the vertices from which these edges emanate.

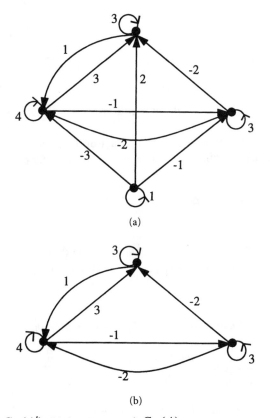

(a)

(b)

8.4   (a) The Mason graph $G_m(A')$; (b) the Mason graph $G_m(A)$.

Note that to obtain the Coates graph $G_c(A)$ from the Mason graph $G_m(A)$ we simply subtract one from the weight of each self-loop. Equivalently, we may add at each vertex of the Mason graph a self-loop of weight -1. Let S denote the set of all such loops of weight -1 added to construct the Coates graph G_c from the Mason graph $G_m(A)$.

---

[2]In network and systems theory literature Mason graphs are usually referred to as **signal flow graphs**.

Consider now the Coates graph G_c constructed as above and a 1-factor H in G_c having j self-loops from the set S. If H has a total of $L_Q + j$ directed circuits, then removing the j self-loops from H will result in a subgraph Q of $G_m(A)$ which is a collection of L_Q vertex disjoint directed circuits. Also,

$$w(H) = (-1)^j w(Q)$$

Then, from Theorem 8.3 we get

$$
\begin{aligned}
\det A &= (-1)^n \sum_H (-1)^{L_Q + j} w(H) \\
&= (-1)^n \sum_Q (-1)^{L_Q} w(Q) \\
&= (-1)^n \left[1 + \sum_Q (-1)^{L_Q} w(Q) \right]
\end{aligned}
\tag{8.11}
$$

We can rewrite the above as:

$$\det A = (-1)^n \left[1 - \sum_j Q_{j1} + \sum_j Q_{j2} - \sum_j Q_{j3} \cdots \right] \tag{8.12}$$

where each term in $\sum_j Q_{ji}$ is the weight of a collection of i vertex-disjoint directed circuits in $G_m(A)$.

Suppose we refer to $(-1)^n \det A$ as the determinant of the graph $G_m(A)$. Then, starting from $H_{n+1,k}$ and reasoning exactly as above we can express the numerator of (8.3) as

$$\sum_{i=1}^n b_i \Delta_{ik} = (-1)^n \sum_j w(P_{n+1,k}^j) \Delta_j \tag{8.13}$$

where $P_{n+1,k}^j$ is a directed path from x_{n+1} to x_k of $G_m(A')$ and Δ_j is the determinant of the subgraph of $G_m(A')$ which is vertex-disjoint from the path $P_{n+1,k}^j$. From (8.12) and (8.13) we get the following theorem.

THEOREM 8.6   *If the coefficient matrix A is (8.2) is nonsingular, then*

$$\frac{x_k}{x_{n+1}} = \frac{\sum_j w(P_{n+1,k}^j) \Delta_j}{\Delta}, \qquad k = 1, 2, \ldots, n \tag{8.14}$$

where $P_{n+1,k}^j$ is the jth directed path from x_{n+1} to x_k of $G_m(A')$, Δ_j is the determinant of the subgraph of $G_m(A')$ which is vertex-disjoint from the jth directed path $P_{n+1,k}^j$, and Δ is the determinant of the graph $G_m(A)$.

Equation (8.14) is known as **Mason's gain formula**. In network and systems theory $P_{n+1,k}^j$ is referred to as a **forward path** from vertex x_{n+1} to vertex x_k. The directed circuits of $G_m(A')$ are called the **feedback loops**.

We now illustrate Mason's method by solving the system (8.4) for x_2/x_4. To compute the denominator in (8.14) we determine the different collections of vertex-disjoint directed circuits of the Mason graph $G_m(A)$ shown in Fig. 8.4(b). They are listed below along with their weights.

Collection of Vertex-Disjoint Directed Circuits of $G_m(A)$	Weight	No. of Directed Circuits
(x_1)	04	1
(x_2)	03	1
(x_3)	03	1
(x_1, x_2)	02	1
(x_1, x_3)	03	1
(x_1, x_2, x_3)	02	1
$(x_1)(x_2)$	12	2
$(x_1)(x_3)$	12	2
$(x_2)(x_3)$	09	2
$(x_2)(x_1, x_3)$	09	2
$(x_3)(x_1, x_2)$	06	2
$(x_1)(x_2)(x_3)$	36	3

From the above we obtain the denominator in (8.14)

$$\Delta = 1 + (-1)^1[4 + 3 + 3 + 2 + 3 + 2]$$
$$+ (-1)^2[12 + 12 + 9 + 9 + 6] + (-1)^3 36 = -4$$

To compute the numerator in (8.14) we need the forward paths in $G_m(A')$ from x_4 to x_2. They are listed below with their weights.

j	$P_{4,2}^j$	Weight
1	(x_4, x_2)	-1
2	(x_4, x_1, x_2)	3
3	(x_4, x_3, x_1, x_2)	-2

The directed circuits which are vertex-disjoint from $P_{4,2}^1$ are (x_1), (x_3), (x_1, x_3). Thus

$$\Delta_1 = 1 - (4 + 3 + 3) + 12 = 1 - 10 + 12 = 3.$$

(x_3) is the only directed circuit which is vertex-disjoint from $P_{4,2}^2$. So,

$$\Delta_2 = 1 - 3 = -2.$$

No directed circuit is vertex-disjoint from $P_{4,2}^3$, so $\Delta_3 = 1$. Thus, the numerator in (8.14) is

$$P_{4,2}^1 \Delta_1 + P_{4,2}^1 \Delta_2 + P_{4,3}^1 \Delta_3 = -3 - 6 - 2 = -11$$

and

$$\frac{x_2}{x_4} = \frac{11}{4}$$

References

[1] Coates, C.L., "Flow graph solutions of linear algebraic equations," *IRE Trans. Circuit Theory*, CT-6, 170–187, 1959.

[2] Mason, S.J., "Feedback theory: some properties of signal flow graphs," *Proc. IRE*, 41, 1144–1156, 1953.

[3] Mason, S.J., "Feedback theory: further properties of signal flow graphs," *Proc. IRE*, 44, 920–926, 1956.

[4] Thulasiraman, K. and Swamy, M.N.S., *Graphs: Theory and Algorithms*, New York, Wiley Interscience, 1992.

[5] Chen, W.K., *Applied Graph Theory*, Amsterdam, North Holland, 1971.

[6] Harary, F., "The determinant of the adjacency matrix of a graph," *SIAM Rev.*, 4, 202–210, 1962.

[7] Hohn, F.E., *Elementary Matrix Algebra*, New York, Macmillan, 1958.

Index